P9-EDU-795

Recent Progress in

HORMONE RESEARCH

Proceedings of the Laurentian Hormone Conference

VOLUME 47

RECENT PROGRESS IN
HORMONE RESEARCH

Proceedings of the
1990 Laurentian Hormone Conference

Edited by
C. WAYNE BARDIN

VOLUME 47

PROGRAM COMMITTEE

C. W. Bardin	A. R. Means
H. G. Friesen	M. New
D. K. Granner	D. Orth
P. A. Kelly	G. Ringold
I. A. Kourides	N. B. Schwartz
S. McKnight	J. L. Vaitukaitis

W. Vale

ACADEMIC PRESS, INC.
Harcourt Brace Jovanovich, Publishers
San Diego   New York   Boston
London   Sydney   Tokyo   Toronto

This book is printed on acid-free paper. ∞

Copyright © 1991 by ACADEMIC PRESS, INC.

All Rights Reserved.

No part of this publication may be reproduced or transmitted in any form or by any means, electronic or mechanical, including photocopy, recording, or any information storage and retrieval system, without permission in writing from the publisher.

Academic Press, Inc.
San Diego, California 92101

69685

United Kingdom Edition published by
Academic Press Limited
24–28 Oval Road, London NW1 7DX

Library of Congress Catalog Number: Med. 47-38

International Standard Book Number: 0-12-571147-6

PRINTED IN THE UNITED STATES OF AMERICA
91 92 93 94   9 8 7 6 5 4 3 2 1

CONTENTS

LIST OF CONTRIBUTORS AND DISCUSSANTS

L. Aguilar-Bryan
M. Bagchi
A. C. Bauer-Dantoin
C. W. Bardin
J. D. Baxter
G. I. Bell
L. M. Besecke
A. E. Boyd III
J. Bryan
C. F. Burant
W. Chu
J. Clark
L. A. Conaghan
W. F. Crowley, Jr.
A. C. Dalkin
F. M. DeNoto
A. Dunaif
K. Duncan
P. Epstein
J. S. Finkelstein
C. Forest
H. Fukumoto
J. W. Funder
D. Granner
M. A. Haidar
D. J. Haisenleder
R. Hazelwood
R. Hertz
W. Hseuh
E. Imai
M. N. G. James
J. L. Jameson
T. Kayano
R. P. Kelch
P. Kelly
J. Kirkland
D. L. Kunze
V. Lakshmi
S. J. Legan
J. E. Levine
P. Lucas
J. C. Marshall
S. McKnight

A. Means
P. L. Mellon
J. M. Meredith
J. Mitchell
L. Moss
F. Murdoch
S. Nagamatsu
D. A. Nelson
M. New
R. O'Brien
L. St. L. O'Dea
B. W. O'Malley
G. A. Ortolano
S. J. Paul
S. Pavlou
J. E. Pessin
A. Powers
H. Raef
A. S. Rajan
T. L. Reudelhuber
R. B. Russell
W. T. Schrader
N. Schwartz
S. Seino
W. I. Sivitz
F. J. Strobl
G. Teitelman
M.-J. Tsai
S. Y. Tsai
J. H. Urban
W. Vale
I. Vermes
K. M. Vogelsong
N. L. Weigel
R. I. Weiner
M. Weirman
J. Weiss
R. W. Whitcomb
J. J. Windle
A. M. Wolfe
H. Xiang
G. C. Yaney

PREFACE

Tools of modern science have led to enhanced understanding of biology. Nowhere is this more apparent than in the field of endocrinology and metabolism as reflected in this volume of *Recent Progress in Hormone Research*. The topics covered range from the molecular regulation of genes to current topics in clinical endocrinology. For example, studies of progesterone receptor activation of model genes emphasize that steroid hormone receptors are part of a larger class of transcription factors. As a consequence, a better understanding of how steroid receptors act can now be approached in cell-free systems. A series of manuscripts describe the complex regulation of LH and FSH secretion through studies of the LHRH pulse generator in intact animals. Abnormalities of this regulator in humans are also defined, and a newly developed LHRH-secreting neuronal cell line produced by targeted oncogenes is described that will provide new information on the pulsatile secretion of LHRH *in vitro*. The latter studies were not thought possible a few months ago. Insights into hormone-induced hypertension are now possible through the understanding of how the specificities of cortisone and aldosterone for the glucocorticoid and mineralocorticoid receptors are determined. In addition, the structure of renin suggests new approaches for the treatment of hypertension. Finally, a series of manuscripts relating to carbohydrate metabolism illustrate the origin of insulin-secreting cells of the pancreas, define the sulfonylurea receptor, and demonstrate the diversity of glucose transporters. Genetic analysis of the phosphoenolpyruvate carboxylkinase gene illustrates how complex hormonal regulators control intermediary metabolism.

The 1990 meeting of the Laurentian Hormone Conference, which is the basis for this volume of *Recent Progress in Hormone Research,* marks the end of Dr. James Clark's tenure as President. The Board of the Conference thanks Dr. Clark for his years of service.

C. Wayne Bardin

Molecular Mechanism of Action of a Steroid Hormone Receptor[1]

BERT W. O'MALLEY, SOPHIA Y. TSAI, MILAN BAGCHI, NANCY L. WEIGEL, WILLIAM T. SCHRADER, AND MING-JER TSAI

Department of Cell Biology, Baylor College of Medicine, Houston, Texas 77030

I. Background Review

In the early 1960s, predominant theories existed that steroid hormones acted at the level of the cell membrane to facilitate ion or substrate transport or to catalyze energy exchange. When labeled estradiol was synthesized by Jensen's laboratory, a new series of experiments was initiated that led to the observation that estradiol bound with specificity and high affinity to an intracellular protein in target cells, termed a receptor (Jensen *et al.*, 1966, 1968; O'Malley *et al.*, 1979; Gorski and Cannon, 1976; Gorski *et al.*, 1968; Yamamoto, 1985). A flurry of experiments extended this concept to progestins, androgens, glucocorticoids, and other steroid hormones. These receptors were noted to be capable of binding to DNA and were thought to be concentrated in the nucleus following hormone administration. In separate experiments during this period, steroid hormones were shown also to increase the incorporation of radiolabeled nucleotides and amino acids into precipitable macromolecules. Heretical voices suggested that steroid hormones may activate genes via the concurrent theories of Jacob and Monod generated from bacterial studies. For the most part, such investigators were correct in their interpretations, but critics were entitled to point out that their results could emanate from increased transport of labeled precursors into cellular pools rather than from a net increase in macromolecular synthesis.

The next important observations were qualitative and involved experiments which showed that steroid hormones (1) caused appearance and accumulation of new species of nuclear (hybridizable) RNAs, (2) caused stimulation of *de novo* synthesis of new specific proteins, (3) caused a corresponding increase in the net cellular level of specific mRNAs, and (4) stimulated the rate of transcription of certain nuclear genes (O'Malley *et al.*, 1969, 1979; O'Malley and Means, 1974). These observations argued for a nuclear role for steroid hormone and were accompanied by studies which revealed that receptors had an inherent affinity for

---

[1] The Gregory Pincus Memorial Lecture.

1

DNA (O'Malley, 1984; Yamamoto, 1985). At this point in the early 1970s, the circle was closed and the primary pathway for steroid hormone action was defined as follows: steroid → (steroid–receptor) → (steroid–receptor–DNA) → mRNA → protein → functional response (Fig. 1).

Steroid receptors were purified then to single species and characterized as to size, charge, etc. Antibodies were developed and structural domains were postulated by proteolytic analyses. After an initial description by Jensen *et al.* (1966, 1968), assays of sex steroid receptors became commonplace in the diagnosis and assignment of therapy for breast cancer (Jensen and DeSombre, 1973). Investigators isolated specific target genes for steroid hormones, defined their structure, and proved that cis-acting regulatory sequences were located near such genes, usually in the 5′ flanking sequences (Payvar *et al.*, 1983; Renkawitz *et al.*, 1984; Yamamoto, 1985;). When such sequences, termed steroid response elements (SREs), were occupied by receptors these genes came under the control of their respective hormones.

The molecular biological community became increasingly interested in these receptors for steroid hormones as they came to realize that they were the most highly studied and purified trans-activation factors for control of eukaryotic transcription. Also, they were the specific activators of an emerging and fascinating genetic cis element, the enhancer. In the past 5 years, a great deal more has been learned of the structure–function relationships of steroid receptors and the mechanisms by which they interact with DNA. Biochemical studies in the late 1970s suggested that steroid receptors, thyroid receptors, and receptors for vi-

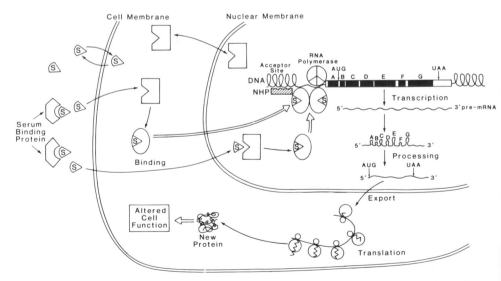

FIG. 1.   Pathway for steroid hormone regulation of a target gene (ovalbumin). S, Steroid; NHP, nonhistone proteins.

tamins such as vitamin D_3 were likely to belong to a family of gene regulatory proteins. Furthermore, it was known that these functional proteins were organized into domains which contained the functions of (1) specific and high-affinity ligand binding, (2) specific DNA binding, and (3) "transcriptional modulation."

Molecular cloning of the steroid/thyroid and vitamin receptors proved this "superfamily" concept (Giguere et al., 1987; Evans, 1988; Beato, 1989; O'Malley, 1990). A surprising observation has been that certain oncogenes such as v-erbA are members of this receptor gene family. The avian erythroblastosis virus appears to have captured the cellular gene coding for thyroid receptor and this retrovirus uses this mutated molecule for its own oncogenic purposes. Further sequence analyses not only substantiated the existence of functional domains within receptors but showed that they could be rearranged as independent cassettes within their own molecules or as hybrid molecules with other regulatory peptides (Green and Chambon, 1987). Perhaps the most intensively studied domain has been that responsible for DNA binding. This domain has been shown to be a cysteine-rich region which is capable of binding zinc in a manner which creates two peptide projections referred to as "zinc fingers" (Evans, 1988). These zinc fingers promote the interactions of receptors with target enhancers and clearly mark each as a member of this evolutionarily conserved family. A broad region of the C-terminal domain was shown to comprise the ligand-binding site and a transcriptional activation region; the heterogeneous N-terminal domains of receptors were revealed to contain an additional transcriptional modulation domain (for review, see Beato, 1989).

The precise sequences of the steroid response elements (SREs) which are regulated by steroid hormones have been described for all members of this family to date. In general they are 15-base pair (bp) core sequences, composed of two half-sites of 5 or 6 bp arranged in a dyad axis of symmetry (inverted repeats); the half-sites are split by a few central base pairs of random composition (Strahle et al., 1989). The SREs for various receptors share similarities in sequence and, in fact, the identical sequence allows activation by glucocorticoid, progesterone, and androgen receptors. One copy of such an SRE is sufficient usually to bring a promoter under moderate hormonal control and two copies often provide a synergistic response to the cognate hormone.

The precise mechanism of interaction of receptors with their target SREs has come under close scrutiny of late. After cytoplasmic synthesis, many steroid receptors form complexes with heat shock proteins, such as hsp90 and hsp70 (Schuh et al., 1985; Catelli et al., 1985; Pratt et al., 1988). Extracts of receptors for glucocorticoid, progesterone, and estrogen exist in aggregate form with these heat shock proteins when prepared from cells; thyroid and vitamin D_3 receptors appear not to be extracted as such complexes. In cells, this interaction may promote proper folding and stability of the molecule; in complex with hsp90, receptors cannot bind to DNA. When a receptor is complexed with heat shock proteins in vitro and in vivo, binding of hormone causes dissociation of the complex (Denis *et al.*, 1988; Kost *et al.*, 1989). The exact physiologic meaning of this association of certain receptors with heat shock proteins is unclear at present but it is thought that heat shock proteins aid posttranslational folding of the molecules, prevent denaturation, and chaperone the receptors to their appropriate cellular compartments.

Evidence shows that glucocorticoid, progesterone, and estrogen receptors bind to their SREs as dimers, one molecule to each half-site (Tsai *et al.*, 1988; Kumar *et al.*, 1988; Klein-Hitpass *et al.*, 1990; Fawell *et al.*, 1990). This interaction appears to be cooperative, at least for glucocorticoid and progesterone receptors (Tsai *et al.*, 1988). In this manner, receptor dimers bind with greater affinity and stability to their SREs (Fig. 1). Interactions between receptor dimers at separate SREs also allow a higher order cooperative interaction which stabilizes the two sets of dimers into a tetrameric structure which has a 100-fold greater affinity for its SRE sequences (Tsai *et al.*, 1989; Schmid *et al.*, 1989). Such protein–protein interactions may occur among homologous receptor complexes, or receptor–promoter/TATA complexes. Multimeric interactions among receptors at their target DNA sites facilitate the initiation of transcription at nearby genes. Protein–protein interaction is the currently popular watchword for formation of stable transcription complexes at regulated genes.

The precise role of ligand in receptor activation remains still a bit of a mystery (O'Malley, 1990). Certainly receptors are functionally inactive in cells in the absence of ligand. Hormone administration *in vivo* leads to formation of a bound protein complex at SREs of hormone-regulated genes, as evidenced by *in vivo* footprint analyses (Becker *et al.*, 1986). Crude receptor extracts of cells bind DNA poorly until they undergo a temperature- and salt-aided "activation" which is thought to be driven by hormone in the intact cell. This activation is accompanied by disaggregation of receptors from heat shock and other inhibitory proteins. Anti-hormones promote disaggregation but do not activate target genes, indicating an additional level of ligand-induced allosteric control. Purified receptors on the other hand, show good binding preference for their SREs whether or not they are complexed with hormone (Willmann and Beato, 1986; Klein-Hitpass *et al.*, 1990). Obviously, the hormone plays a pivotal role in activating the intracellular receptor but to date its exact role can only be investigated using *in vitro* studies.

II.   Molecular Mechanisms of Steroid Hormone Action

A.   RECEPTOR ACTIONS IN A CELL-FREE SYSTEM

In this setting, we will summarize our experimental approaches to elucidate the precise molecular mechanism by which steroid receptors regulate the initiation of target gene transcription and the role of the hormonal ligand in this process. For determining the direct actions of receptor on DNA transcription, it

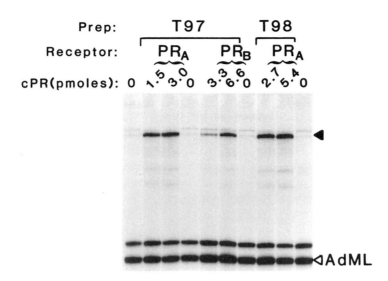

FIG. 2. Stimulation of *in vitro* transcription of progesterone response element 2 (PRE₂)–TATA promoter template by different chicken progesterone receptor (cPR) preparations. AdML, Adenovicus major late promoter.

was obvious to us that a cell-free (reconstituted) transcription system was required. In such a system the concentration of receptor, ligand, general transcription factors, and target genes could be manipulated. For a series of reasons beyond the scope of this chapter and having to do primarily with the complexity of the components and reaction conditions, it took us nearly a decade to develop a reliable, sensitive, and reproducible system using native chicken progesterone receptor (cPR). In this chapter we present a selection of our results to date. The primary data, certain controls, and precise methodological details can be found in the original referred publications (Klein-Hitpass *et al.*, 1990; Bagchi *et al.*, 1990a).

In our initial experiments we measured the effect of purified cPR (50–80% pure) or a purified cPR derivative expressed in *Escherichia coli* on transcription of templates that lacked or contained progesterone response elements (PREs). A HeLa cell nuclear extract which does not contain measurable amounts of progesterone receptor was used to provide basal transcriptional machinery (Reinberg *et al.*, 1987). Our templates contained a TATA box, which is essential for correct transcriptional initiation (Grosschedl and Birnstiel, 1980; Zarucki *et al.*, 1982), and a "G-free cassette" (Sawadogo and Roeder, 1985a), which serves as a "reporter gene." Two copies of the PRE of the tyrosine aminotransferase (*TAT*) gene, which have been shown to be able to confer progesterone inducibility to promoters *in vivo* (Strahle *et al.*, 1987), were inserted about 20 bp upstream of the TATA box. A similar construction in which the PREs were located relatively close to the TATA box has been shown previously to be functional in transfection experiments (Bradshaw *et al.*, 1988). A template containing the adenovirus major late (AdML) promoter (-400 to $+10$) linked to a shorter G-free cassette was included in the reaction mixture as a noninducible internal control gene.

As indicated in Fig. 2, transcripts correctly initiated at the control AdML promoter are 190 nucleotides (nt) in length, whereas correct initiation at the target promoter (about 30 bp downstream of the TATA box) yields transcripts of 360 nt in length. Transcripts slightly longer than both of these correctly initiated products represent "read-through" transcripts initiated inappropriately on the reporter plasmids upstream of the G-free cassette and which can be monitored; they are trimmed also by the G-specific ribonuclease T1, which is included in the reactions (Sawadogo and Roeder, 1985a).

B. cPR STIMULATES *in Vitro* TRANSCRIPTION

First we analyzed the effect of cPR_A on transcripts initiated from our template. Progesterone receptor was purified from chick oviducts as a hormone–receptor complex. Salt steps in the purification yielded an "activated" receptor expressing DNA-binding activity. Templates either lacking or containing two progesterone receptor binding sites (PREs) were used for this assay. Addition of 1.5 pmol cPR stimulated *in vitro* transcription of the template containing PREs (Fig. 2). This amount of cPR represents a final *in vitro* concentration of ~ 50 nM, a level that closely approximates the physiologic concentration in nuclei of progesterone-stimulated oviduct cells. At this concentration the number of receptor molecules is approximately 10-fold higher than the PRE_2 template, which binds four molecules of the receptor. Using several separate receptor preparations, reproducible inductions were observed (Fig. 2). As expected, the activity of the AdML promoter was unaltered by cPR.

In chicken oviduct cytosol two forms (cPR_A and cPR_B) of progesterone receptor exist which we believe to arise from differential initiation of translation from two AUGs (Conneely *et al.*, 1989). Both forms can bind specifically to the PRE of the *TAT* gene *in vitro* and stimulate transcription of the mouse mammary tumor virus (MMTV) promoter *in vivo*. We tested the different receptor forms in our *in vitro* transcription assay and found that both cPR_A and cPR_B preparations induced transcription of the test promoter containing PREs (Fig. 2).

C. COOPERATIVE BINDING OF PROGESTERONE RECEPTORS TO A PRE DIMER CONTRIBUTES TO SYNERGISTIC INDUCTION OF TRANSCRIPTION *in Vitro*

We have demonstrated previously that when progesterone receptor dimers interact with a DNA fragment containing two PREs, binding of a dimer to the first PRE site enhances the binding of receptor to the second PRE (Tsai *et al.*,

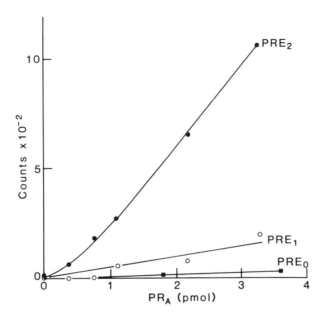

FIG. 3. Synergistic induction of minimal promoter by progesterone response element (*in vitro* transcription).

1989). This result suggests that cooperative binding of receptors to the multiple PREs which reside in the control regions of many target genes contributes to the synergistic induction of target gene expression observed *in vivo*. To directly test the relationship of cooperative binding to functional cooperativity, we used test genes containing zero, one, and two copies of a PRE in our *in vitro* transcription assay. As shown in Fig. 3, the level of cPR-dependent RNA synthesis from the test gene lacking a PRE was extremely low, even at high levels of PR_A (3.6 pmol). This result confirmed a requirement for PRE elements for cPR-mediated transcription. A test gene containing one PRE (PRE_1) yielded significantly higher levels of receptor-dependent transcription when 2.1 and 3.2 pmol cPR was added. The levels of transcription with receptor were at least fivefold higher than that of the construct lacking PRE (PRE_0). In contrast, a test gene containing two PREs (PRE_2) was very efficiently transcribed. At 0.72 pmol cPR, the level of transcription was already equal to that observed with PRE_1 at 3.2 pmol cPR; at 3.2 pmol cPR, it was 5.5-fold higher than that of PRE_1 and ~30-fold higher than that of PRE_0. We conclude that the two PREs act cooperatively to induce functional synergism in this *in vitro* transcription assay.

D. THE MECHANISM OF PROGESTERONE RECEPTOR ACTION IS TO ENHANCE THE ASSEMBLY OF A STABLE PREINITIATION TRANSCRIPTION COMPLEX

To examine the mechanism by which progesterone receptor interacts with general transcription factors to stimulate RNA synthesis *in vitro*, we investigated whether the receptor participates in formation of a stable preinitiation (rapid start) complex of transcription factors (Fig. 4). For these studies, two test templates having different length "G-free" cassette sequences were employed. The assay involves template competition for a limiting concentration of general transcription factors in the presence or absence of receptor.

To attempt to investigate the mechanism by which regulatory proteins, such as steroid hormone receptors, interact with core promoters to enhance transcription, we turned to our cell-free transcription system. Using a series of preincubation and template competition analyses combined with kinetic analyses, we were able to dissect the mechanism of action of cPR in our *in vitro* transcription system. The details of these experiments are reported elsewhere (Klein-Hitpass *et al.*, 1990) and will be summarized in concept only. In short, the steroid receptor enhances the formation of a rapid start complex by RNA polymerases (Fig. 4) (Bagchi *et al.*, 1990b). It appears to do this by enhancing the assembly of a template-committed complex of transcription factors at the TATA element. This stable complex is now poised for rapid and repeated initiation of transcription by RNA polymerase.

Multiple factors must interact at the proximal promoter (TATA box) of a gene to form a committed complex and allow initiation of transcription (Buratowski *et al.*, 1989). This sequence of events is thought to be dependent on the reversible interaction of the TF-IID protein with the TATA sequence itself. This interaction is subsequently stabilized by sequential binding of TF-IIA, TF-IIB, RNA polymerase, and TF-IIE/F but the precise roles of these factors in the initiation complex are only partially known (Buratowski *et al.*, 1989). Nevertheless, together they form a three-dimensional surface conformation which directs RNA polymerase to begin the transcriptional process. This entire complex is relatively unstable and the individual factors are all in low concentrations in cells. RNA polymerase itself may be in a concentration as low as $<100,000$ molecules/cell.

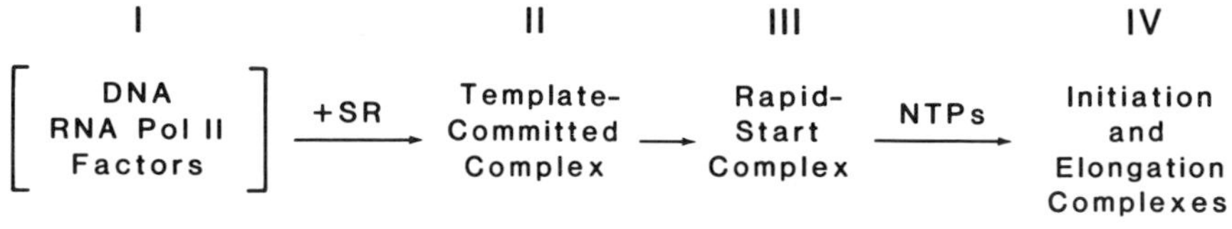

FIG. 4.   Steroid receptor enhances formation of a stable preinitiation (rapid-start) complex.

In a given cell, as many as 10,000 genes may be in competition for these factors. These factors must assemble and remain on an active gene for it to be repetitively transcribed at a high rate. Any additional specific transcription factors which would aid in recruiting or stabilizing these general factors should lead to enhanced expression of a given gene. Following assembly of the transcription factors they interact to form a rapid-start complex which is poised to initiate transcription as soon as nucleotide triphosphates are available (Fig. 5).

In our cell-free system, cPR acts to enhance formation of a stable template-committed complex of these transcription factors, thereby promoting formation of the rapid-start complex and stimulating transcriptional initiation (Fig. 5). Receptor may do this by facilitating recognition of the promoter by such factors, simply by stabilizing the promoter DNA–protein complex once it is formed, or perhaps by influencing both reactions. In either case, it is likely that the *in vivo* effect is modulated via local protein–protein interactions between the ligand-activated receptor and the promoter complex (Klein-Hitpass *et al.*, 1990; O'Malley, 1990). Additional questions to be asked in this context relate to the precise contacts between the receptor amino acids and the surface of individual promoter factors. Also, we must understand whether there is a continuing role for bound ligand in this process. Finally, the continued development of cell-free transcription systems to test the function of steroid hormone receptors should permit a precise definition of the transcription activation domains of receptors, the role of hormone in receptor activation, and other interactions of receptor with enhancer and silencer binding proteins in eukaryotic cells.

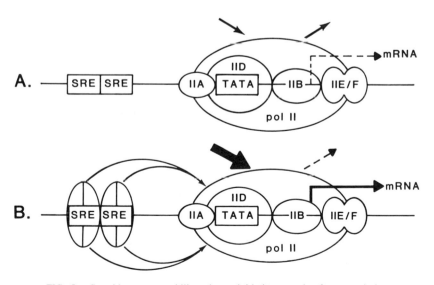

FIG. 5. Steroid receptor stabilizes the preinitiation complex for transcription.

E. HORMONE-FREE cPR CAN ACTIVATE *in Vitro* TRANSCRIPTION WHEN PURIFIED

To investigate whether receptor function in our assay is dependent absolutely on the presence of hormone, we employed hormone-free cPR (aporeceptor) purified from chicken oviduct cytosol after salt activation. Previously we showed that chicken progesterone aporeceptor is able to form a stable PRE–receptor complex and the addition of hormone changes the relative affinity of the protein for PRE by only a few fold (Rodriguez *et al.*, 1989). On a molar basis aporeceptor was as effective as receptor preparations incubated *in vitro* with hormone (Klein-Hitpass *et al.*, 1990). Preincubation of the aporeceptor with progesterone $(10^{-7} M)$ for 2 hours at 4°C (sufficient to label more than 90% of the receptors present in the preparation) did not alter the magnitude of the response. We conclude from this result that bound hormone is not required for activation of transcription by "purified cPR" *in vitro*.

Our result, that purified chicken progesterone aporeceptor can activate *in vitro* transcription as efficiently as holoreceptor and that aporeceptor cannot be stimulated further by hormone treatment *in vitro*, was a rather surprising result to us. Domain-switching experiments have indicated that the activation function localized within the hormone-binding domains of human estrogen receptor (hER) (Webster *et al.*, 1988), rat glucocorticoid receptor (GR) (Godowski *et al.*, 1988), and hGR (Hollenberg and Evans, 1988) is absolutely dependent on hormone *in vivo*. One might speculate that *in vivo* hormone is required only to convert the receptor to the DNA-binding form; this process can be accomplished *in vitro* by salt activation alone, which dissociates non-hormone-binding proteins such as hsp90 (Groyer *et al.*, 1987; Denis *et al.*, 1988) from the aporeceptor. Once the receptor is able to bind to DNA, it can utilize its inherent transcriptional potential. Alternatively, we can speculate that hormone is required both to dissociate inhibitory (heat shock) proteins and to convert the receptor allosterically to a form which will trans-activate target genes. During purification, we may denature or alter the receptor so that a subpopulation of molecules is in the proper allosteric conformation to interact with the transcriptional machinery and stimulate gene expression. In any event, our results indicate that purified cPR does not require bound hormone for subsequent productive interaction with the transcription machinery. They do not, however, indicate that steroid hormone lacks the capacity to allosterically maintain its cognate receptor in a conformation which maximizes its covalent modifications and/or transcriptional potential *in vivo*.

To summarize the results described above, we have developed a cell-free transcription system in which the function of purified native steroid receptor can be studied directly *in vitro*. We have reported previously that another member of the steroid receptor superfamily, chicken ovalbumin upstream promoter (COUP) transcription factor, could stimulate transcription *in vitro* using a natural promoter (Sagami *et al.*, 1986). Estrogen-dependent transcription of an estrogen target gene has been reported in whole nuclear extracts of *Xenopus* liver (Corthesy *et al.*, 1988); also, transcription mediated by the DNA-binding fragment of the glucocorticoid receptor has been reported previously (Freedman *et al.*, 1989). Reconstituted cell-free systems have the advantage that the concentration of all regulatory factors can be manipulated for diverse experimental protocols.

F.   MULTIPLE STEROID RECEPTORS ACT BY A COMMON MECHANISM TO REGULATE INITIATION OF TRANSCRIPTION AT TARGET GENES

We wondered next whether all steroid receptors regulate DNA transcription via a conserved mechanism. To test the generality of our cell-free response system, we obtained samples of human glucocorticoid receptor from Dr. Brad Thompson's laboratory (Srinivasan and Thompson, 1990) and mouse estrogen receptor from Dr. Malcolm Parker's laboratory; both were produced from recombinant cDNAs in baculoviral expression systems. Target genes containing G-free cassettes were constructed using the cognate steroid response element for each receptor and the tests were carried out as described above in collaboration with the Thompson and Parker laboratories.

Recombinant full-length human glucocorticoid receptor stimulated transcription *in vitro* of test genes containing synthetic GRE/PRE sites or an MMTV promoter (Tsai *et al.*, 1990). The receptor expressed in a baculoviral vector was highly active, enhancing transcription of hormone response genes greater than 30-fold even at a receptor concentration of 1.2 nM (Fig. 6). The enhancement of transcription is glucocorticoid/progesterone response element (GRE/PRE) dependent, suggesting that it is a receptor-mediated event. *In vitro* and *in vivo* treatment with the agonist dexamethasone or with the antagonist Ru486 did not alter significantly the functional activity of partially purified receptor. In the presence of general HeLa cell transcription factors, kinetic studies suggested that glucocorticoid receptor is required for formation of a stable committed transcriptional complex. These results indicate that the action of glucocorticoid receptor on gene transcription is similar to that defined for the progesterone receptor and suggests that this molecular mechanism may be a general characteristic for steroid receptors (Fig. 5).

Preparations of mouse recombinant estrogen receptor were shown also to stimulate RNA synthesis directly in an *in vitro* transcription assay (Elliston *et al.*, 1990). Stimulation of transcription by receptor again required the presence of the cognate estrogen response elements in the template and could be specifically inhibited by addition of competitor oligonucleotides containing estrogen response element, (EREs). Moreover, polyclonal antibodies directed against the DNA-binding region of ER inhibit transcriptional stimulation by estrogen recep-

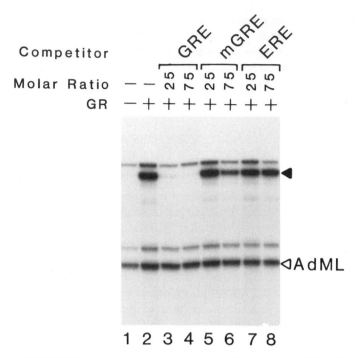

FIG. 6. GRE/PRE-Dependent transcription by glucocorticoid receptor. GRE, Glucocorticoid response element; ERE, estrogen response element.

tor. Finally, we found that estrogen receptor acted by facilitating the formation of a stable preinitiation complex of trans-acting factors at the target gene promoter, thus augmenting the initiation of transcription by RNA polymerase II. Again, these observations were very similar to what we found using the progesterone receptor in its own response system and lend strong support to our hypothesis as to the mechanism of steroid receptor-regulated gene expression (Fig. 5). Considered as a whole, our conclusions apply to at least three different hormone receptors and suggest strong conservation of molecular function among all members of the steroid receptor superfamily.

G. A SYSTEM FOR STUDY OF LIGAND ACTIVATION OF RECEPTOR-MEDIATED TRANSCRIPTION

T47D cells are enriched in human progesterone receptor (hPR) and considerable amounts of receptor are present in the crude nuclear fractions. We tested crude nuclear extracts of T47D cells, grown under hormone-deprived conditions, for DNA binding and transcriptional activity in the presence or absence of added hormone. In a gel retardation experiment, unoccupied receptors did not generate any specific DNA–protein complex with PRE. Incubation of the extract with hormone induced the formation of specific DNA–protein complexes. In a previous study using monospecific antibodies against hPR, we have established that these complexes were indeed formed by binding of A and B forms of hPR to PRE sequences (Bagchi *et al.*, 1988).

To study the effects of hormone on the transcriptional activities of these extracts, we used the same synthetic promoter containing two copies of a PRE and the ovalbumin TATA box as described above. Treatment of nuclear extracts with 10^{-7} M progesterone resulted in a 10-fold induction in the synthesis of the correctly initiated (360 nt) transcript (Fig. 7). Very little accurately initiated transcription occurred in the absence of hormone. This hormonal induction of transcription was observed over a wide range of concentrations of the nuclear extract (Bagchi *et al.*, 1990a). The hormone-induced transcriptional enhancement was entirely promoter specific since the activity of the adenovirus promoter remained unchanged on hormone treatment. These results demonstrated clearly

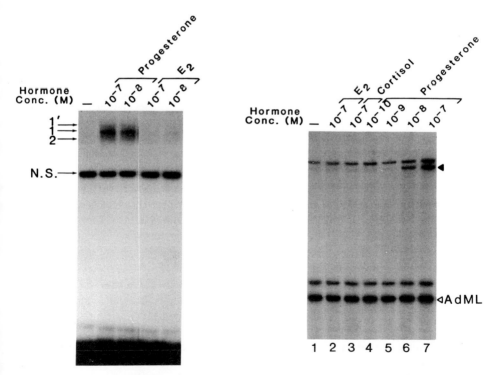

FIG. 7. Specificity of hormonal activation of DNA binding and transcriptional enhancement by PR. E_2, Estradiol; N.S., nonspecific protein–DNA complex.

that treatment of hPR with progesterone alone results in the activation of RNA synthesis from a PRE-driven promoter. This event occurred concurrently with hormonal induction of high-affinity binding of hPR to PREs and was mediated by PR itself. Progesterone failed to induce the synthesis of any correctly initiated transcripts from a promoter that did not contain PREs whereas it stimulated transcription from the PRE-containing promoter.

The hormonal specificity of the transcriptional enhancement was examined. Progesterone, but not 17β-estradiol or cortisol, enhanced transcription *in vitro* (Bagchi *et al.*, 1990a). A dose–response experiment with progesterone indicated that maximal transcriptional stimulation was achieved at a progesterone concentration slightly greater than 10 nM (Fig. 7). At this concentration, maximal hormone-dependent binding of receptor was noted. These results substantiate further the involvement of hPR in mediating hormonal induction of transcription.

The compound Ru486 has strong antiprogestin and antiglucocorticoid activity *in vivo*. The molecular basis of these antagonistic activities is not currently understood. We tested the transcriptional activity of Ru486–PR in our *in vitro* system and compared it with that of the agonist R5020–PR. Addition of 10–100 nM Ru486 led to some transcriptional activation of the test promoter. The extent of this transcriptional enhancement however, was only about 25–30% (an average of three experiments) of that elicited by the progestin agonist R5020. The much weaker transcriptional activation by the Ru486–PR complex may be the result of an allosterically altered receptor structure which affects its interaction with the transcriptional machinery. The fact that Ru486 treatment can induce a weak transcriptional activation is consistent with *in vivo* data showing it to be a weak agonist.

To ascertain the structural status of the hPR in the salt extracts of nuclei of hormone-free and hormone-stimulated T47D cells, we analyzed the sedimentation behavior of this unfractionated PR in various conditions (Denis *et al.*, 1988). Even in the presence of molybdate, which is known to stabilize an 8S form of receptor, the salt-treated R5020–PR complex sedimented in a 4S form. These results suggested that in our nuclear salt extract, the 8S form of receptor had already undergone transformation to a 4S form. Samples of 4S receptor were assayed using antibody to hsp90 and shown to be free of this inhibitory protein. At this stage p59, another associated protein, has been removed also. Our results are consistent with the results of numerous laboratories which reported that the high salt-mediated dissociation of hsp90 from receptors leads to a conversion from an 8S to a 4S structural complex (Pratt *et al.*, 1988).

The observation that a 4S form of PR requires hormone to be transcriptionally active provides a novel and interesting insight into the catalytic role of hormone in receptor-mediated gene activation. Although the 4S PR is apparently free of hsp90, it remains transcriptionally inactive (Fig. 8). The receptor may, however, still be associated with other lower molecular mass inhibitory proteins which

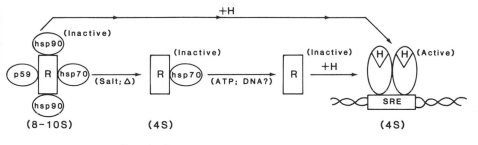

FIG. 8.   Hypothetical kinetic intermediates in steroid hormone activation of receptor.

mask its dimerization, DNA binding, and/or trans-activation properties. For example, in contrast to hsp90, hsp70 remains associated with glucocorticoid and progesterone receptors even in its 4S form. In such a scenario, the role of the hormonal ligand could be to bind to the receptor, induce a structural change, and trigger the release of hsp70 inhibitory protein(s). The free receptor could then acquire the proper allosteric conformation to bind to DNA and effect gene trans-activation.

To examine this possibility, we have subjected the salt-dissociated receptor to DNA column purification and then to a procedure which employs ATP to dissociate hsp70 from receptor (Kost *et al.*, 1989). Greater than 90% of hsp70 is removed from the receptor at this point. When tested in our cell-free transcription assay, the receptor still requires hormone for trans-activation of target gene expression. Again, Ru486 did not cause appropriate activation of transcription. These results are the strongest to date that reveal an allosteric (or covalent) modification of a steroid receptor which is induced by its cognate hormone independent of dissociation from heat shock proteins. This has been a frequent and controversial question in our field which is not possible to answer using whole-cell experiments.

As described above, we have shown that highly purified chicken PR can bind to target DNA and enhance transcription in a hormone-independent manner. Our current opinion is that during purification, various *in vitro* manipulations may have accomplished an irreversible conformational change in the purified receptor and converted it to a form which can cause trans-activation without ligand binding. *In vivo* and in crude extracts, the receptor will not have undergone such a conformational change and hormone would be absolutely necessary to transform it into a form that is active in DNA binding and trans-activation.

Our results are consistent with the following model for ligand-induced activation of the receptor (Fig. 8). Before addition of hormone, PR is present in the cell in an inactive 8S complex with hsp90, hsp70, and perhaps other inhibitory proteins such as p59. Hormone treatment results in the release of these chaperone proteins and in the conversion of the 8S to a 4S form. In our experimental setup, we have bypassed the first part of this hormone-induced transformation event as salt treatment *in vitro* has converted the 8S to the 4S form. This cell-free system has allowed us to trap a previously unidentified intermediate form of the receptor—a salt-dissociated 4S form which is not competent to bind specifically to DNA and induce initiation of transcription (Fig. 8). It is free of hsp90 but not hsp70. Subsequent ATP treatment removes hsp70 but the receptor is still inactive in transcription. Only addition of authentic progestin causes appropriate activation of the receptor and it now stimulates target gene transcription. As dissociation of hsp90, hsp70, etc., is not sufficient to make the receptor competent for trans-activation, we propose that an additional conformation alteration(s) in the receptor molecule must precede the activation of target genes by PR (Fig. 8). This alteration includes dimerization, an allosteric change, and perhaps a covalent modification such as phosphorylation.

H. RECEPTOR-MEDIATED REGULATION OF COMPLEX PROMOTERS

The expression of a typical eukaryotic gene is controlled by multiple sequence elements within its 5' flanking region. Each of these elements represents a binding site for one or more regulatory factors. Classically, promoter-binding factors have been viewed as capable of action only when positioned adjacent to the initiation site. Enhancer-binding proteins, on the other hand, function whether bound adjacent to or distant from the initiation site. It is the individual and combinatorial activities of these binding sites which are thought to control the developmental and tissue-specific expression of myriad target genes (Serfling *et al.*, 1985; Ptashne, 1988). Enhancer elements, located at a significant distance from the cap site of a target gene, may be unable to efficiently influence binding of transcription factors to the TATA box without the aid of interactions with other factors bound at the upstream promoter (Fig. 9). This cooperation would be reflected in functional synergism between enhancer and upstream promoter regulatory elements.

As examples of this type of cooperative interaction, activation via the simian virus 40 (SV40) enhancer requires Sp1-binding sites in the early promoter (Everett *et al.*, 1983) while activation of the heterologous β-globin gene by this enhancer (Treisman and Maniatis, 1985) or multiple heat shock elements (Bienz and Pelham, 1986) is also dependent on distinct promoter sequences. Likewise, κ enhancer-induced expression from the heterologous thymidine kinase gene (*TK*) promoter is down regulated by the removal of a linked immunoglobulin octamer sequence (Parslow *et al.*, 1987). Steroid hormone receptors interact with downstream transcription factors to influence cellular gene expression. Steroid

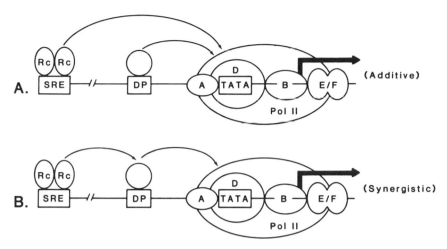

FIG. 9. Mechanisms of action of steroid-responsive enhancer.

response elements (SREs) can cooperate with a variety of heterologous binding sites, including those for nuclear factor 1 (NF-1), Sp1, and octanucleotide transcription factor (OTF), to stimulate the *TK* promoter (Schule *et al.*, 1988; Strahle *et al.*, 1988) and the β-globin promoter. Furthermore, glucocorticoid-dependent activation of the mouse mammary tumor virus (MMTV) long terminal repeat (LTR) is abolished by removal of an NF-1 binding site in the distal promoter (Buetti and Kuhnel, 1986).

Previous experimental analysis of the interactions between gene regulatory factors has relied upon transient transfection analysis of recombinant DNA constructs in living cells. Such studies are usually carried out by gross mutagenesis of regulatory sequences, which may remove overlapping binding sites or alter spatial relationships. Furthermore, the level of a transcription factor varies widely in different cell types and cannot be controlled. However, precise functional dissection and quantitation of these interactions can be carried out in cell-free assays using purified components. Since we have already established a receptor-dependent *in vitro* transcription system using a minimal promoter containing tandem glucocorticoid/progesterone response elements (GRE/PREs) adjacent to the TATA box of the chicken ovalbumin gene (Klein-Hitpass *et al.*, 1990), we can now use this same system to construct more complex cis and trans interactions. To this end, we have investigated the requirements of steroid receptors for distal promoter elements and their binding proteins during *in vitro* activation of two complex promoters: the MMTV LTR and the intact ovalbumin promoter. We posed the question as to whether synergistic cooperation between GRE/PREs and distal promoter elements is dependent on the identity of the transcription factor that binds to the distal promoter.

I. EFFICIENT STEROID RECEPTOR-DEPENDENT TRANSCRIPTION OCCURS FROM COMPLEX PROMOTERS

To examine enhancer–promoter interactions in a situation where the response elements are in their natural sequence context, the region of the MMTV LTR from -400 to -20 was cloned upstream of the G-free cassette (pMMTV-400. Steroid responsiveness *in vivo* requires LTR sequences between -180 and -170 and sequences centered at -120, both of which bind purified glucocorticoid receptor (GR) and progesterone receptor (PR) *in vitro* (Scheidereit and Beato, 1984). A third site centered at -70 is reported to be essential for steroid induction and binds the ubiquitous nuclear factor NF-1 (Nowock *et al.*, 1985).

pMMTV-400 was transcribed in the presence of a HeLa nuclear extract and human GR prepared using a baculoviral expression system (Srinivasan and Thompson, 1990). Twenty femtomoles of receptor induced a greater than 10-fold increase in specific transcription without a concomitant effect on activity of the adenovirus major late promoter (AdML) used as an internal control. In the absence of added GR, the low basal activity of the promoter can be eliminated by competition with an oligonucleotide containing the adenovirus NF-1 binding site. GR-dependent transcription in the presence of this oligonucleotide is severely curtailed, being reduced by more than fivefold compared to activity in its absence. These data clearly show that the NF-1 trans-activator is necessary for maximal induction of the MMTV LTR by GR under these *in vitro* assay conditions. Furthermore, quantitation of the specific transcripts produced from pMMTV-400 under each condition allows calculation of the strength of the resultant synergism between receptor and NF-1 transcription factor.

In contrast to these studies, substitution of another distal promoter such as COUP revealed no synergism with upstream GRE/PREs. In this construction, the distal promoter element is dispensable for gene induction and suggests that COUP–TF may be inherently unable to interact with GR in this context. Most importantly, when an NF-1 site is substituted for the COUP element in this construct, a marked synergism is again noted. Thus, the transcriptional response to receptor is dependent quantitatively on the nature of the distal promoter in complex genes.

Because regulatory factors such as adenomajor late transcription factor (MLTF), activating transcription factor (ATF), PR, ER, and GR can enhance preinitiation complex formation when bound close to the cap site (Sawadogo and Roeder, 1985b; Hai *et al.*, 1988; Klein-Hitpass *et al.*, 1990), it appears likely that TATA-proximal elements can modulate gene activity independently of other control sequences. Basal transcription of the MMTV and ovalbumin genes may be mediated in this way by NF-1 and COUP-TF, respectively. Enhancers, on the other hand, are likely to be less capable of directly influencing initiation complex stability due, according to one model, to the greater energy required for DNA loop formation between widely separated binding sites (Ptashne, 1988).

Our results with NF-1 in the MMTV LTR show that distal promoter elements can have dual functions: determination of the level of uninduced gene expression and mediation of the effects of upstream regulatory sequences (Fig. 9B). Interactions between enhancer- and distal promoter-bound factors may affect complex formation by bringing the upstream factors into closer proximity to the cap site or by enhancing the inherent ability of the promoter factor to stabilize assembly of general transcription factors at the TATA element. Alternatively, cooperation could manifest itself by an increase in binding to the distal promoter element.

Interestingly, we have also described a situation where the enhancer (tandem GRE/PREs) activates independently of downstream control elements (COUP) (Fig. 9A). Substitution of COUP by the binding site for another transcription factor such as NF-1 substantially increased cooperativity. We feel that, depending upon the circumstances, natural SREs in steroid target genes will be shown to work by either of these mechanisms (Fig. 9). The strength of the cooperative interaction may depend not only on the identity of the distal promoter-binding factor but also on its availability in the cell and on the presence of additional proteins needed to mediate an interaction. A complex control mechanism such as this would lend extraordinary sensitivity and versatility to steroid-regulated gene expression.

J. FUTURE APPLICATIONS OF CELL-FREE TRANSCRIPTION

In conclusion, we have established a simple model system which we have employed for initial analyses of the molecular details of gene regulation by steroid receptors. Complete reconstitution outside the cell of the events controlled by these molecules at cellular promoters may soon be possible. For example, (1) by reconstituting the transcriptional machinery from known purified components of nuclear extracts, factors essential for steroid receptor-mediated transcriptional activation can be identified and their roles and stoichiometric interactions can be determined; (2) the function of purified receptor derivatives or mutants can be assayed under controlled and receptor-free conditions; this should be particularly useful to study receptor mutants which are defective for nuclear translocation and thus cannot be studied *in vivo*; (3) the mechanisms by which receptor modifications (e.g., phosphorylation/dephosphorylation) modulate receptor can be assessed; (4) it should be quite useful to study "transcriptional interference" (Meyer *et al.*, 1989), a phenomenon which is related to "squelching" (Gill and Ptashne, 1988) and represents the competition of different steroid receptors (or gene regulatory proteins) for common transcription factors which are present at limiting concentrations in cells; and, finally, (5) the combination of *in vitro* transcription and chromatin reconstitution technology makes it now feasible to study the role of higher order structures in receptor-dependent gene expression.

III. Orphan Receptors

COUP-TF AND ORPHAN RECEPTORS

Finally, some comment should be made on one of the most fascinating observations to evolve from the cloning of cDNAs for steroid receptors; it is the unexpected large size of the steroid receptor superfamily of related genes. After elucidation of the receptors for the more traditional members of this family (glucomineralocorticoids, sex steroids, thyroid hormone, vitamin D_3, and retinoic acid), a large number of cloned receptoroids have been discovered (Evans, 1988). These molecules can be considered to be "orphan receptors" in search of a function and a ligand. Since they were cloned by cDNA cross-hybridization screening using cDNA probes, we have little clue as to their cellular physiology. A function is implicit, however, since they are expressed in cells as fully processed cytoplasmic mRNAs.

The first report of two such molecules was by the Evans laboratory; they were termed ERR-1 (estrogen–receptor related) and ERR-2 (Giguere *et al.*, 1988). Their function remains unknown to date. A number of orphan receptor sequences have been published (deThé *et al.*, 1987; Nauber *et al.*, 1988; Oro *et al.*, 1988; Lazar *et al.*, 1989). They are recognized easily by their extensive homology in the DNA-binding region and their absolute conservation of type II zinc fingers. We estimate that more than 20 additional related molecules may have been cloned in the combined laboratories in this field. Over eight of these molecules have been cloned from *Drosophila*. Until recently, all of these putative receptors were without designated functions.

The assignment of COUP-TF to the steroid receptor superfamily of proteins has a number of apparent implications (Wang *et al.*, 1989). It was originally noted to be an orphan receptor. It designates promoter regulatory proteins as a legitimate subtype in this family. The family had been thought previously to include only enhancer regulatory proteins. Second, it provides information which is useful for understanding the evolution of this family of transcriptional regulators. Finally, it raises the question as to whether COUP-TF, and other promoter activators for that matter, may be ligand-activated gene regulators. This latter query remains to be answered experimentally. Deductive reasoning should permit us to conclude that if one of these orphan receptors has been adopted for a function, then in time, others will follow suit.

Although admittedly speculative, the question as to whether the orphan receptors have endogenous ligands is clearly the most exciting to be considered

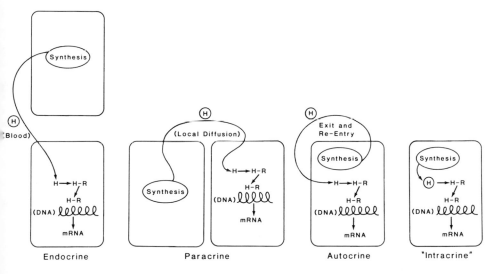

FIG. 10. System variations in hormonal regulation of endocrine target cells.

(O'Malley, 1989). If they do have ligands, we would predict that the ligands are indigenous to the cells of origin since transfection of these molecules into most cultured cells shows them to be active in an apparently constitutive manner. At present, our best guess is that these molecules do bind ligands and that the ligands are produced within the same cell. This hypothesis is substantiated to some degree by the fact that C-terminal mutational analysis of COUP-TF provides a response pattern in cells which is identical to that observed with authentic steroid receptors.

If existent, we would predict that many of these putative ligands will be hydrophobic in nature and many may be nutritionally or metabolically derived. In fact, this could be the tip of a new "intracrine" iceberg wherein a series of yet to be described hormones are discovered to be indigenous to the specific target cells expressing these orphan receptors. This intracrine system (Fig. 10) would of course now be subject to exogenous regulation by new drugs using pharmacologically derived agonists and antagonists of these response systems. If true, the elucidation of such a new intracrine regulatory system would inject a great deal of interest and excitement into the fields of hormone action and molecular endocrinology.

REFERENCES

Bagchi, M., Elliston, J. F., Tsai, S. Y., Edwards, D. P., Tsai, M.-J., and O'Malley, B. W. (1988), *Mol. Endocrinol.* **2,** 1221–1229.

Bagchi, M., Tsai, S. Y., Tsai, M.-J., and O'Malley, B. W. (1990a). *Nature (London)* **345,** 547–550.
Bagchi, M., Tsai, S. Y., Weigel, N. L., Tsai, M.-J., and O'Malley, B. W. (1990b). *J. Biol. Chem.* **265,** 5129–5134.
Beato, M. (1989).*Cell (Cambridge, Mass.)* **56,** 335–344.
Becker, P. B., Gloss, B., Schmid, W., Strähle, V., and Schultz, G. (1986). *Nature (London)* **324,** 686–688.
Bienz, M. and Pelham, H. R. B. (1986). *Cell (Cambridge, Mass.)* **45,** 753–760.
Bradshaw, M. S., Tsai, M.-J., and O'Malley, B. W. (1988). *Mol. Endocrinol.* **2,** 1286–1293.
Buetti, E., and Kuhnel, B. (1986). *J. Mol. Biol.* **190,** 379–389.
Buratowski, S., Hahn, S., Guarente, L., and Sharp, P. A. (1989). *Cell (Cambridge, Mass.)* **56,** 549–561.
Catelli, M.-G., Binart, N., Jung-Testas, I., Renoir, J. M., Baulieu, E.-E., Feramisco, J. R., and Welch, W. J. (1985). *EMBO J.* **4,** 3131.
Conneely, O. M., Kettelberger, D. M., Tsai, M.-J., Schrader, W. T., and O'Malley, B. W. (1989). *J. Biol. Chem.* **264,** 14062–14064.
Corthesy, B., Hipskind, R., Theulaz, I., and Wahli, W. (1988). *Science* **239,** 1137–1139.
Denis, M. Poellinger, L., Wikstom, A. C., and Gustafsson, J. A. (1988). *Nature (London)* **333,** 686–688.
deThé, H., Marchio, A., Tiollois, P., and Dejean, A. (1987). *Nature (London)* **330,** 667–670.
Elliston, J. R., Fawell, S. E. Klein-Hitpass, L., Tsai, S. Y., Tsai, M.-J. Parker, M. G., and O'Malley, B. W. (1991). *Mol. Cell. Biol.* **10,** 6607–6612.
Evans, R. M. (1988). *Science* **24,** 889–895.
Everett, R. D., Baty, D., and Chambon, P. (1983). *Nucleic Acids Res.* **11,** 2447–2464.
Fawell, S. E., Lees, J. A., White, R., and Parker, M. G. (1990). *Cell (Cambridge, Mass.)* **60,** 963–962.
Freedman, L. P., Yoshinaga, S. K., Vanderbilt, J. N., and Yamamoto, K. R. (1989). *Science* **245,** 298–301.
Giguere, V., Ong, E. S., Sequi, P., and Evans, R. M. (1987). *Nature (London)* **330,** 624–629.
Giguere, V., Yang, N., Segui, P., and Evans, R. M. (1988). *Nature (London)* **331,** 91–94.
Gill, G., and Ptashne, M. (1988). *Nature (London)* **334,** 721–724.
Godowski, R. J., Picard, D., and Yamamoto, K. R. (1988). *Science* **241,** 812–816.
Gorski, J., Gannon, F. (1976). *Annu. Rev. Biochem.* **28,** 425.
Gorski, J., Toft, D., Shyamala, G., Smith, D., and Notides, A. (1968). *Recent Prog. Horm. Res.* **24,** 45–80.
Green, S., and Chambon, P. (1987). *Nature (London)* **325,** 75–78.
Grosschedl, R., and Birnstiel, M. L. (1980). *Proc. Natl. Acad. Sci. U.S.A.* **77,** 1432–1436.
Groyer, A., Schweizer-Groyer, G., Cadepond, F., Mariller, M., and Baulieu, E. E. (1987). *Nature (London)* **328,** 624–626.
Hai, T., Horikoshi, M., Roeder, R. G., and Green, M. R. (1988). *Cell (Cambridge, Mass.)* **54,** 1043–1051.
Hollenberg, S. M., and Evans, R. M. (1988). *Cell (Cambridge, Mass.)* **55,** 899–906.
Jensen, E. V., and DeSombre, E. R. (1973). *Proc. Int. Congr. Endocrinol., 4th, 1972* pp. 1227–1231.
Jensen, E. V., Jacobsen, H. I., Flesher, J. W., Saha, N. N., Gupta, G. N., Smith, S., Colucci, V., Shiplacoff, D., Neuman H. G., DeSombre, E. R., and Jungblut, P. W. (1966). *In* "Steroid Dynamics" (G. Pincus, T. Nakao, and J. R. Tait, eds.), pp. 133–156. Academic Press, New York.
Jensen, E. V., Suzuki, T., Kawashima, T., Stumpf, W. E., Jungblut, P. W., and DeSombre, E. R. (1968). *Proc. Natl. Acad. Sci. U.S.A.* **59,** 632–638.
Klein-Hitpass, L., Tsai, S. Y., Greene, G. L., Clark, J. H., Tsai, M.-J., and O'Malley, B. W. (1989). *Mol. Cell. Biol.* **9,** 43–49.

Klein-Hitpass, L., Tsai, S. Y., Weigel, N. L., Allan, G. F., Riley, D., Rodriguez, R., Schrader, W. T., Tsai, M.-J., and O'Malley, B. W. (1990). *Cell (Cambridge, Mass.)* **60,** 247–257.

Kost, S. L., Smith, D. F., Sullivan, W. P., Welch, W. J., and Toft, D. O. (1989). *Mol. Cell. Biol.* **9,** 3829–3838.

Kumar, V., Green, S., Stack, G., Berry, M., Jin, J.-R., and Chambon, P. (1988). *Cell (Cambridge, Mass.)* **55,** 145–156.

Lazar, M. A., Hodin, R. A., Darling, D. S., and Chin, W. W. (1989). *Mol. Cell. Biol.* **9,** 1128–1136.

Meyer, M.-E., Gronemeyer, H., Tucotte, B., Bocquel, M.-T., Tasset, D., and Chambon, P. (1989). *Cell (Cambridge, Mass.)* **57,** 433–442.

Nauber, U., Pankratz, M. J., Kienlin, A., Seifert, E., Klemm, U., and Jackle, H. (1988). *Nature (London)* **336,** 489–492.

Nowock, J., Borgmeyer, U., Puschel, A. W., Rupp, R. A. W., and Sippel, A. E. (1985). *Nucleic Acids Res.* **13,** 2045–2061.

O'Malley, B. W. (1984). *J. Clin. Invest.* **74,** 207.

O'Malley, B. W. (1989). *Endocrinology (Baltimore)* **125**(3), 1119–1120.

O'Malley, B. W. (1990). *Mol. Endocrinol.* **4,** 363–369.

O'Malley, B. W., and Means, A. R. (1974). *Science* **183,** 610–620.

O'Malley, B. W., McGuire, W. L., Kohler, P. O., Korenman, S. (1969). *Recent Prog. Horm. Res.* **25,** 105.

O'Malley, B. W., Roop, D. R., Lai, E. C., Nordstrom, J. L., Catterall, J. F., Swaneck, G. E., Colbert, D. A., Tsai, M.-J., Dugaiczyk, A., and Woo, S. L. C. (1979). *Recent Prog. Horm. Res.* **35,** 1–46.

Oro, A. E., Org, E. S., Margolis, J. S., Posakony, J. W., McKeown, M., and Evans, R. M. (1988). *Nature (London)* **336,** 493–496.

Parslow, T. G., Jones, S. D., Bond, B., and Yamamoto, K. R. (1987). *Science* **235,** 1498–1501.

Payvar, F., DeFranco, D., Firestone, G. L., Edgar, B., Wrange, O., Okret, S., Gustafsson, J. A., and Yamamoto, K. R. (1983). *Cell (Cambridge, Mass.)* **35,** 381–392.

Pratt, W. B., Jolly, D. J., Pratt, D. V., Hollenberg, S. M., Giguere, V., Cadepond, F. M., Schweizer-Groyer, G., Catelli, M.-G., Evans, R. M., and Baulieu, E.-E. (1988). *J. Biol. Chem.* **263,** 267–273.

Ptashne, M. (1988). *Nature (London)* **335,** 683–689.

Reinberg, D., Horikoshi, M., and Roeder, R. G. (1987). *J. Biol. Chem.* **262,** 3322–3330.

Renkawitz, R., Schütz, G., von der Ahe, D., and Beato, M. (1984). *Cell (Cambridge, Mass.)* **37,** 503–510.

Rodriguez, R., Carson, M. A., Weigel, N. L., O'Malley, B. W., and Schrader, W. T. (1989). *Mol. Endocrinol.* **3,** 356–362.

Sagami, I., Tsai, S. Y., Wang, H., Tsai, M.-J., and O'Malley, B. W. (1986). *Mol. Cell. Biol.* **6,** 4259–4267.

Sawadogo, M., and Roeder, R. G. (1985a). *Proc. Natl. Acad. Sci. U.S.A.* **82,** 4394–4398.

Sawadogo, M., and Roeder, R. G. (1985b). *Cell (Cambridge, Mass.)* **43,** 165–175.

Scheidereit, C., and Beato, M. (1984). *Proc. Natl. Acad. Sci. U.S.A.* **81,** 3029–3033.

Schmid, W., Strahle, U., Schutz, G., Schmitt, J., and Stunnenberg, H. (1989). *EMBO J.* **8,** 2257–2263.

Schuh, S. W., Yonemoto, J., Brugge, V. J., Bauer, R. M., Riehl, W. P., Sullivan, W. P., and Toft, D. O. (1985). *J. Biol. Chem.* **260,** 14292–14296.

Schule, R., Muller, M., Kaltschmidt, C., and Renkawitz, R. (1988). *Science* **242,** 1416–1420.

Serfling, E., Jasin, M., and Schaffner, W. (1985). *Trends Genet.* **1,** 224–230.

Srinivasan, G., and Thompson, E. B. (1990). *Mol. Endocrinol.* **4,** 209–216.

Strahle, U., Klock, G., and Schutz, G. (1987). *Proc. Natl. Acad. Sci. U.S.A.* **84,** 7871–7875.

Strahle, U., Schmid, W., and Schutz, G. (1988). *EMBO J.* **7,** 3389–3395.

Strahle, U., Boshart, M., Klock, G., Stewart, F., and Schutz, G. (1989). *Nature (London)* **339,** 629–632.

Treisman, R., and Maniatis, T. (1985). *Nature (London)* **315,** 72–75.

Tsai, S. Y., Carlstedt-Duke J.-A., Weigel, N. L., Dahlman, K., Gustafsson, J.-A., Tsai, M.-J., and O'Malley, B. W. (1988). *Cell (Cambridge, Mass.)* **55,** 361–369.

Tsai, S. Y., Tsai, M.-J., and O'Malley, B. W. (1989). *Cell (Cambridge, Mass.)* **57,** 443–448.

Tsai, S. Y., Srinivasan, G., Allan, G. F., Thompson, E. B., O'Malley, B. W., and Tsai, M. J. (1990). *J. Biol. Chem.* **265,** 17055–17061.

Wang, L.-H., Tsai, S. Y., Cook, R. G., Beattie, W. G., Tsai, M.-J., and O'Malley, B. W. (1989). *Nature (London)* **340,** 163–166.

Webster, N. J. G., Green, S., Jin, J.-R., and Chambon, P. (1988). *Cell (Cambridge, Mass.)* **54,** 199–207.

Willmann, T., and Beato, M. (1986). *Nature (London)* **324,** 688–691.

Yamamoto, K. R. (1985). *Annu. Rev. Genet.* **19,** 209–252.

Zarucki, T., Tsai, S. Y., Itakura, K., Soberon, X., Wallace, R. B., Tsai, M.-J., Woo, S. L. C., and O'Malley, B. W. (1982). *J. Biol. Chem.* **257,** 11070–11077.

DISCUSSION

F. Murdoch. I am very interested in all the work you discussed on the hormone dependence of transcription. Certainly it is good to see a system that replicates *in vivo* transcription in that manner. We have done quite a bit of work on DNA binding and the hormone dependence of this binding. In Jack Gorski's laboratory we have examined the hormone dependence of the estrogen receptor binding to DNA. We have tried a number of different approaches and have never seen any hormone dependence. The one piece of data I want to discuss stems from an experiment in which we used the estrogen response element from the vitellogenin gene and rat uterine cytosol as our source of estrogen receptor; this is unpurified crude cytosols. In lanes 2 and 3 of Fig. 11 cytosol is present in the absence of any hormone. A protein–DNA complex is formed. In lane 3 an antibody that specifically recognizes the estrogen receptor was added to the reaction and a supershift is seen suggesting that the receptor is present in the complex. In lanes 4 and 5 a similar experiment was performed, this time in the presence of estradiol. Qualitatively and quantitatively the complexes are the same. On the right-hand side of the panel a similar experiment was performed with the mutated response element, and there are no complexes formed showing that complex formation was specific. We have done this type of experiment with whole cell extracts from MCF-7 cells as well. In the case of rat uterine cytosols we do go through a heating step to see the DNA binding which is the first step you described, but we have not seen the hormone dependence. You discussed at great length why you think you now see a hormone dependence. Why do you think you did not see this before?

B. O'Malley. I would say that we do not have a complete answer to this as yet. When we purify progesterone receptor, it binds DNA well and is active in transcription. It is not as active on a molar basis as is the more crude receptor form partially purified from T-47 D cells. It is our best guess at this time that we are activating the molecule with purification. Through purification we are removing heat shock proteins, but we may also be allosterically activating the receptor or permitting its enzymatic modification.

J. Baxter. Under the conditions in which you demonstrated hormone-dependent transcriptional stimulation, did you also have steroid-dependent receptor binding to DNA?

B. O'Malley. In our system, we retain clear hormone-dependent binding of progesterone to DNA also.

J. Funder. My question relates to the orphan receptors. If you do mutations in the hormone binding domain and they are inactive, do you also affect the internal economy of the receptors in terms of dimerization or the binding of other proteins?

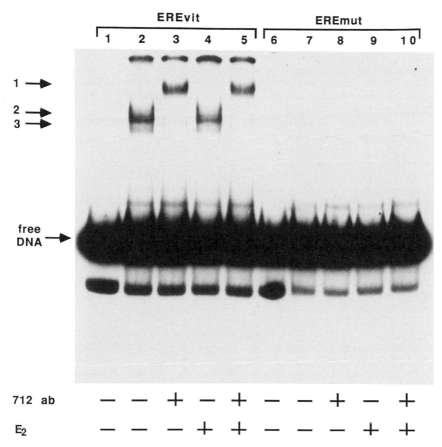

FIG. 11. Effect of estradiol on gel mobility shift assay of uterine cytosol with restriction fragments containing the *Xenopus* vitellogenin A2 gene estrogen response element (EREvit) or a 2-bp mutant (EREmut). Cytosol from immature rat uteri was heated in the presence of 10 n*M* 17β-estradiol or ethanol as indicated. Samples containing 0.4 n*M* estrogen receptor were incubated with 3 n*M* of either the ^{32}P-labeled EREvit or the ^{32}P-labeled EREmut for 19.5 hours at 4°C. Lanes 3,5,8, and 10 had antisera 712 against the estrogen receptor included during this incubation. Lanes 1 and 6 are controls without cytosol. Free and protein-bound DNA was separated by nondenaturing 4% polyacrylamide gel electrophoresis, and DNA is visualized by autoradiography. From Murdoch, F. E., Meier, D. A., Furlow, J. D., Grunwald, K. A. A., and Gorski, J. (1990). Estrogen receptor binding to a DNA response element *in vitro* is not dependent upon estradiol. *Biochemistry* **29**(36), 8377–8385.

R. O'Malley. We have not explored intracellular localization or dimerization with the orphan receptors. Since these appear to be activated in cells, I would expect that they would be in the nucleus. I would also expect that for the most part they would be free of heat shock proteins.

J. Funder. Has anybody formally studied ligands such as butyrate, which has extraordinary effects on transcription and cell replication?

B. O'Malley.   Give me your list of compounds to test and we will test them. We have about a million on our list at present!

W. F. Crowley, Jr.   What you discussed about these "orphan" receptors in terms of their being an "intracrine" system is an interesting speculation. It strikes me that there is a great deal of parallelism between this thinking and the oncogenic concept for polypeptide proteins. These receptors might be the steroid equivalent of oncogenes. Have you had the opportunity to study any hormone-dependent tumors for overexpression of any of these orphan receptors?

B. O'Malley.   Minor overexpression of COUP-TF has been noted in some breast tumors.

J. Baxter.   How far can you move the PRE or GRE away from the TATA box in your target gene constructs for cell-free transcription?

B. O'Malley.   It depends on the construct. In the case of a minimal promoter (TATA), we cannot move the PRE elements very far. If the construct also contains an upstream promoter (e.g., COUP, NF1, etc.), we can move the PRE elements some hundreds of bases upstream and still maintain steroid responsiveness. In this case, it is relatively position independent.

J. Baxter.   Can you place one PRE adjacent to the TATA box and move the other one farther away and still obtain receptor-dependent stimulation of transcription?

B. O'Malley.   We have not split the PRE elements to date.

J. Baxter.   You implied that the receptor stimulates the binding of the TATA box factors. Do you have any data that would differentiate between receptor-mediated stimulation of the binding of factors and activation of factors that were already bound?

B. O'Malley.   I think the competition studies would distinguish stabilization from simple activation. It is difficult, however, to separate recruitment and stabilization of factors. If we study the kinetics of this reaction, our best guess is that the predominant effect is via stabilization rather than recruitment of transcription factors.

W. Bardin.   You hypothesized that the steroid ligand either dissociates heat shock proteins or modifies the receptor so that it becomes an active complex. You favored the latter. Have you excluded that the ligand does both?

B. O'Malley.   No, my message was that the ligand does do both. Let me restate my conclusion. In the cell, the ligand induces both dissociation of heat shock proteins and an allosteric change in the molecule, which may be accompanied by covalent modification (e.g., phosphorylation). In other words, there is clearly life after heat shock dissociation.

Neuroendocrine Control of Human Reproduction in the Male

William F. Crowley, Jr.,* Randall W. Whitcomb,* J. Larry Jameson,† Jeffrey Weiss,‡ Joel S. Finkelstein,‡ and Louis St. L. O'Dea§

*Reproductive Endocrine and †Thyroid Units, ‡Massachusetts General Hospital, Boston, Massachusetts 02114 and §Division of Endocrinology and Metabolism, Montreal General Hospital, Montreal, Quebec, Canada

I. Introduction

The fundamental obstacles to the study of the physiology of gonadotropin-releasing hormone (GnRH) in the human still exist. The rapid metabolism of GnRH continues to make its direct measurement in the peripheral circulation limited in its utility in defining the neuroendocrine control of reproduction in the human (Handelsman et al., 1984; Spratt et al., 1985). The inaccessibility of the hypophyseal–portal blood supply to direct sampling also means that a series of indirect approaches must be undertaken with cross-roughing of several strategies being required to piece together a complete story of the hypothalamic control of gonadotropin function. Thus, we have continued to use complementary approaches involving the tandem study of GnRH-deficient men as well as normal men with intact hypothalamic pituitary axes as outlined (cf. Fig. 1) in a previous publication (Crowley et al., 1985).

What has changed considerably in the past few years has been the validity of using pulsatile lutinizing hormone (LH) secretion as a surrogate for endogenous GnRH secretion in the human. Its role as an alternative marker of hypothalamic secretion has been dramatically improved by two advances. The first is the widespread agreement that increasing the intensity of sampling of peripheral blood for LH levels to 10-minute intervals has harvested considerably more precise information about patterns of LH release, permitted a sharper discernment of each individual LH pulse, and consequently allowed more precise estimates of LH and consequently GnRH pulse frequency (Filicori et al., 1987). Second, a growing body of experimental evidence in several species has now confirmed the concordancy of GnRH secretion from the hypothalamus with subsequent bursts of pulsatile LH release from the anterior pituitary (Levine et al., 1982; Caraty et al., 1982; Clarke et al., 1984; Karsch et al., 1984). Taken together, both of these advances have yielded considerably more information as well as enhanced the certitude of this information.

27

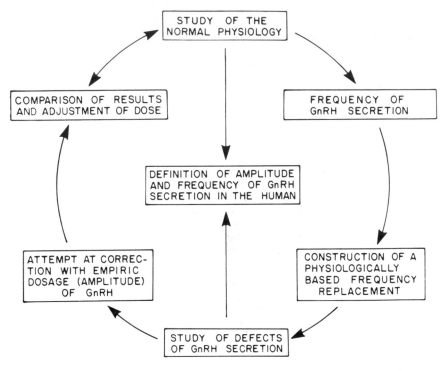

FIG. 1.   A schematic diagram of a physiological approach to the study of GnRH in the human. A normative database is generated from which the frequency of GnRH secretion is inferred. This frequency program is then applied to individuals with defects of endogenous GnRH secretion, utilizing an empiric but individualized dose of GnRH to recreate normal LH secretory dynamics including mean level, amplitude, and frequency. Comparison of the results of the exogenous program with endogenous patterns then allows the investigator to refine the optimal replacement program and make inferences regarding endogenous GnRH secretion (Crowley et al., 1985).

II.   Models for the Study of GnRH Physiology

Each of the various models chosen for the study of GnRH physiology has its unique strength; however, this strength is invariably combined with a limitation such that interpretations of GnRH physiology on the basis of any single model is flawed in some fundamental way. It is thus critical to understand the unique contribution each model can bring to the study of GnRH physiology while bearing in mind its limitation, for it is that very limitation that usually requires the tandem study of a complementary model.

A.   NORMAL MALES

The first model is that of the normal male (cf. Figs. 2 and 3). Its advantages are numerous, including the relative abundance of subjects and their ease of study (given the enormous limitations inherent in any study of humans). They also permit a visualization of the entire hypothalamic–pituitary–gonadal axis, allowing the investigator to monitor the free-running rhythms of the fully integrated neuroendocrine system, at least over the short periods of time involved in 24–48 hours of study. Such studies also afford an understanding of the interindividual variability inherent in the normal male reproductive system. Thus, when studied in sufficient numbers to overcome this variance, examination of normal men permits assembly of a robust normative database. This precision of normative information can then be used to establish statistical limits of confidence that can be turned to the ever-refined definition of clinical abnormalities of the reproductive system.

The study of the normal male possesses certain limitations. The variance inherent in this model demands the study of large numbers of patients. In addition, since their hypothalamic–pituitary axes are fully integrated, compensatory adaptations can occur in response to any experimental manipulation that might be attempted of this system. For example, when the endogenous steroid hormone levels are altered experimentally, the resulting changes in the hypothalamic–pituitary axis are ambiguous as to the relative contributions of the hypothalamic or pituitary components. Consequently, the normal male, when studied alone, does not permit a dissection of the hypothalamic from the pituitary components

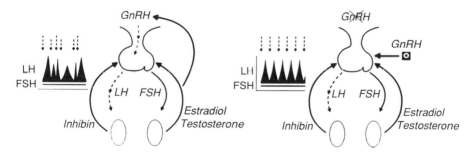

FIG. 2.   Hypothalamic–pituitary–gonadal axes in normal (left) and GnRH-deficient (right) men. The free-running nature of the normal male GnRH secretory system results in GnRH-induced LH pulses of differing amplitudes and frequencies with evidence of feedback of gonadal steroids and/or inhibin potentially occurring at both the hypothalamus and/or the pituitary. In contrast, GnRH-deficient men receiving an experimentally definable regimen of exogenous GnRH have gonadotropin pulse amplitudes and frequencies that can be fixed. Hence they are capable of responding to gonadal feedback only at the level of the pituitary.

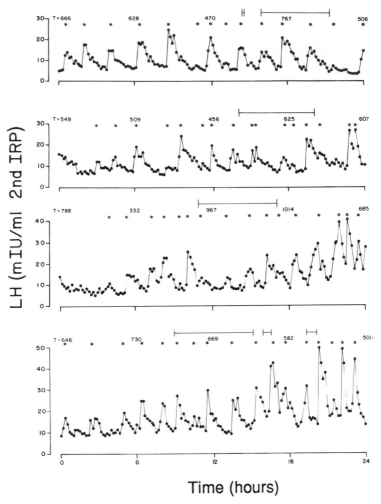

FIG. 3. LH secretory patterns of four normal men sampled at 10-minute intervals for 24 hours. Identified pulses are marked with an asterisk. Note the variability in pulse amplitudes within individuals. T, Serum testosterone levels (in ng/dl); IRP, International Reference Preparation. Sleep periods are bracketed at the top of each panel.

of the reproductive system and thus requires the parallel study of a complementary model.

B. GnRH-DEFICIENT MALES

GnRH-deficient men represent a unique "experiment of nature" that in many ways is invaluable to study in parallel with the intact male (cf. Figs. 2 and 4).

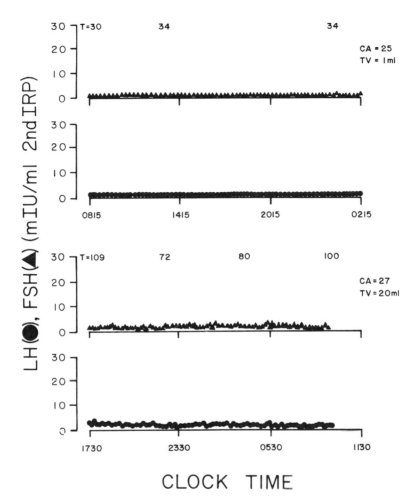

FIG. 4.   Serum LH (●) and FSH (▲) concentrations determined at 10-minute intervals in two representative men with complete GnRH deficiency. The patient shown in the upper panel had testicular volumes (TV) of 1 m, anosmia, and no history of spontaneous puberty. The patient shown in the lower panel had testicular volumes of 20 m, a normal sense of smell, and history of partial puberty with subsequent regression of sexual function. CA, Chronological age; T, serum testosterone levels (in ng/dl). From Spratt *et al.* (1987a).

Since these patients completely lack functional secretion of GnRH (Spratt *et al.*, 1987a), their reproductive competency can be totally restored with the proper regimen of pulsatile GnRH (Valk *et al.*, 1980; Hoffman and Crowley, 1982; Whitcomb and Crowley, 1990). When competency of their pituitary–gonadal axes has been restored to normal with a physiologic program of exogenous

GnRH administration, these men permit a rather clear distinction in the responsiveness of the hypothalamic from the pituitary component of this system. In the face of a physiologic GnRH regimen that can be experimentally "clamped," all modifications of their gonadotropin responses to any physiologic manipulation must occur at the level of the anterior pituitary. Thus, when any change in the system is observed in normal men in response to any experimental pertubation (for example, a change in gonadotropin amplitude and/or frequency), GnRH-deficient subjects receiving a fixed regimen of exogenous GnRH permit direct inferences to be made about the hypothalamic contribution to these adaptations when viewed in parallel with normal men (cf. Fig. 2).

Several significant limitations of this human model also exist. It is a rare disorder. Estimates of the disease frequency range from between 1 in 100,000 to 1 in 1,000,000. Additionally, great care mut be taken to define the "physiologic nature" of the exogenous GnRH regimen used in their replacement in terms of GnRH dose, frequency, variability of interpulse interval, etc. Finally, longitudinal studies are required since the administration of exogenous GnRH to these patients for the first 2–3 months elicits priming of the pituitary responsiveness and hence a changing pattern of pituitary responsiveness during this period (Hoffman and Crowley, 1982). Thus, changes of the pituitary responsiveness during this developmental window render interpretation of experimental manipulations problematic. Once the competency of their pituitary–gonadal axis has been restored, their period of GnRH administration has emerged from this developmental window, and their pituitary–gonadal axes have become acclimated to long-term exogenous GnRH replacement, these patients serve as excellent models for the study of the physiology of GnRH in the human.

C.   DISPERSED PITUICYTE PERIFUSION SYSTEMS

Both of the previous models, however, share the combined limitation of examining GnRH–pituitary interactions in the presence of an intact gonad and hence limit studies to the secretory level. To overcome these obstacles, *in vitro* pituitary cells isolated from experimental animals can be perifused with physiologic regimens of GnRH (Loughlin *et al.*, 1981; Badger *et al.*, 1983; Weiss *et al.*, 1990a,b). Once these pituicytes have been isolated, they were dispersed into single cells and attached to various support matrices. These cells retain both their ability to synthesize and secrete gonadotropins in response to varying regimens of exogenous GnRH. This system also has the advantage of being widely applicable to various species. Thus, their study, free from the influences of gonadal feedback, can be useful when combined with the above models.

There are a few additional advantages of these perifused pituicyte systems that have not been previously stressed yet are particularly attractive. The first is that, given that the "metabolic clearance" of gonadotropins in these systems is equal to the perifusion rate of the columns, this model permits the direct study of secretory events free from the confounding influence of metabolic clearance. For example, the "clearance" of LH, follicle-stimulating hormone (FSH), and free α subunit in this *in vitro* system are equivalent, as opposed to the *in vivo* circumstances, wherein they vary by more than a log order from α subunit to FSH. In addition, reagents for assay of these glycoproteins, at least in the rat, are considerably purer than those employed in the human. Rat LH and free α subunit standards are greater than 95% pure (Parlow, personal communication, 1989) whereas those for FSH are at least 50% pure. Therefore, an initial attempt at the relative quantification of the secretion of these three glycoproteins can be made in this *in vitro* system—a feature not possible given the current impurity of human gonadotropin standards. Consequently, examining the relative molar amounts of LH vs free α subunit vs FSH released in response to varying doses and frequencies of GnRH administration is uniquely possible in this system (Weiss *et al.*, 1990b). Finally, and most importantly in bringing the advances of molecular biology to bear on the study of GnRH control of gonadotropin bio-synthesis, mRNA levels can be recovered in these perifused systems to examine both the differential control by GnRH of the biosynthesis of each of the anterior pituitary glycoproteins as well as the relationship between their synthesis and secretion (Weiss *et al.*, 1990a,b). As the technologies evolve in this area, tran-scriptional regulation of anterior pituitary glycoproteins can also be approached using these dispersed animal pituicytes (Weiss *et al.*, 1990a).

The disadvantages of these *in vitro* systems, however, are that they are highly species specific, critically dependent on prior development of the appropriate assay systems with the requisite sensitivity, and somewhat influenced by the sex and physiologic status of the animal from which these cells are harvested. More-over, while gonadal peptides and protein hormones have proven ready models for study in the perifused pituitary cell system, to date, sex steroid feedback has been more limited in its demonstrability in this system. Dispersed pituicytes are also useful for only short periods of time since they have a tendency to become infected or to undergo necrosis. This latter feature limits the utility of these systems from hours to days. Finally, this model also tends to be expensive, depending on the animal species and physiologic state under investigation.

Nonetheless, when viewed in the aggregate, normal men, GnRH-deficient patients receiving experimentally definable regimens of exogenous GnRH, and the *in vitro* perifused pituitary cell system each provide unique and complemen-tary information regarding the physiology of GnRH. Each is of critical impor-tance in assembling a piece of the picture; only the combined study of all of these models reaps the greatest harvest of physiologic information.

III. The Role of the Bolus Dose of GnRH in Gonadotrope Secretion

Even the most casual review of the 24-hour pattern of pulsatile LH release in the normal male as outlined in Fig. 3 reveals a striking variability of the LH pulse amplitude over a 24-hour period. The question thus arises, is this variation in LH pulse amplitude due to (1) varying bolus doses of endogenous GnRH secretion, (2) varying intervals of pituitary stimulation by a fixed dose of GnRH, (3) short-term feedback effects of gonadal steroids and/or proteins at the anterior pituitary, or (4) a combination of these factors?

Of course, it is not possible to examine this issue farther in an intact male other than to document its presence. Therefore, this problem can be approached using both GnRH-deficient men as well as the *in vitro* perifused rat pituitary system. Both these systems share the advantage of being able to control the dose and frequency of GnRH stimulation experimentally. *In vivo,* this frequency of GnRH stimulation was set to the mean frequency of that documented to occur in normal males of approximately 2 hours, a frequency deduced from sampling a large series of normal adult males at 10-minute intervals for 24 hours (Spratt *et al.,* 1988). *In vitro,* this frequency was set at once hourly, a frequency documented to occur *in vivo* in the rat (Ellis and Desjardins, 1981). Prior to exogenous GnRH administration, all GnRH-deficient men exhibited a complete absence of endogenous GnRH activity as attested to by their apulsatile LH secretion during baseline studies (cf. Fig. 4). Each man subsequently received exogenous GnRH administered at 2-hour intervals until normalization of their pituitary-gonadal axis had been documented for at least 3 months. Subsequently, they were admitted to the General Clinic Research Center of the Massachusetts General Hospital, where the frequency of their GnRH regimen was held constant at 2-hour intervals, their gonadal steroid levels were determined to be within the physiologic range for normal adult males, and the intravenously administered doses of GnRH were varied to span two log orders (2.5–250 ng/kg/bolus). As can be seen in Fig. 5A and B, a remarkably linear relationship exists between the LH amplitude (or area under the curve) and the log of the GnRH dose. This dose–response relationship is 10-fold higher for LH than for FSH (cf. Fig. 6) and exists for both biologically and immunologically active LH (cf. Fig. 7). When the biologic activity of the LH is measured in the dispersed rat Leydig cell assay and regressed on the immunoreactive LH dose, a predictable linear relationship can be demonstrated between these two variables (cf. Fig. 7). However, somewhat surprisingly, the

---

→

FIG. 5.   (A) Serum LH and FSH responses to varying doses but a fixed frequency (q2 h) of GnRH in a man with GnRH deficiency. Note the progressive increases in amplitude and area of the LH pulses in response to intravenous GnRH doses of 2.5–75 ng/kg in the presence of a physiologic level of testosterone. The LH response to 250 ng/kg GnRH demonstrates a further increase in area under the curve, but no increase in pulse amplitude. T, Serum testosterone levels (in ng/dl). (B) Log-linear transformation of LH data from (A). Note the linear relationship between the log of the GnRH dose and the resultant LH pulse amplitude. The shaded area indicates the mean ± 2 SD of LH pulse amplitudes from 30 normal men sampled at 10-minute intervals for 24 hours. Using these data, it is possible to select and individualize a dose of GnRH that will yield an LH pulse amplitude within the normal range for further physiological studies (Spratt *et al.,* 1986).

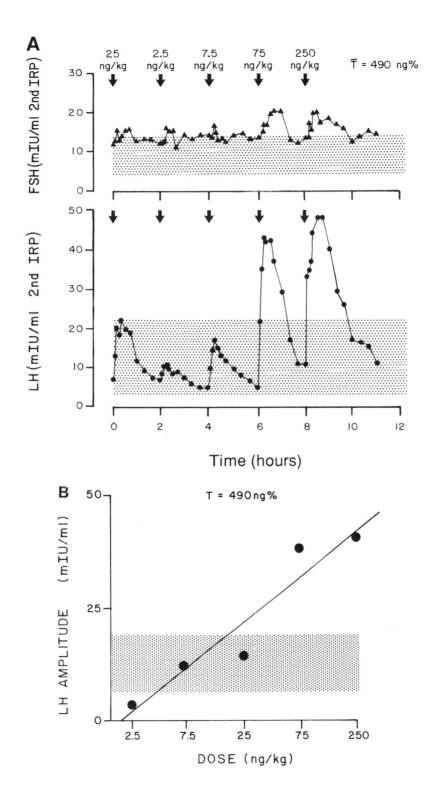

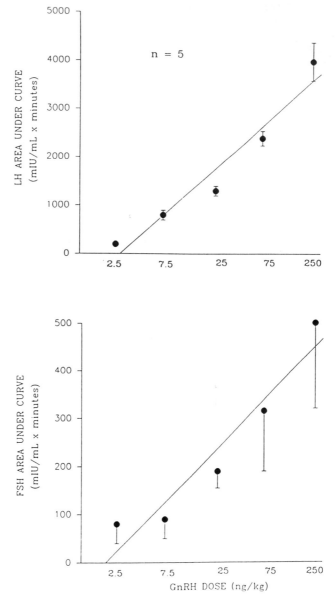

FIG. 6. Mean (± SEM) immunoactive dose–response curves for LH (top) and FSH (bottom) constructed from the area under the secretory curve of pulses after varying doses of GnRH in five men with GnRH deficiency. Note that this relationship is 10-fold higher for LH than for FSH in these men with intact gonads.

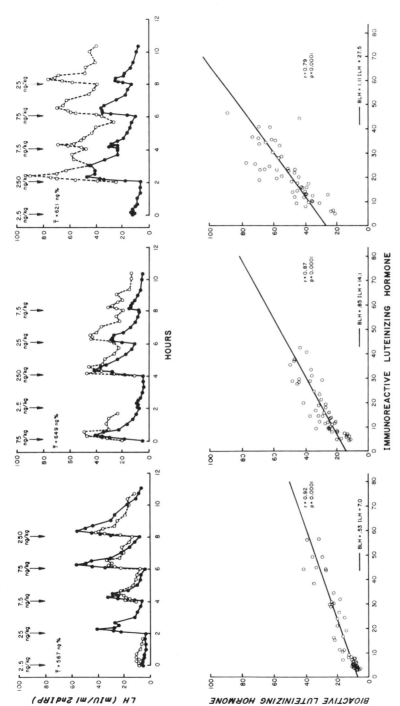

FIG 7. *Top:* Bioactive (○) and immunoactive (●) LH responses to varying doses of intravenous GnRH in three men with GnRH deficiency. Note the variation in relationships of bio- to immunoactivity of LH between the patients. T, Serum testosterone levels (in ng/dl). *Bottom:* Each bioactive LH value (BLH) is graphed on the y axis against the corresponding immunoactive values (ILH) on the x axis. Note the variation in both slopes and y intercepts between patients. Also note that the y intercept is positive in all patients, i.e., by extrapolation of all detectable values, there is bioactive LH-like activity when immunoactive LH becomes undetectable (Spratt *et al.*, 1986).

slope of this relationship of biologic to immunologic LH is unique to each patient (Fig. 7) and varies remarkably from patient to patient. In some subjects there was approximate equality of both bio- and immunoactive LH, whereas in others there appeared to be a preferential enrichment in either biologically active LH or immunologically active LH. Another unexpected finding of these dose–response relationships was that there was always a positive intercept to the bio vs immunoreactive LH curves (cf. Fig. 7) such that even when the immunoactive LH was undetectable, there was always a discernable degree of bioactive LH present in the rat Leydig cell assays. Once again, this "blank" in the LH bioassay varied from individual to individual but was uniformly present (Spratt *et al.*, 1986).

Examining this same issue in the absence of the gonad via the dispersed pituicyte system, it was also quite clear that a dose–response relationship existed *in vitro* between LH, FSH, and free α subunit (FAS) vs the log of the GnRH dose quite similar to that documented *in vivo* in the presence of gonadal feedback (cf. Fig. 8). Moreover, as the "metabolic clearance" (i.e., the perifusion rate as all three glycoproteins are identical in this system), the dose–response relationships for each glycoprotein were found to plateau (unlike the *in vivo* circumstance) and true ED_{50} doses for LH, FSH, and free α subunit could be calculated that were all in the nanomolar range and within close proximity to each other (cf. Fig. 8) (Weiss *et al.*, 1990b). However, when the absolute quantity of each glycoprotein secreted was compared using the relatively pure rat standards (acknowledging that the FSH was the least pure standard and therefore set to unity in this system), it appeared that the relative quantities of anterior pituitary glycoprotein secreted following each dose of GnRH were not equal. As can be seen in Fig. 8, the molar ratio of LH:FAS:FSH was 3.0:1.7:1.0. Recognizing that the FSH standards might be only 50% pure, these data suggest that there is an excess of LH and FAS over FSH secreted after each pulse of GnRH from the rat gonadotroph. This quantitative diversity of gonadotrope response is most interesting and suggests further differences in second messenger, transcriptional, or posttranscriptional events for the three anterior pituitary glycoproteins.

Taken together, these studies confirm a quantitative relationship between the amount of GnRH administered exogenously (or presumably secreted endogenously) and all three of the ensuing gonadotrope glycoprotein secretory responses. Moreover, these dose–response relationships can now be quantitated in relation to the data obtained from extensive sampling of our normal population (cf. Fig. 5), such that an individual dose of exogenous GnRH can be selected in each GnRH-deficient subject that will predictably evoke an LH secretory response the amplitude of which is within the midnormal range, when applied at a physiologic frequency and in a physiologic gonadal hormonal milieu. This ability to select an ED_{50} or physiologic dose of GnRH (i.e., that dose capable of producing a midphysiologic amplitude of an LH pulse), is critically important in the subsequent examination of the role of GnRH frequency, the effects of gonadal

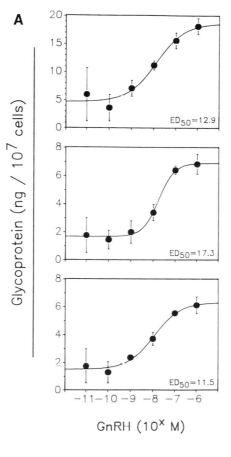

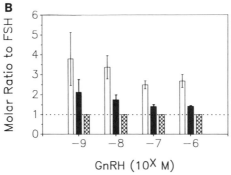

FIG. 8.   (A) Dose–response relationship between GnRH and LH (top), FSH (middle), and free α subunit (FAS; bottom) in the *in vitro* dispersed rat pituicyte system. (B) The molar ratio of LH (open bars) and FAS (solid bars) to that of FSH (hatched bars) for all of the doses (Weiss *et al.*, 1990b).

steroids, and variations in the interpulse interval. However, this selection of GnRH dosage depends on the blending of both normative data with experiments conducted in GnRH-deficient men. The tandem study of these two human models thus makes it possible to select and individualize a dose of GnRH with some degree of certitude that will reconstruct the normal dynamics of hypothalamic–pituitary interaction in each GnRH-deficient man.

IV.   The Role of Frequency of GnRH Stimulation in Gonadotrope Secretion

Although there is a clear-cut linear dose–response relationship between the dose of exogenous GnRH and the subsequent pituitary response, this information is not sufficient to answer the question first posited regarding the LH amplitudes that occur in the normal male, i.e., whether all their variability is attributable to individual differences of endogenous GnRH doses secreted by the hypothalamus. Merely because a similar variability of LH amplitude could be produced by varying the dose of GnRH does not *a priori* prove that all variations in the LH amplitudes observed in normal men represent changing endogenous doses of GnRH secretion from the hypothalamus. Thus, the next series of experiments systematically examined the role of increasing and decreasing the frequency of a fixed and "physiologic" bolus of GnRH input to the gonadotrope. Having defined the dose of GnRH that produces a midphysiologic response of LH in subjects with a normal gonadal hormonal milieu, this dose can be held constant during serial manipulations of the GnRH frequency. These studies were designed to examine the alternative hypothesis that variations in frequency of stimulation by a fixed (and "physiologic") dose of GnRH could also result in varied pituitary responsiveness per se. This is a particularly important hypothesis to examine since the hypothalamus is known to vary the frequency of endogenous GnRH secretion and hence gonadotroph stimulation widely across puberty (Boyar *et al.*, 1972; Jakacki *et al.*, 1982), the menstrual cycle (Filicori *et al.*, 1986), and after removal of gonadal influences (Yen *et al.*, 1972).

A.   INCREASING FREQUENCY OF GnRH STIMULATION

An individualized GnRH dosage was administered intravenously at the mean normal adult male frequency of every 2 hours to a series of GnRH-deficient patients (Spratt *et al.*, 1987b). After a 12-hour period of baseline monitoring, the frequency of this fixed GnRH dose was progressively increased at weekly intervals from 120- to 60- to 30- to 15-minute frequencies. The patients were admitted every 7 days for an equivalent period of monitoring of these intravenous GnRH doses. The increase to the next frequency was then made in the middle of each admission. As shown in representative subjects on Fig. 9, a progressive rise

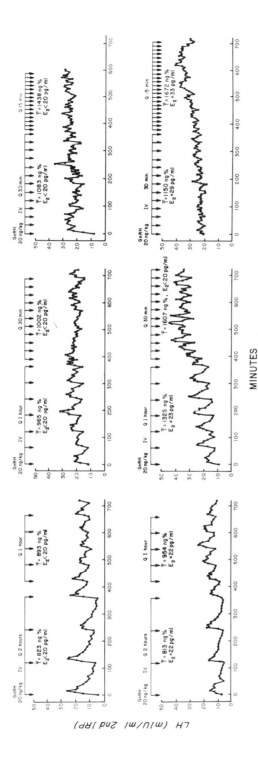

FIG. 9. Serum LH concentrations determined at 5- and 10-minute intervals during three sequential admissions of a "midphysiologic dose" of intravenous GnRH to two GnRH-deficient subjects as the GnRH frequencies were progressively increased in the middle of each 12-hour study from 2-hourly to hourly to half hourly. Testosterone concentrations (T) were determined in serum pools formed from samples from each 6-hour period and expressed in nanograms per deciliter. Note that at higher frequencies, GnRH injections were not uniformly followed by LH pulses (Spratt *et al.*, 1987b).

of the mean LH level occurred with increasing frequencies of GnRH stimulation whereas FSH levels did not change significantly (Spratt *et al.*, 1987b). This increase in mean LH levels was, however, accompanied by a decreasing amplitude of individual LH responses at each level of increased frequency, a phenomenon that is considerably more apparent for FSH than LH, since the mean FSH level did not increase in spite of the progressive shortening of the interpulse interval of GnRH stimulation (cf. Fig. 10). This failure of LH, and especially FSH, to rise in parallel with the expectations that had been established by the earlier dose–response relationships represents the earliest example of "desensitization" of pituitary gonadotropin secretion induced by frequency changes. This term signifies the shortfall in pituitary gonadotropin responsiveness from that which would be expected accompanying increasing GnRH stimulation. Moreover, this earliest form of desensitization associated with increasing the frequency of GnRH stimulation appears to be an unstable process, as can be seen in Fig. 11, wherein both striking rises and falls in gonadotrope response to fixed doses and intervals of frequent GnRH stimulation occur as the mean frequency is increased. The resemblance of these rises and falls to those occurring during both the onset and termination of the LH surge in the normal menstrual cycle is quite striking. This observation then raises the question of whether this frequency-

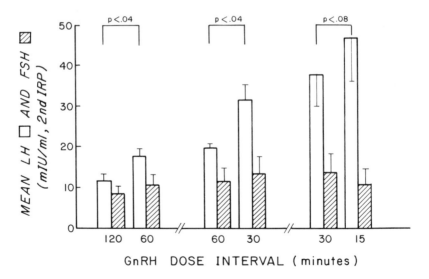

FIG. 10.   Mean (± SEM) serum LH and FSH concentrations during the three weekly 12-hour admissions described in Fig. 9 while the frequency of GnRH administration of a physiologic dose of GnRH was progressively increased to five GnRH-deficient men as shown in Figs. 9 and 11. Mean LH concentrations increased progressively as GnRH frequency increased, whereas no significant change in FSH levels occurred. Thus, LH was preferentially secreted at higher frequencies of GnRH whereas FSH secretion is preferentially desensitized by increasing GnRH frequency (Spratt *et al.*, 1987b).

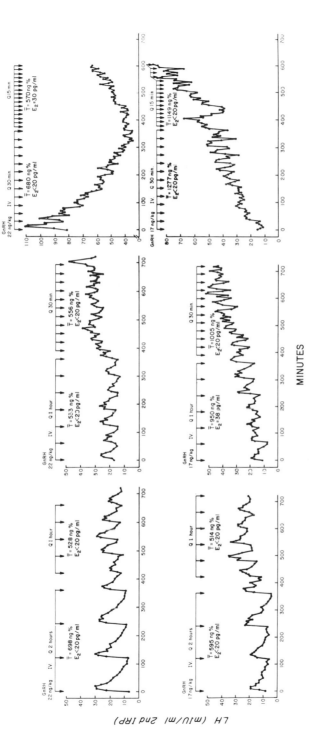

FIG. 11. Serial serum LH concentrations determined in two additional GnRH-deficient subjects as described in Fig. 9 as GnRH frequency of a mid-physiologic dose was progressively increased from q 2 h to q 30 minutes. The instability of LH responses to these fixed and midphysiologic doses of GnRH administered at q 30 minute frequencies seen in the panels at the far right demonstrate the unstable nature of the desensitization process. The rise in LH levels in response to a fixed dose and rapid frequency (30 minute) GnRH frequency shown in the right panels is strikingly similar to that which occurs in normal women at the midcycle LH surge (Spratt et al., 1987b).

mediated form of desensitization of gonadotropin secretion might play a role in these other physiologic states. Additionally, this progressive rise in mean LH levels with a fixed FSH level results in a high LH:FSH ratio associated with these rapid GnRH pulse frequencies (cf. Fig. 10). This high ratio of LH to FSH is quite analogous to those gonadotropin ratios that have been documented to occur in the clinical entity of polycystic ovarian disease (Rebar *et al.*, 1976). Subsequently, other studies from our group have now demonstrated that, in fact, patients with polycystic ovarian disease do exhibit a similar high frequency and relatively fixed pattern of GnRH secretion over time (Waldstreicher *et al.*, 1988).

B. DECREASING FREQUENCY OF GnRH STIMULATION

Having examined the ability of increasing frequencies of fixed GnRH doses to alter the quantity and ratios of gonadotropins secreted, the effects of decreasing the frequency of the ED_{50} or physiologic dosage of GnRH progressively were determined. As can be seen in Fig. 12, mild decreases in the frequency of GnRH stimulation from 2 to 3 to 4 hours, i.e., in line with those intervals that can occur in normal adult males (cf. Fig. 3) (Spratt *et al.*, 1988), are associated with a

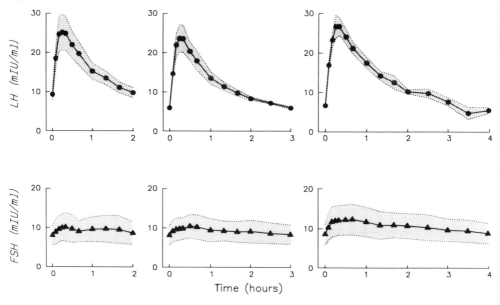

FIG. 12.   Mean (± SEM) serum LH and FSH concentrations of four GnRH-induced pulses in four men with GnRH deficiency when the frequency of intravenous GnRH administration was progressively decreased at weekly intervals from 2 hours (left) to 3 hours (middle) to 4 hours (right) (Finkelstein *et al.*, 1988).

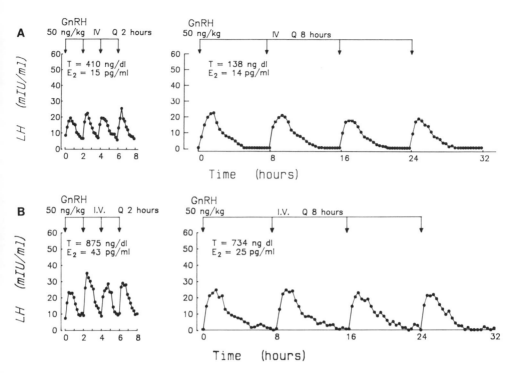

FIG. 13. Serum LH concentrations determined at frequent intervals in two men with isolated GnRH deficiency as the frequency of GnRH administration was decreased from every 2 to every 8 hours at weekly intervals without testosterone, T, replacement (A) or with T replacement (B). Note that the increase in pituitary responsiveness to the slower frequencies of GnRH administration is independent of changes in ambient gonadal steroid levels (Finkelstein *et al.*, 1988).

relatively fixed peak of the LH pulse, but progressive increases in LH amplitudes and area under the secretory curve are associated with a more prolonged interval between pulses of GnRH stimulation. These longer intervals permit mean gonadotropin levels to fall to a lower baseline level by allowing increased time for metabolic clearance to occur (Finkelstein *et al.*, 1988). This phenomenon is especially dramatic as the frequency of GnRH stimulation is decreased from 2 to 8 hours, at which point the interval between the episodes of GnRH stimulation is so long that the pretreatment apulsatile baseline state of the patient's GnRH deficiency can be observed (cf. Fig. 13). When groups of patients with GnRH deficiency are pooled during decreases of GnRH stimulation from 2 to 8 hours, it is clear once again that the LH amplitude and area under the secretory curve increase even more strikingly, primarily due to a further fall in the baseline before each pulse. Once again, however, there appears to be a relative preservation of the peak of each LH pulse in response to the fixed bolus of GnRH stimulation (Finkelstein *et al.*, 1988).

However, as the pulse frequencies are reduced to 8 hours a supervening complication arises in that the serum testosterone levels fall due to the long absence of LH stimulation from the Leydig cells (cf. Fig. 13). While there had been milder falls of the mean testosterone levels as GnRH frequency was reduced from 2 to 3 to 4 hours, these became more dramatic and achieved statistical significance only at 8-hour frequencies. Consequently, the increased amplitude and area under the LH secretory curve observed during these declining GnRH frequencies might well have been due to the decreasing negative feedback of testosterone on the pituitary. To control for this variable, a companion series of patients received exogenous testosterone as their GnRH frequency was decreased to 8 hours, as indicated in Fig. 14, and showed identical changes in gonadotropin secretory dynamics. This finding indicates that the increasing responsiveness of LH secretion observed during decreasing frequencies of GnRH stimulation is unrelated to any changes in ambient serum testosterone levels. Alternatively stated, the frequency of GnRH stimulation, when isolated from the impact of its dosage and ambient gonadal steroid levels, encodes information regarding the

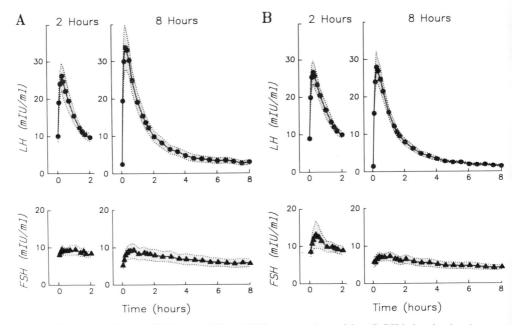

FIG. 14.   Mean (± SEM) serum LH and FSH concentrations of four GnRH-induced pulses in four GnRH-deficient men when GnRH was administered every 2 or 8 hours (A) without T replacement or (B) with T replacement (Finkelstein *et al.*, 1988).

quantity and pattern of gonadotropins to be secreted from the gonadotropin. Somewhat in contrast to previous findings (Gross *et al.*, 1987) the serum FSH levels failed to demonstrate any changes in these subjects with intact gonads over a wide range of increasing and decreasing GnRH frequencies. These results indicate that the overriding control of circulating FSH levels in the adult male may well not be GnRH dosage, frequency, or ambient sex steroid levels. Rather, gonadal peptides such as inhibin and activin may be the dominant influence on FSH.

Finally, to examine whether these relationships between the frequency of GnRH stimulation and the ensuing pituitary responses occur in the absence of gonadal sex steroids, the perifusion model in which all gonadal peptide and steroid influences are absent was examined and demonstrated once again that the frequency of GnRH stimulation per se is a variable independent from that of GnRH dosage on the quantity of gonadotropin secretion. Of interest in the

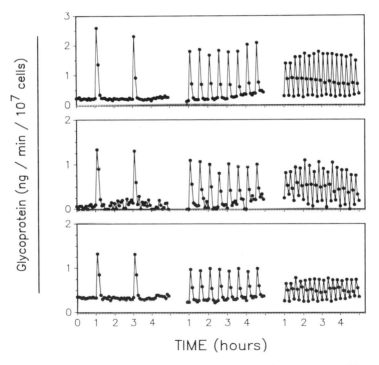

FIG. 15.   The effects of varying the frequency of GnRH stimulation on LH (top), FSH (middle), and FAS (bottom) secretion in perifused male rate pituitary cells *in vitro*. In the absence of gonadal sex steroids, effects of decreasing GnRH frequency on pituitary responsiveness are apparently similar to those seen *in vivo*.

perifusion system is that the previously noted ratio of LH:FAS:FSH secretion during the dose–response studies was maintained over a wide range of frequencies as well (cf. Fig. 15).

V. The Role of Variable Interpulse Interval in Determining Gonadotrope Function

Having addressed the dose responsiveness of this system and systematically examined the impact of altering the GnRH frequency, we then sought to determine the role of the variability of the interpulse interval around this mean frequency of GnRH stimulation. The spectrum of amplitudes and interpulse intervals of LH pulses observed in normal men was first established and this permitted construction of frequency histograms of both of these variables (cf. Fig. 16) (Butler *et al.*, 1986; O'Dea *et al.*, 1989). The variability in LH amplitudes and intervals encountered in subjects with intact hypothalamic–pituitary–gonadal axes is readily apparent. Thus, the role of the variable interpulse interval of GnRH stimulation about the mean adult male frequency in determining the LH amplitudes could now be examined. The individually determined ED_{50} physiologic dose of GnRH was therefore administered intravenously to GnRH-deficient men at a series of interpulse intervals in a random order selected by a computer program that modeled the administered intervals to those which occurred in normal subjects (cf. Fig. 16). If the spectrum of LH amplitudes encountered in normals could be reconstituted in our GnRH-deficient patients by merely varying the interpulse interval of a fixed GnRH dose, then this finding would indicate that all of the variability of LH amplitudes in normal men must be related solely to the variability of their interpulse intervals of GnRH stimulation. On the other hand, should the variable interpulse interval of these fixed GnRH doses fail to reconstruct the spectrum of normal amplitudes, then any residual variability of LH amplitude must have its origin in something other than varying the interpulse intervals of fixed doses of GnRH stimulation, such as varying the endogenous dose of GnRH secreted in the intact men.

As can be seen in Fig. 16, our computer program quite faithfully reconstructed the spectrum of interpulse intervals of normal men. It also successfully recapitulated the changes in LH amplitude, most of which occur following long interpulse intervals (cf. Fig. 16), by showing that the high-amplitude LH pulses (which frequently follow long interpulse intervals) were faithfully reconstructed in the GnRH-deficient men administered a fixed dose of GnRH. Of great interest, however, was the fact that the high percentage of small-amplitude LH pulses characteristic of the normal male were not reconstructed by the shorter interpulse intervals of fixed GnRH doses in our GnRH-deficient patients. Rather, larger than normal LH amplitudes (most of which occur at the shorter interpulse intervals) were evident in the GnRH-deficient men.

Taken together, this experiment suggests that the normal male hypothalamus

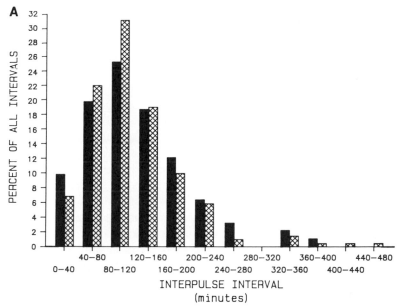

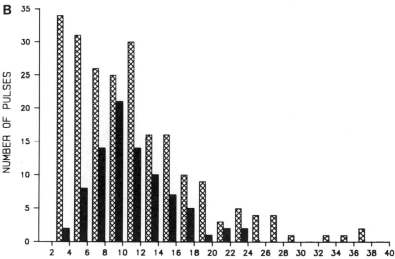

LH AMPLITUDE (mIU/ml, 2nd IRP)

FIG. 16. (A) Frequency distribution of interpulse intervals in 20 normal (cross-hatched bars) and in 8 GnRH-deficient (solid bars) men. Data are expressed as percentage of all LH interpulse intervals. The computer model used to generate the experimental interpulse intervals in the GnRH men was clearly successful in recapitulating the spectrum of interpulse intervals found in normal men. (B) Frequency distribution of LH pulse amplitudes of GnRH-deficient men (solid bars) after administration of GnRH at the intervals demonstrated in the upper panel compared to the LH pulse amplitudes occurring in normal men (hatched bars). Data are expressed as total number of pulses in each group. The failure of the experimental group to recapitulate the high percentage of small-amplitude LH pulses occurring in normals signifies the ability of the normal hypothalamus to alter the bolus dose of GnRH, whereas the GnRH dose for GnRH-deficient men was held constant (O'Dea et al., 1989).

often produces larger amplitude LH pulses by merely lengthening the interval of GnRH stimulation, i.e., decreasing the frequency of a bolus of GnRH secretion from the hypothalamus transiently. On the other hand, the smaller amplitude LH pulses could not be reconstructed by merely shortening the GnRH interpulse interval (i.e., increasing the frequency of GnRH stimulation) of a fixed dose of GnRH in the GnRH-deficient subjects. Thus, the normal subjects are likely to produce the higher percentage of their low-amplitude LH pulses by another mechanism, such as being able to decrease the dose of endogenous GnRH secreted. The significance of this finding is that the interpulse interval, independent of GnRH dose, mean frequency of GnRH stimulation, or sex steroids imparts important information to the pituitary concerning its gonadotropin responsiveness. The other important conclusion here is that the endogenous hypothalamic secretion of GnRH must be variable in terms of both dose and frequency since there is no way to reconstitute the spectrum of amplitudes of LH pulses occurring in the normal without such variations. The fact that the GnRH-deficient model cannot vary the dose of GnRH, and thus fails to recreate the smaller amplitude LH pulses observed in the normal, allows a clear visualization of this role of variable dose of endogenous GnRH secretion for the first time.

VI. The Role of Sex Steroid Modulation in GnRH-Induced Gonadotropin Secretion

To define the site of action of sex steroid feedback in the human male, the models of the intact and GnRH-deficient men were again studied in tandem. By infusing sex steroids into normal men with intact hypothalamic–pituitary axes, the net effect of sex steroids on both the hypothalamus and anterior pituitary can be examined. By combining these studies with those of the GnRH-deficient men, who can experience sex steroid feedback only at the level of the anterior pituitary as their GnRH regimen is fixed, the hypothalamic component of sex steroid feedback can be dissected from that of the pituitary. The responses of the GnRH-deficient men represent the isolated pituitary component, which can then be "subtracted" form that of normal men to deduce the role of the hypothalamus in steroid hormone feedback. The assumptions involved in the parallel use of these models require that the same immunoassays, same experimental protocol, and the same regimens of sex steroid infusion be applied in both circumstances.

All the GnRH-deficient men selected for these studies (1) were determined to be apulsatile in their pretreatment status; (2) had been receiving exogenous GnRH with normalization of their pituitary–gonadal axis for at least 3 months, (3) had previously undergone an individualized dose–response curve such that a dose of GnRH was determined that produced LH pulse amplitudes within the midphysiologic range when compared with normal adult males (Spratt et al., 1988), and (4) were converted to intravenous GnRH administration during these studies.

A. ESTRADIOL

Estradiol was infused intravenously at a dose of 90 μg/24 hours for a period of 96 hours. This dose of estradiol represents twice the endogenous production rate of the normal male (Baird *et al.*, 1969) and had been used by previous investigators in similar studies (Sherins and Loriaux, 1973). The normal men were sampled at 10-minute intervals for 12 hours prior to and during the last day of the estradiol infusion. The GnRH-deficient men were sampled similarly, during which three intravenous doses of the ED_{50} physiologic dose of GnRH were administered followed by a GnRH dose–response curve spanning from 7.5 to 250 ng/kg. All blood sampling was performed prior to and during the last day of the estradiol infusion in both normal and GnRH-deficient men.

As can be seen in Figs. 17 and 18, estradiol suppressed LH pulse amplitude in both normal and GnRH-deficient men (Finkelstein *et al.*, 1991a). The data in the individual cases in Fig. 17 were quite typical of those seen in the population of six men studied in each subgroup (cf. Fig. 18) and revealed that there was equivalent suppression of gonadotropins in both normal and GnRH-deficient men in terms of both percentage of mean LH levels (cf. Fig. 23) and amplitudes of response (cf. Fig. 18). Additionally, there was a slight decrease in the apparent frequency of LH pulses in the normal men; however, most of this change in frequency was due to the difficulty of visualizing low-amplitude LH pulses in some of the normal men and thus was largely artifactual. Thus, the percentage decreases in mean LH and FSH levels in both normal and GnRH deficient men were identical (cf. Fig. 19). Consequently, it is apparent that the majority, if not all, of the effects of estradiol are exerted at the level of the anterior pituitary. In addition, there is a uniform blunting of the dose–response curves in the estradiol-treated GnRH-deficient men, indicating a global suppression of the responsiveness of the anterior pituitary to GnRH by estradiol.

B. TESTOSTERONE

Unlike the findings with estradiol, testosterone administered at a dose of 15 mg/24 hours (i.e., at similar twice-production rate levels) (Horton *et al.*, 1965) exhibited marked differences between the GnRH-deficient and normal men, thus indicating testosterone has a dual site of action. In the normal men (cf. Fig. 20), the predominant effect of testosterone was to reduce the endogenous GnRH pulse frequency dramatically (Finkelstein *et al.*, 1991b). In some cases, it virtually eradicated all evidence of GnRH secretion. In contrast, in the GnRH-deficient men in whom the frequency of GnRH stimulation was fixed (Fig. 20), there was a mild but discernible effect of testosterone at the level of the anterior pituitary with decreases in mean LH levels and LH pulse amplitudes following its administration (cf. Fig. 21). When viewed as a group, the advantages of being able to control the GnRH pulse frequency and dosage in the GnRH-deficient men becomes readily apparent (Fig. 21). Since these patients can exhibit no decrease in

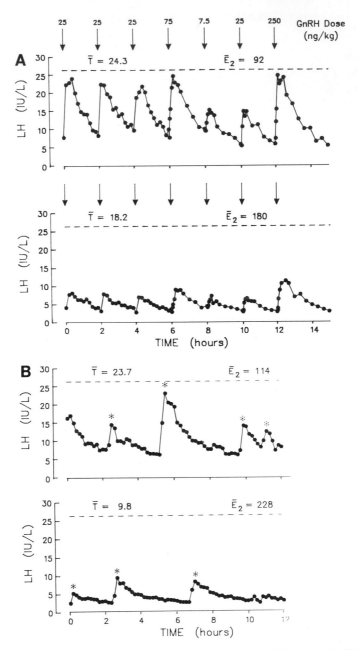

FIG. 17. (A) Serum LH levels determined at frequent intervals for 15 hours in a GnRH-deficient man before (top) and during (bottom) intravenous infusion of 90 μg of estradiol for 96 hours. The arrows indicate each GnRH bolus and the GnRH dose administered is shown in nanograms per kilogram. The dashed line denotes the upper 95% confidence limit for peak LH levels in our normal men. (B) Serum LH levels determined at frequent intervals for 12 hours in a normal man before (top) and during (bottom) intravenous infusion of 90 μg of estradiol. LH pulses are indicated by asterisks. Mean serum testosterone (nmol/liter) and estradiol (pmol/liter) are indicated on each panel (Finkelstein *et al.*, 1991a).

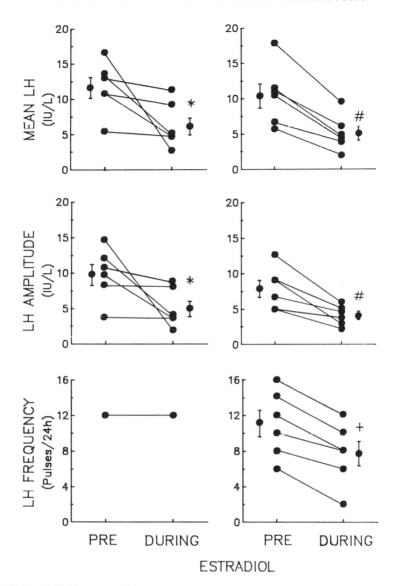

FIG. 18. Individual mean LH levels, LH amplitude, and LH frequency in six GnRH-deficient (left) and six normal (right) men before and during intravenous infusion of 90 μg of estradiol. Group means ± SE for each parameter are indicated to the side of each graph.(*) $p < 0.02$; (+) $p < 0.05$; (#) $p < 0.01$ (Finkelstein et $al.$, 1991a).

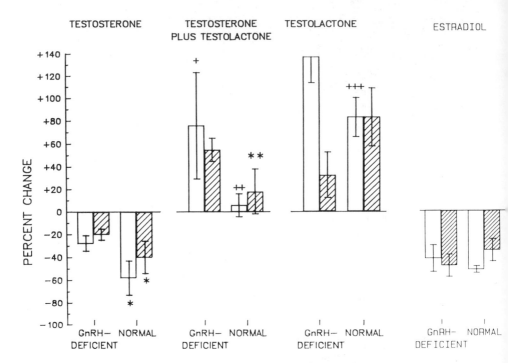

FIG. 19.   Percentage change in mean LH (open bars) and FSH (hatched bars) levels in GnRH-deficient and normal men during administration of testosterone (T), testosterone plus testolactone (TL), TL alone, or estradiol. (*) $p < 0.05$ vs mean LH and FSH levels in GnRH-deficient men receiving T alone; (**) $p < 0.02$ vs mean FSH levels in GnRH-deficient men receiving T plus TL; (+) $p < 0.05$ vs mean LH levels in GnRH-deficient men receiving T alone; (++) $p < 0.01$ vs mean LH levels in normal men receiving TL alone; (+++) $p < 0.01$ vs mean LH levels in normal men receiving T plus TL (Finkelstein *et al.*, 1991a,b).

mean GnRH pulse frequency, all decreases in their mean levels must be solely attributable to the isolated impact of testosterone on the amplitude of the pituitary LH response to GnRH. It is also noteworthy that the majority of this blunting effect of testosterone on the anterior pituitary was manifest in response to those doses of GnRH that produced LH pulse amplitudes within the normal adult male

$\longrightarrow$

FIG. 20.   (A) Serum LH levels determined at frequent intervals for 15 hours in a GnRH-deficient man before (top) and during (bottom) intravenous infusion of 15 mg of testosterone. The arrows indicate each GnRH bolus and the GnRH dose administered. The dashed line denotes the upper 95% confidence limit for peak LH levels in our normal men. (B) Serum LH levels determined at frequent intervals for 12 hours in a normal man before (top) and during (bottom) intravenous T infusion. LH pulses are indicated by asterisks. Mean serum T (nmol/liter) and E_2 (pmol/liter) are indicated on each panel (Finkelstein *et al.*, 1991b).

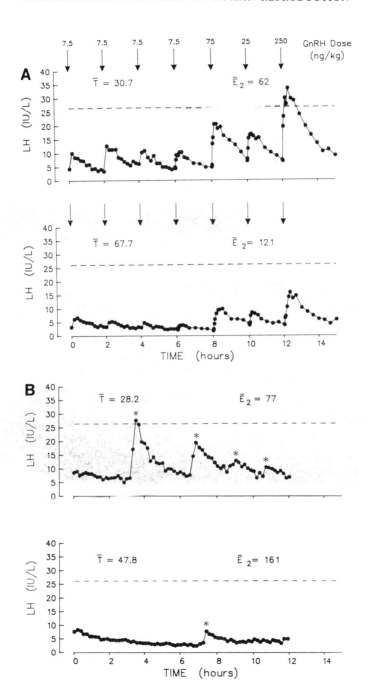

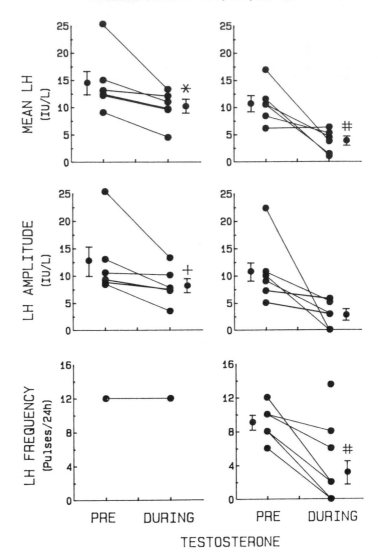

FIG. 21. Individual mean LH levels, LH amplitude, and LH frequency in six GnRH-deficient (left) and six normal (right) men before and during intravenous testosterone infusion. Group means ± SE for each parameter are indicated to the side of each graph. (*) $p < 0.02$; (+) $p < 0.05$; (#) $p < 0.01$ (Finkelstein et al., 1991b).

range. In contrast, the testosterone infusions had no significant effect on the LH pulse amplitudes that occurred after GnRH doses at 250 ng/kg, i.e., a dose of GnRH that routinely produces LH pulse amplitudes above the normal range and hence represents a pharmacologic dose of GnRH. This finding may well explain previous studies that have demonstrated that the responsiveness to a dose of exogenous GnRH in the pharmacologic range used for GnRH testing is un-affected by testosterone infusion. The final point of interest during the testosterone studies is that, given that testosterone induces striking decreases in the endogenous GnRH pulse frequency in normal men and our previous studies, where controlled decreases in GnRH frequency demonstrated an increased amplitude in the ensuing LH pulse, the fact that the normal men did not exhibit accompanying rises in LH pulse amplitude associated with their decreased frequency of endogenous GnRH frequency can now be discerned to represent the direct pituitary effect of testosterone feedback in the normal men. Only by the combined study of these two groups of subjects, however, could such an effect become apparent.

C. TESTOSTERONE PLUS TESTOLACTONE

To determine the degree to which the effects of testosterone on the hypo-thalamic–pituitary axis of the normal male and on the pituitary responses of GnRH-deficient males might be related to aromatization, a parallel series of studies were undertaken in which the testosterone infusions were repeated with the addition of 2 g of testolactone administered orally in both normal and GnRH-deficient men (Finkelstein *et al.*, 1991b). A companion series of steroid produc-tion rate studies, performed in collaboration with Dr. Chris Longcope, confirmed that aromatization had been inhibited by 67 to 80% in the two subjects as studied with blood and urinary steroid production rates. The results of blockade of aromatization produced strikingly different results in both normal and GnRH-deficient (cf. Fig. 19) men. In the normal males, inhibition of aromatization completely blocked the effect of testosterone on GnRH pulse frequency as well as mean LH levels and amplitude. Thus, blockade of aromatase activity rendered normal men virtually immune to the effects of exogenous testosterone infusion (cf. Figs. 19, 22, and 23). In contrast, GnRH-deficient men exhibited a striking increase in both LH pulse amplitude and mean LH level in view of their con-trolled frequency of GnRH administration. This "overshoot" of the GnRH-deficient men during aromatase blockade was taken to signify the liberation of the pituitary from estradiol feedback (previously documented in the above stud-ies) combined with the fact that the frequency of GnRH stimulation was con-trolled. Consequently, the fact that the normal men did not exhibit this overshoot seen in the GnRH-deficient men suggests that testosterone itself, independent of its aromatization to estradiol, has an additional effect on the hypothalamus in

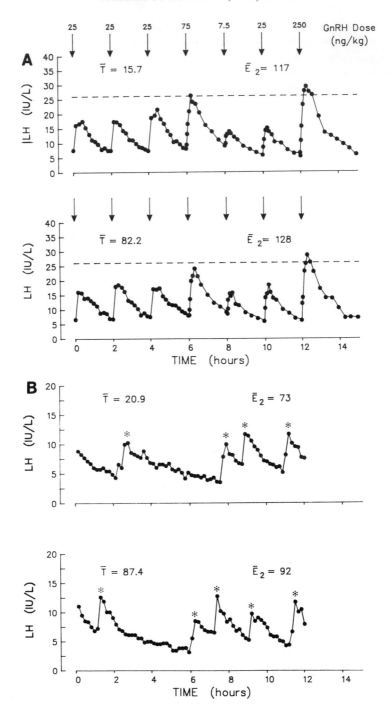

normal men to slow GnRH frequency and/or reduce its dosage of secretion and thus prevent the overshoot seen in GnRH-deficient men, whose dose and frequency of GnRH stimulation remained unchanged.

When the mean LH and FSH levels are compared for all these studies (cf. Fig. 19), one can see that the differences between the ability of estradiol to suppress FSH and LH in normal and GnRH-deficient men are quite small, indicating that the predominant effect of this steroid is at the level of the anterior pituitary (cf. Figs. 18 and 19). On the other hand, testosterone suppresses gonadotropins to a greater degree in the normal than the GnRH-deficient men, indicating that it has both a pituitary and a hypothalamic effect. Quantitating these effects would indicate that approximately half of the effect of testosterone is due to its anterior pituitary site of action, whereas half appears due to its hypothalamic effect on GnRH frequency as judged by suppression of mean LH levels (cf. Fig. 19). Finally, the effects of testosterone plus testolactone represent merely a restoration of the gonadotropins to pretreatment levels in normals. The GnRH-deficient subjects, however, exhibit the above-mentioned overshoot, suggesting that freedom from inhibition of the anterior pituitary by blockade of aromatase as seen in the GnRH-deficient subjects is somehow further restrained in the normal men, thus targeting an additional hypothalamic site to the direct action of testosterone.

VII.   Summary and Conclusions

The traditional difficulty in studying the neuroendocrine control of reproduction in the human male has been the inability to tease out the hypothalamic from the pituitary component of this neuroendocrine system. The use of multiple models, each with its own strength and weakness, represents an overlapping approach that has permitted further insights to be gained into the hypothalamic control of the neuroendocrine regulation of gonadotropin secretion in the human. Such an insight is an important prerequisite to the understanding of the pathophysiology of various disease states, the unraveling of a control of FSH secretion by GnRH vs other modulators, and the subsequent design of rational therapies for male reproductive disorders.

FIG. 22.   (A) Serum LH levels determined at frequent intervals for 15 hours in a GnRH-deficient man before (top) and during (bottom) intravenous testosterone (15 mg/24 hours) plus oral testolactone (2 g/day) administration. The arrows indicate each GnRH bolus and the GnRH dose administered. The dashed line denotes the upper 95% confidence limit for peak LH levels in normal men. (B) Serum LH levels determined at frequent intervals for 12 hours in a normal man before (top) and during (bottom) testosterone plus testolactone administration. LH pulses are indicated by asterisks. Mean serum T (nmol/liter) and E_2 (pmol/liter) are indicated on each panel (Finkelstein *et al.*, 1991b).

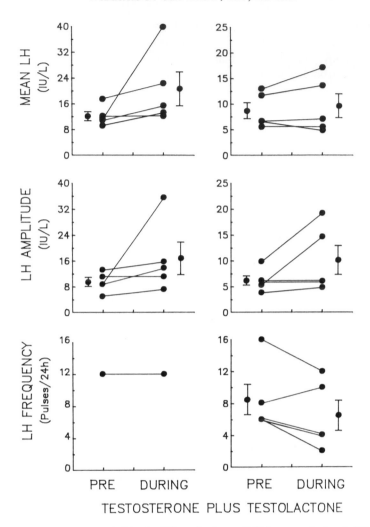

FIG. 23.   Individual mean LH levels, LH amplitude, and LH frequency in five GnRH-deficient (left) and five normal (right) men before and during testosterone plus testoalactone administration. Group means ± SE for each parameter are indicated to the side of each graph (Finkelstein *et al.*, 1991b).

ACKNOWLEDGMENT

Supported in part by National Institutes of Health Grants RO1-HD15788 and RR 1066 and Food and Drug Administration FD-U-000523.

REFERENCES

Badger, T. M., Loughlin, J. S., and Naddaff, P. G. (1983). *Endocrinology (Baltimore)* **112**, 793–799.

Baird, D. T., Horton, R., Longcope, C., and Tait, J. F. (1969). *Recent Prog. Horm. Res.* **25**, 611–664.

Boyar, R., Finkelstein, J., Roffwarg, H., Kapen, S., Weitzman, E., and Hellman, L. (1972). *N. Engl. J. Med.* **287**, 582–586.

Butler, J. P., Spratt, D. I., O'Dea, L. St. L., and Crowley, W. F. (1986). *Am. J. Physiol.* **250**, E338–E340.

Caraty, A., Orgeur, P., and Thiery, J. C. (1982). *C. R. Hebd. Seances Acad. Sci.* **295**, 103–106.

Clarke, I. J., Cummins, J. T., Findlay, J. K., Burman, K. J., and Doughton, B. W. (1984). *Neuroendocrinology* **39**, 214–221.

Crowley, W. F., Filicori, M., Spratt, D. I., and Santoro, N. (1985). *Recent Prog. Horm. Res.* **41**, 473–531.

Ellis, G. B., and Desjardins, C. (1981). *Endocrinology (Baltimore)* **110**, 1618–1627.

Filicori, M., Santoro, N. F., Merriam, G. R., and Crowley, W. F. (1986). *J. Clin. Endocrinol. Metab.* **62**, 1136–1144.

Filicori, M., Flamigni, C., and Crowley, W. F. (1987). *In* "The Episodic Secretion of Hormones" (W. F. Crowley and J. G. Hoefler, eds.), pp. 5–13. Churchill-Livingstone, New York.

Finkelstein, J. S., Badger, T. M. O'Dea, L.St. L., Spratt, D. I., and Crowley, W. F. (1988) *J. Clin. Invest.* **81**, 1725–1733.

Finkelstein, J. S., O'Dea, L. St. L., Whitcomb, R. W., and Crowley, W. F. (1991a). *J. Clin. Endocrinol. Metab.* (in press).

Finkelstein, J. S., Whitcomb, R. W., O'Dea, L. S., Longcope, C., Schoenfeld, D. A., and Crowley, W. F. (1991b). *J. Clin. Endocrinol. Metab.* (in press).

Gross, K. M., Matsumoto, A. M., and Bremner, W. J. (1987). *J. Clin. Endocrinol. Metab.* **64**, 675–680.

Handelsman, D. J., Jansen, R. P. S., Boylan, L. M., Spaliviero, J. A., and Turtle, J. R. (1984). *J. Clin. Endocrinol. Metab.* **59**, 739–744.

Hoffman, A. R., and Crowley, W. F. (1982). *N. Engl. J. Med.* **307**, 1237–1241.

Horton, R., Shinsako, J., and Forsham, P. H. (1965). *Acta Endocrinol. (Copenhagen)* **48**, 446–458.

Jakacki, R. I., Kelch, R. P., Sauder, S. E., Lloyd, J. S., Hopwood, N. J., and Marshall, J. C. (1982). *J. Clin. Endocrinol. Metab.* **55**, 453–458.

Karsch, F. J., Bittman, E. L., Foster, D. L., Goodman, R. L., Legan, S. J., and Robinson, J. E. (1984). *Recent Prog. Horm. Res.* **40**, 185–232.

Levine, J. E., Pau, K.-Y. F., Ramirez, V. D., and Jackson, G. L. (1982). *Endocrinology (Baltimore)* **111**, 1449–1455.

Loughlin, J. S., Badger, T. M., and Crowley, W. F. (1981). *Am. J. Physiol.* **240**, E591–E596.

O'Dea, L. St. L., Finkelstein, J. S., Schoenfeld, D. A., Butler, J. P., and Crowley, W. F. (1989). *Am. J. Physiol.* **256**, E510–E515.

Rebar, R., Judd, H. L., Yen, S. S. C., Rakoff, J., Vandenberg, G., and Naftolin, F. (1976). *J. Clin. Invest.* **57**, 1320–1325.

Sherins, R. J., and Loriaux, D. L. (1973). *J. Clin. Endocrinol. Metab.* **36**, 886–893.

Spratt, D. I., Crowley, W. F., Butler, J. P., Hoffman, A. R., Conn, P. M., and Badger, T. M. (1985). *J. Clin. Endocrinol. Metab.* **61**, 890–895.

Spratt, D. I., Finkelstein, J. S., Badger, T. M., Butler, J. P., and Crowley, W. F. (1986). *J. Clin. Endocrinol. Metab.* **63**, 143–150.

Spratt, D. I., Carr, D. B., Merriam, G. R., Scully, R. E., Rao, P. N., and Crowley, W. F. (1987a). *J. Clin. Endocrinol. Metab.* **64**, 283–291.

Spratt, D. I., Finkelstein, J. S., Butler, J. P., Badger, T. M., and Crowley, W. F. (1987b). *J. Clin. Endocrinol. Metab.* **64**, 1179–1186.

Spratt, D. I., O'Dea, L. St. L., Schoenfeld, D. A., Butler, J., Rao, P. N., and Crowley, W. F. (1988). *Am. J. Physiol.* **254**, E658–E666.

Valk, T. W., Corley, K. P., Kelch, R. P., and Marshall, J. C. (1980). *J. Clin. Endocrinol. Metab.* **51**, 730–738.

Waldstreicher, J., Santoro, N. F., Hall, J. E., Filicori, M., and Crowley, W. F. (1988). *J. Clin. Endocrinol. Metab.* **66**, 165–172.

Weiss, J., Jameson, J. L., and Burrin, J. M. (1990a). *Mol. Endocrinol.* **4**, 557–564.

Weiss, J., Duca, K. A., and Crowley, W. F. (1990b). *Endocrinology (Baltimore)* **127**, 2364–2371.

Whitcomb, R. W., and Crowley, W. F. (1990). *J. Clin. Endocrinol. Metab.* **70**, 3–7.

Yen, S. S. C., Tsai, C. C., Naftolin, F., Vandenberg, G., and Ajabor, L. (1972). *J. Clin. Endocrinol. Metab.* **34**, 671–675.

DISCUSSION

J. Levine. It is very gratifying to see the mix of unequivocal hypothalamic and pituitary feedback mechanisms in the human. We have noted almost exactly the same things in the rat. Our rat version of the GnRH-deficient human is the hypophysectomized rat bearing an ectopic pituitary transplant. Using this model in the female rat, we have also seen that there is amplitude regulation of GnRH-stimulated LH secretion by steroids. The direct suppressive effect of estrogen is a very rapid one. It also appears that a similar mechanism operates in the male rat, with respect to testosterone. There is a rapid release from direct pituitary feedback following castration. We have also seen that there is a frequency regulation of GnRH release at the hypothalamic level, so for once the data seem to really agree very well between the rat and the human. Have you studied the respective time courses of the feedback actions at the two levels? Is there a difference between the rapidity with which changing levels of steroids can influence events at the hypothalamic versus pituitary levels? We have seen that GnRH frequency regulation is a relatively sluggish process, whereas direct pituitary effects of steroids can be manifest very rapidly. In the case of estrogen action on the pituitary, this can occur literally within minutes. Do you see any difference?

W. F. Crowley, Jr. We have not had the opportunity to examine the time course of these effects. We have had to restrict our examinations to fixed points 4 days apart in the steroid studies to date because of the problems of stabilizing all of the variables in the human. Basically, each individual study involves a separate 5-day CRC (Clinical Research Center) admission, two separate 12 hours of frequent sampling, and intravenous administration of all medications. Consequently, we presently have no information on the time course. In the female, we do have some information on the time course of positive sex steroid feedback as we have dissected that process out a little more carefully; however, in the male we have not yet done this.

N. Schwartz. As you know we have been using a perifusion set-up with pituitary fragments from the rat. We feel that the fragments which we use within 15 minutes of removal from the animal have given us data which much better reflect the *in vivo* situation than the dispersed cells which we have also tried. I would like to remind you of some sex differences that really surprised us. It has been an assumption among neuroendocrinologists that all the sex differences are in the hypothalamus, not in the pituitary. We have published a number of papers, including the very recent paper in *Endocrinology* (August, 1990), which show that the female pituitary always gives us what we expect. We can see cycle differences; estrogen *in vivo* makes a real difference in terms of GnRH responses *in vitro*. All this is exactly what we expected. What we did not expect to find was that in the normal male, when we remove the pituitary, the secretion rates for LH and FSH rise fantastically right at the beginning. It is as though there is an unleashing from a negative feedback of a tremendous nature in the male, so that the FSH secretion is so high that we think we are at the maximum the pituitary can achieve. It is kind of all downhill for the next 8 hours. The basal LH and FSH secretion rate in the absence of GnRH starts out very high in the male and falls continuously for 8 hours. In the female it levels off and then, regardless of the cycle stage, FSH starts to rise. We think this is an escape from an *in vivo* inhibin effect, so we are seeing a clear sex difference there. In GnRH responsiveness, then, the male has a very mild response to GnRH in that fragment; we do not yet see some of the responses that you have shown *in vitro*. We thought that if we put testosterone in the medium we would enhance GnRH responses, but that has not worked. The only thing we have gotten to work in the male pituitary to increase the GnRH responsibility is to give the animal estrogen *in vivo*. Do you have any insight from female and male patients that would help?

W. F. Crowley, Jr. Yes. We have followed the sex differences your group obtained quite closely, and you are right. They are very striking, but I do not understand them well. Of course, we all have a big segment of the picture missing here, which is the role of inhibin and activin in the physiology of gonadotropin control, especially the role that these gonadal proteins may play in this dimorphism. There is, however, one major sex difference in this system in the human that is quite striking, and almost nobody has paid any attention to it yet since very few groups are actively working on both male and female models in parallel. This difference involves the GnRH dose required to reconstitute the physiologic integrity of the system in GnRH-deficient humans. Approximately 25 ng/kg of intravenously administered GnRH will restore midphysiological ranges of gonadotropins in a GnRH-deficient male. On the other hand, to restore a normal ovulatory cycle to a GnRH-deficient woman, the mean GnRH dose is 75 ng/kg, i.e., about three times the average dosage of GnRH required by the human male. So I think one of the sexually dimorphic variables is the dose of GnRH. Now, if the dose of GnRH has an effect of partially desensitizing FSH secretion in the intact animal, then potentially the time to escape from such a partial desensitization may be different between the two sexes. How that speculation translates to the rat, I do not know, but it would not surprise me that some of the variables required to produce LH levels in the normal range, such as the amount of GnRH secreted, might not prove to be partially desensitizing to FSH.

The second area in which we see sex differences in the human occurs when you start to examine the doses of GnRH antagonists required to block endogenous GnRH secretion. Assuming that GnRH antagonists as receptor blockers give some information about endogenous levels of GnRH secretion, just as naloxone does for endogenous opiates, you can again see some differences in the inhibition of various cycle stages in the female. So this is another way of evaluating the quantity of endogenous GnRH secreted, and, again, a sexual dimorphism is apparent.

A. Dunaif. I may have misunderstood, but it seemed to me that in your intact males, when you administered testolactone with testosterone, the estradiol levels increased, whereas in your hypogonadotropic males the estradiol levels remained low. We observed the same thing in our intact women on testolactone; this was secondary to increased secretion of estradiol, despite the fact that aromatization was blocked. This would suggest that you did change some parameter of GnRH secretion and that you might not have detected it. What do you think?

W. F. Crowley, Jr. We employed steroid production rates measured with isotopic infusions, in collaboration with Chris Longcope, to determine that a blockade of aromatization did occur and that estradiol production rates fell by 58–70%. We believe that testolactone or its metabolites interfere in the serum estradiol radioimmunoassay, where we had somewhat puzzling results. Dick Santen and Howard Judd have pointed out that some of the derivatives of testolactone interfere in some of the assays for estradiol. That is why we used production rate determinations to measure aromatase inhibition.

A. Dunaif. Why don't you think that you found this assay artifact?

W. F. Crowley, Jr. It did happen in some of the IHH males. It was a variable finding in both study populations.

S. Pavlou. Does LH bioactivity of the LH B/I ratio change when the frequency of administration of GnRH pulses is varied? We have seen in studies we have done, especially using a GnRH antagonist, that bioactivity changes within the same individual and also changes depending on which part of the LH pulse is sampled. LH is characterized by microheterogeneity, and some species of LH are more bioactive than others. We have seen that if you analyze LH by chromatofocusing the ratio of B/I changes so much between fractions that when you go from basic to more acidic forms the B/I increases from 2 to 3 to as high as 25 or 30.

W. F. Crowley, Jr. For these human studies, we feel that some of the gonadotropin bioassays, as presently constructed, have problems with other modulators in the circulation for the following reasons. For the LH bioassay, there are known modulators of LH action in the circulation (such as EGF) which increase the Leydig cell responsiveness to LH markedly. For the FSH bioassay, there is a whole family of circulating compounds that compete for FSH receptor binding. One of the existing FSH bioassays uses the preprecipitation of serum with polyethylene glycol before actual assay for FSH. Such a step serves to remove some of these competitors. I am concerned about these methodological problems. I think the best way to use these bioassays is when you separate the gonadotropins from the serum as you have done and perform, for example, chromatofocusing. However, one has to recognize that even these steps do not always separate all forms. It just separates them out, relatively speaking, and then assays them. However, we have not done any of these experiments and we have not examined the frequency change experiments with bioassay ratios, partly stemming from our lack of complete confidence in these methods, especially for the subtler determinations that would be necessary to study that question with the requisite detail.

S. Pavlou. When you take just only one random blood sample, you never know at which part of the LH pulse you are: whether you are at the rising portion of the LH peak, the peak, or in between pulses. This is why there are many differences in LH bioactivity, not only between individuals but also within individuals.

W. F. Crowley, Jr. This is why we took all the points in the LH pulse for bio- and immunoassay. In fact, that line of the regression of bio- vs immunoassayable LH levels I discussed made use of all points from all pulses: nadir, ascending limb, descending limb, and peak. Thus, that positive y intercept of the regression line is quite solid within all patients. Between patients the slope of it varies, but there is a very consistent positive bioassayable LH when immunoassay levels are zero according to the lines of regression. Hence, we have some ongoing concern about the LH bioassay and its use in making some of the subtler distinctions you alluded to in your questions.

I. Vermes I would like to discuss a simple problem, namely, standardization of LH and FSH immunoassays. You presented a huge number of LH and FSH results measured during the past 5 years. I am not convinced that what you showed was the same LH assay.

W. F. Crowley, Jr. All these studies used the same assay. We did not change anything in the assay during the time period of these studies. When any antisera change is made in our lab (e.g., from one bleed to the next), extensive cross-validation of old bloods are performed to assure us of equivalency.

I. Vermes. I think it might be dangerous to draw conclusions on molar ratios based on immunological measurements knowing the problems of standardization of these assays.

W. F. Crowley, Jr. I would like to relate some details of our human gonadotropin assays because we have taken some care on this issue. If you are going to put in place a normative database that is continually contrasting physiologic and pathologic events, you must know your assay specificity.

The LH and FSH antibodies were raised by me and a colleague of mine, Dr. Marco Filocori. The

LH assay is β directed; it sees the conserved and homologous region of the β subunit shared by LH and hCG; it has no α cross-reactivity; and therefore, of course, cannot be used to measure LH in pregnancy or in other circumstances in which hCG circulates. It does not see the unique sequences of the free form of the LH β subunit and, therefore, it is the type of assay that certainly cannot distinguish between the free β subunit of LH (for which, if secreted, its changes over time are not yet charted). The FSH assay is quite specific for the FSH dimer; it does not have significant cross-reactivity with either the α or the β subunit and it only sees intact FSH. The free α subunit assay sees the 21 to 29-amino acid sequence of the free α subunit which is the region obscured following combination with the β subunits in the various dimers, so it only sees the α subunit in its uncombined or secreted form.

The rat assays are similarly specific, but have the added advantage of having their standards > 95% pure for LH and α subunit and > 50% pure for FSH. Thus, *relative* quantitation can be undertaken in the rat system and, hence, that is the only part of our studies that we did make such inferences about relative amounts of secreted glycoproteins.

J. Clark. Why don't you use DHT in the androgen infusion experiment? Is there some problem with using it?

W. F. Crowley, Jr. We have given a lot of consideration to this, but the serum binding globulin, testosterone estrogen binding globulin (TEBG), has a very high and preferential affinity for DHT over T and E_2. Consequently, it soaks up an enormous mount of DHT before the free fraction is increased. When you study the literature on DHT effects, you find that the results have been relatively inconstant. In the doses of DHT that have had to be used (i.e., up into the milligram quantities as opposed to its low microgram production rate), the results are variable, whereas when you look at the T and E_2 effects, all these studies have used only twice production rate levels—pharmacologic levels to be sure—giving you levels which are not so far out of the physiologic range, whereas the DHT doses required have been clearly well out of the physiologic range.

Now this introduces a second feature and that is it is not just the dose of steroid you have to give but what that pertubation does in terms of unloading TEBG of its other steroids. In other words, you suddenly convert all the estradiol into the free form when you give DHT; you also convert all of the testosterone into the free form, and then you have problems of what these adaptive changes from the unloading of TEBG by DHT do to the system. What you really want to do is get something that is a nonaromatizable androgen at the hypothalamic level that does not bind to TEBG, that will tcase out the direct hypothalamic component of androgen feedback. That is the very next step in these studies. We have been thinking more of synthetic androgens which do not affect the binding globulin abnormalities.

J. Clark. Do you know whether the steroid receptors are stable in the pituitary pieces *in vitro*?

W. F. Crowley, Jr. I do not know, but Neena Schwartz would probably be the expert on that.

J. Clark. When you take the uterus out of an animal and place it *in vitro*, estrogen receptors are lost within the first hour of the incubation in the absence of estrogen.

N. Schwartz. We do find that estrogen in the medium does cause good suppression of LH and a small suppression of FSH over an 8-hour period, so there are at least some receptors there for that long.

In your *in vitro* set-up have you tried altering GnRH frequency and looking at the FSH and LH ratio? I think it is a crucial experiment in terms of theory.

W. F. Crowley, Jr. Yes, I agree. Of course, one of the problems with the perifusion system is that you do not see much data in the literature on FSH secretion in this system. This is because by the time you disperse the pituitary cells, the FSH levels used are undetectable in many of these systems or at best, running at very, very low levels when you initially start to use male pituitary cells. That is in sharp contradistinction to what you see in the serum of these animals, underscoring the role of circulating $T_{1/2}$ in FSH dynamics *in vivo* vs secretory phenomenon *in vitro*. Consequently, we had to improve the sensitivity of the FSH assay, increase the number of cells on the column, and tinker with the assay conditions to be able to read all the FSH levels consistently in the first place. We have done some *in vitro* studies with increasing the GnRH frequency, but have not measured the FSH levels. We still have some of the samples left and plan to do that.

I can also tell you that one of the newer and interesting things about increasing the GnRH frequency is that the free α subunit is serving as an important alternate marker to LH for endogenous secretion in the human. In circumstances in which very fast GnRH frequencies exist, free α subunit provides an excellent corroborative index of LH secretion. We have used that feature in several circumstances such as polycystic ovarian disease, to back up our previous estimates of the rapid GnRH frequency of that condition. We have also monitored pulsatile free α subunit secretion of the midcycle surge and in postmenopausal women.

J. Kirkland. Those of us who use the GnRH stimulation test in children occasionally find discordance between clinical findings and biochemical findings. I am curious, in view of the data you have discussed, whether a dose–response curve should be constructed for each child before the biochemical results can be interpreted?

W. F. Crowley, Jr. I actually do not use that test much in clinical medicine. The only circumstance in which we perform GnRH tests in our unit is to distinguish centrally mediated precocity from peripherally mediated precocious puberty. I suspect if people "played around" with the test, they could get more out of it, but I doubt that it is worth it because we are finding that other tools, such as gonadal ultrasounds, careful measurements of the sex steroids, and the clinical pictures, can contribute far more to your understanding of most diagnoses than a GnRH test.

P. Kelly. In your presentation using the perfusion studies you showed that the pituitary produces about three molecules of LH for 1.7 molecules of α and one of FSH. Do you have some idea about why the pituitary is producing different ratios?

W. F. Crowley, Jr. I have no idea why there is such stoichometry; this was quite a surprise to us. It could well be that the differential clearance rate is a major factor that we have overlooked. Why is this entire hypothalamic–pituitary–gonadal system so strikingly organized around pulsatility? I am not sure that this is a neuronal phenomenon at all; rather, I wonder if the existence of pulsatility isn't gonadally mediated. The peak hormone levels in the circulation might well be giving us a key tip here. The striking preservation of the LH peaks across a wide range of frequencies may have something to do with the requirements for gonadal stimulation. Therefore, the peak hormone output necessary to achieve maximal receptor occupancy and biologic activity may be a crucial factor. Thus, the sort of half-life accommodations and differential clearance rates are all modifiers designed to keep gonadal receptors occupied in a fashion to optimize gonadal function. That would be my speculation, but I have no direct evidence for this. Somebody really has to examine these variables at the gonadal level in a perifused testis system, which has been done, but not optimized, and study the kinetics of hormone delivery at the "other end of this system" and the effects of stimulating the testes in different ways. I would imagine there would be some pretty striking findings and that this may be the real reason for the "upstairs" pulsatility. It may be a "bottoms-up reason rather than a "top-down" mediated event.

J. Funder. What is the role of folliculostellate cells in all this; for example, between fragments and dispersed cells? Are they different between male and female? What is the state of play on that score?

W. F. Crowley, Jr. We have no histology to address that question. It is quite clear, now that we can block GnRH input to the pituitary in the human with GnRH antagonists, we can suppress LH secretion 80 to 90%, but only suppress α and FSH levels maybe 20 to 30%. For the α subunit, you are not surprised about this finding because you know there is another releasing factor (TRF) that is the other side of its control. For FSH, I think the GnRH antagonists may be revealing to us more clearly the other regulatory system or systems that control FSH secretion. One of the surprises of the GnRH antagonist studies is how modest is the degree of control of FSH secretion by GnRH. The other point is that we may be thinking in too narrow-minded a way about FSH control when we think about a releasing factor, i.e., you give a shot and it induces secretory pulses. However, at the very beginning of the studies on GnRH, Professor Igarashi introduced another concept for thinking about a releasing factor which I think fits very well with some of the recent findings for the putative "FSH releasing factor," that is the concept of a platform function. This is a notion of a diffuse trophic stimulator that causes a rise in hormone synthesis and/or release to a certain level that is constant on which the pulsatile hormones then act. I think there is a lot to recommend that platform notion for "FSH-RF". So the second control system for FSH may be very different from its regulation by GnRH. Some of the early data on activin fit that concept, which is more than 12 years old, of a platform control. How you look for such a factor is, of course, very different from how you look for a pulsatile one. The models, the thinking, and time course are different, and I believe some of this other control system may well have some of these other notions built into it.

S. Pavlou. I would like to comment on the effect of GnRH antagonists on FSH. You are absolutely right that acutely the effects are very minimal, both on α subunit as well as on FSH, whereas LH disappears within 4 or 5 hours. However, continued antagonist administration for 2 weeks renders FSH undetectable. We tried to measure FSH using the newer IRMA assays that have sensitivities of 0.1 units/liter and found that we could not detect any FSH.

W. F. Crowley, Jr. There is a big difference between acute and chronic GnRH antagonism, no question about it. This is a window of opportunity for the antagonist as you well know. On the other hand, there is something different about the acute responses to GnRH which do stabilize acutely.

J. Levine. I would like to reemphasize the point that you made about the acute vs the chronic control of FSH secretion. When we put hypophysectomized rats bearing pituitary transplants on pulsatile GnRH infusions, we see that FSH levels are raised over several days and, as you suggested, reach a platform level as a result of that base of pulsatile GnRH stimulation. However, we can never really discern FSH pulses in response to individual GnRH pulses. This fits also with our data when we have simultaneously measured GnRH and FSH in individual rats. We really see very little correlation between pulses of GnRH and increments in FSH in peripheral blood. So I would agree entirely that it would appear that GnRH is important in maintaining basal FSH secretion, but not in providing an acute signal for pulsatile FSH secretion. GnRH may instead primarily facilitate some other process leading to secretion, such as biosynthesis. Your FSH-β data would certainly be in agreement with this idea.

W. F. Crowley, Jr. Remember one other caveat about these antagonists. We all think that the GnRH antagonist blocks GnRH input and has its impact on the β subunits of LH and FSH. Remember that there is a GnRH response element in the 5' end of the alpha gene, and the way that chronic GnRH antagonism might well work over the long term, where it clearly suppresses FSH secretion, may relate in part to its suppression of the alpha gene. It may well be that inhibition of GnRH input to the gonadotrope may be crucial here.

J. Levine. I was surprised by the strength of your conclusion that alteration of the bolus dose of GnRH must occur normally in pulse generation. Could it not be that what you are seeing are the effects of the waxing and waning of testosterone levels, which would exert varying degrees of feedback suppression at the level of the pituitary gland?

W. F. Crowley, Jr. We have given that idea some thought, but in most of the normal males who have short interpulse intervals, circulating testosterone levels do not vary acutely. Where the variability of circulating testosterone levels comes into play in the normal males is during longer interpulse intervals. Also, the problem we were having in recapitulating the normal male data in GnRH-deficient men is not with the long interpulse intervals; it is with the short ones. So the idea of minute-to-minute feedback is very muted in the human by the presence of testosterone–estrogen binding globulin (TEBG) and is even more muted in the shorter interpulse intervals because the smaller amplitude LH pulses and the shorter interpulse intervals that you were talking about occupy the same domain in GnRH physiology. That is why we have stressed the one (i.e., dose changes) and not the other (steroid feedback). It is a very good thought, though.

Immortalization of Neuroendocrine Cells by Targeted Oncogenesis

PAMELA L. MELLON,* JOLENE J. WINDLE,† AND
RICHARD I. WEINER‡

*The Salk Institute, La Jolla, California 92037, †Division of Cellular Biology, Cancer Therapy
and Research Center, San Antonio, Texas 78229, and ‡Reproductive Endocrinology Center,
University of California, San Francisco, California 94143

I. Introduction

Cultured cell lines which maintain specific differentiated phenotypes have been invaluable in the study of cell biology. For example, the endocrinology and molecular biology of the anterior pituitary hormones, growth hormone (GH), prolactin (PRL), and proopiomelanocortin (POMC), have been substantially advanced due to the development of the somatotropic/lactotropic GH-3 cell line (Nelson et al., 1988; Tashjian et al., 1968) and the corticotropic AtT-20 cell line (Drouin et al., 1987; Nakamura et al., 1978). In contrast, study of the regulation and synthesis of the pituitary gonadotropin hormones and the hypothalamic releasing factors has been impeded by the lack of analogous cell lines representing gonadotropes or hypothalamic neurons.

Study of the molecular and cellular biology of the hypothalamic releasing hormones has been particularly hindered by the cellular complexity of the mammalian brain and the postmitotic nature of mature differentiated neurons. Naturally occurring central nervous system (CNS) neuronal tumors are very rare (Rubinstein et al., 1984) and the experimental transformation or immortalization of highly differentiated CNS neurons quite difficult (Cepko, 1989; Laerum et al., 1984). Several cell lines of peripheral nervous system origin have been established from neuroblastomas and pheochromocytomas [e.g., RT-4 (Imada and Sueoka, 1978) and PC12 (Dichter et al., 1977)]; however, attempts to obtain cell lines of differentiated CNS neurons using a variety of methods have been frustrating. Ronnett et al., 1990) reported the derivation of a cell line from a human cerebral cortical tumor which will differentiate to the neuronal phenotype in the presence of a mixture of neuronal growth factor (NGF), cAMP, and IBMX. This is an important breakthrough, but this approach depends on the rare occurrence of human tumors with the unique capacity to differentiate in this manner. Retroviral transformation of cultured embryonic cerebrocortical cells produced pluripotential or glial cells (Cepko, 1988; Frederiksen et al., 1988; Giotta and Cohn, 1981; Giotta et al., 1980; McKay, 1989). One cell line (Frederiksen et al., 1988),

69

transformed with a temperature-sensitive T antigen, differentiates into a mixed culture with both neural and glial phenotypes upon shift to the nonpermissive temperature. Infection of primary cultures of hypothalamic tissue with simian virus 40 (SV40) resulted in cell lines which were not fully differentiated (de Vitry, 1978; de Vitry *et al.*, 1974).

Transgenic mice have been used successfully to produce tumors in specific tissues (Hanahan, 1989; Jenkins and Copeland, 1989; Rosenfeld *et al.*, 1988), such as the pancreas (Hanahan, 1985) and mammary gland (Stewart *et al.*, 1984), by specifically targeting expression of the potent oncogene producing SV40 T antigen (Tag) with regulatory domains of the insulin gene and mouse mammary tumor virus, respectively. Cell lines derived from transgenic T antigen-induced tumors sometimes maintain the differentiated phenotype of the targeted tissue (Efrat *et al.*, 1988a; Mellon *et al.*, 1990; Ornitz *et al.*, 1987; Windle *et al.*, 1990).

Immortalization by targeted oncogene expression has been less successful in neural cells than in other types of tissues. Transgenic mice bearing Tag under the control of the promoter from the gene producing the hypothalamic peptide hormone, growth hormone-releasing factor (GRF), produced nonspecific thymic hyperplasia but failed to transform cells of the hypothalamus (Botteri *et al.*, 1987). Tissue-specific expression of Tag under the control of the glucagon promoter specifically induced tumors of the α cells of the pancreas in transgenic mice. Although glucagon is also normally produced in CNS neurons, no brain tumors were observed (Efrat *et al.*, 1988b). Tag expression driven by the promoter for the gene producing phenylethanolamine *N*-methyltransferase (PNMT, an enzyme in the catecholamine synthetic pathway), was expressed in the retina but not in the brain (Baetge *et al.*, 1988). Retinal tumors were obtained; however neither the tumors nor cell lines derived from them expressed the endogenous PNMT gene (Hammang *et al.*, 1990).

Using targeted oncogenesis in transgenic mice (Fig. 1), we have transformed specific pituitary and hypothalamic cells and have established clonal cell lines from the resulting tumors (Mellon *et al.*, 1990; Windle *et al.*, 1990). For the immortalization of pituitary cells, T antigen expression was driven by the 5' flanking region of the human glycoprotein hormone α-subunit gene, which is normally expressed in both gonadotropes and thyrotropes. Thus, we anticipated that the resulting cell lines would represent gonadotropes, thyrotropes, or a precursor cell in that lineage. Most of the cell lines expressed the glycoprotein hormone α subunit, but none of these lines expressed any of the β subunits. However, these cells respond to the hypothalamic releasing hormone, gonadotropin-releasing hormone (GnRH), indicating that the cell immortalized represents a developmental precursor of the gonadotrope lineage (Windle *et al.*, 1990). The α-subunit gene is expressed quite early in the development of the anterior pituitary, before the emergence of Rathke's pouch (day e(embryon-

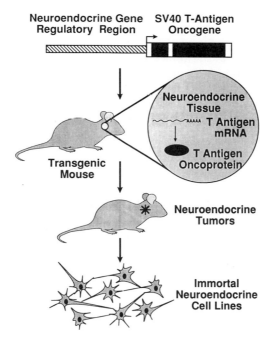

FIG. 1. Strategy for immortalization of neuroendocrine cells by targeted oncogenesis in transgenic mice. A hybrid gene which directs expression of SV40 T antigen using a tissue-specific promoter and regulatory region is introduced into the mouse germ line. Tissue-specifically targeted expression of the T antigen oncogene causes the formation of tumors in the targeted tissue. Tumors are cultured to derive immortal cell lines.

ic)11.5 in the rat; Simmons *et al.*, 1990). Thus, by targeting Tag expression to the early developing anterior pituitary with the α-subunit promoter we have immortalized a developmental precursor of the mature gonadotrope.

Although this approach has proved difficult in the neural cells of the brain, targeting with a gene which is very tightly expressed in a small population of neurons prior to the transition to the postmitotic state may be successful. We have produced specific tumors of GnRH-secreting neurons by introduction of a hybrid gene composed of the GnRH promoter coupled to the coding region for Tag into transgenic mice (Mellon *et al.*, 1990). Clonal cell lines derived from these tumors express GnRH mRNA and secrete GnRH in response to depolarization. Thus, by targeting oncogenesis to a specific, small population of neurons using the regulatory region of a gene which is expressed late in the differentiation of that cell lineage, we have succeeded in immortalizing hypothalamic GnRH neurons which maintain many differentiated functions in culture.

II. Pituitary Gonadotrope Cells

Mammalian reproduction is regulated by the anterior pituitary gonadotropins, luteinizing hormone (LH) and follicle-stimulating hormone (FSH), which are produced in a subpopulation of anterior pituitary cells, the gonadotropes. These hormones are members of a family of glycoprotein hormones which are hetero-dimers, sharing a common α subunit but having unique β subunits (Pierce and Parsons, 1981). This family also includes thyroid-stimulating hormone (TSH), produced in pituitary thyrotropes, and chorionic gonadotropin (CG), produced in the placenta of primates but not rodents.

Studies of the regulation of these hormones have been impeded by the lack of established gonadotrope cell lines. The molecular analysis of LH and FSH regulation has depended on primary pituitary cultures, which have limited viability and progressively lose differentiated function (Tougard and Tixier-Vidal, 1988). Such studies are further complicated by the cellular heterogeneity of the anterior pituitary, which consists of endocrine cells producing growth hormone (GH), prolactin (PRL), and proopiomelanocortin (POMC), in addition to gonadotropes and thyrotropes; as well as nonendocrine endothelial cells, follicular stellate cells, and stromal cells (Farquhar et al., 1975).

The ability to target expression of oncogenes to specific cell types in transgenic mice provides a method for immortalizing rare cell types (Jenkins and Copeland, 1989). We have used this approach to express the simian virus 40 T antigen (SV40 Tag) oncogene in specific pituitary cells by directing its expression with the promoter regulatory region of the human glycoprotein hormone α-subunit gene (Fiddes and Goodman, 1981). Transgenic mice bearing this hybrid gene develop heritable pituitary tumors. These tumors have been cultured to develop immortal clonal cell lines which maintain differentiated functions of gonadotropes, including glycoprotein hormone (α subunit) synthesis and secretion, and responsiveness to gonadotropin-releasing hormone (GnRH; Windle et al., 1990).

A. PITUITARY CELL LINES DERIVED BY TUMORIGENESIS IN TRANSGENIC MICE

Seventeen founder transgenic mice were derived carrying a fusion gene containing the 5' flanking sequences of the human glycoprotein hormone α-subunit gene (1.8 kb) linked to the protein-coding sequences of T antigen. The α-subunit gene is expressed only in anterior pituitary gonadotrope and thyrotrope cells in rodents (Carr and Chin, 1985; Tepper and Roberts, 1984), and thus Tag-driven tumorigenesis was predicted to be anterior pituitary specific (Fox and Solter, 1988). Most of the founder mice developed pituitary tumors with the occurrence of only a few other tumors. Figure 2 demonstrates the tissue-specific expression of Tag in one founder animal carrying a large, encapsulated pituitary tumor. This

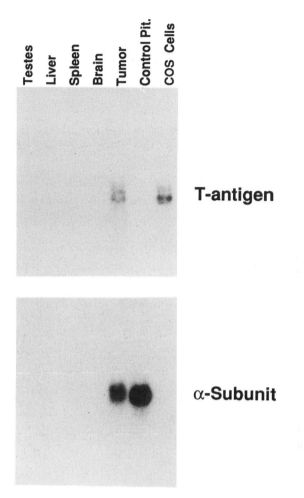

FIG. 2.   Tissue specificity of T-antigen and α-subunit mRNA expression in α–Tag transgenic mice. Northern blot of several tissues from a transgenic mouse demonstrates the tight tissue specificity of expression of the α–Tag transgene to the pituitary tumor. RNA prepared from the tumor and several other tissues from a 49-day-old transgenic male, and control RNAs prepared from normal pituitaries and from a Tag-expressing cell line, COS (Mellon *et al.*, 1981), were run on a Northern blot and hybridized with a Tag-specific probe and a mouse α-subunit cDNA probe (Chin *et al.*, 1981).

tumor was also found to be expressing α-subunit mRNA, consistent with gonadotropic or thyrotropic origin.

Cells from the tumors of several founder mice were dispersed in tissue culture. The initial cultures were very heterogeneous, including endocrine cells, fibroblasts, and blood cells. The endocrine cells attach to plastic dishes more slowly than fibroblasts and were serially passaged to remove fibroblasts. Eventually, stable cultures were established and several clonal cell lines were derived from each by serial dilution (e.g., αT3-1 and αT3-2). The rate of tumor development was directly correlated with the ability to establish cell lines in culture. The αT3-1 and αT4-1 cell lines have been cultured continuously for approximately 2 years without apparent change in phenotype, and can be frozen in liquid nitrogen and thawed successfully. The cells double in culture every 20–30 hours.

B.   ANALYSIS OF PITUITARY HORMONE AND Tag EXPRESSION IN TUMORS AND CELLS

The levels of Tag and α-subunit mRNAs in the uncloned cell populations and in clonal cell lines were compared to those of the original tumors (Fig. 3). A direct relationship between the levels of Tag and α-subunit mRNA was found,

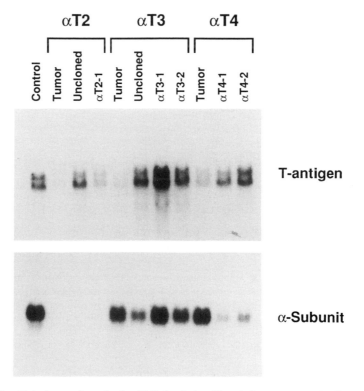

FIG. 3.   T Antigen and α-subunit mRNA levels in α-Tag pituitary tumors and cell lines. Total RNA from the pituitary tumors of three transgenic animals: αT2, αT4 (females), and αT3 (male), from homogeneous populations of cells derived from the tumors, and from representative clonal cell lines were run on a Northern blots with COS RNA and normal pituitary RNA as controls. The filters were hybridized with Tag and α-subunit cDNA probes, respectively, as in Fig. 2.

consistent with the fact that expression of both is driven from an α-subunit gene promoter.

To determine the hormonal characteristics of these cells, poly(A)$^+$ mRNA was hybridized to cDNA probes for each of the pituitary hormones, including α subunit, the β subunits of LH, FSH, and TSH, as well as GH, PRL, and POMC. Since the α subunit is expressed both in gonadotropes (as part of LH and FSH) and thyrotropes (as part of TSH), we expected to obtain cells representing one cell type or the other, or possibly a developmental precursor of the two. The patterns of β subunit expression should distinguish between these possibilities. However, Fig. 4 demonstrates that αT3-1 cells do not express any of the glycoprotein hormone β-subunit genes, consistent with these cells being precursors of gonadotropes and/or thyrotropes which express the α- but none of the β-subunit genes. The cells are also negative for expression of GH, PRL, and POMC, as would be expected since the α-subunit gene is silent in the cell types which produce these hormones.

In addition, α-subunit protein is synthesized and secreted by these cell lines. Secretion of α subunit from the αT4-1 cell line is 140 ± 15.5 ng/plate/24 hours and the cells contain 15 ± 1.73 ng/plate (where one plate contains ~ 2 × 10^6 cells), assayed by radioimmunoassay.

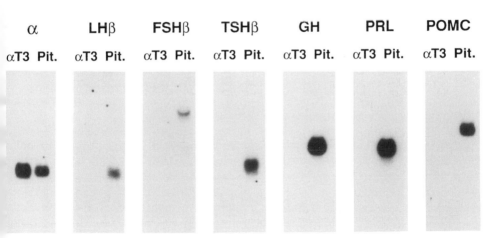

FIG. 4. αT3-1 cells express the α-subunit mRNA but do not express other pituitary hormone mRNAs. Poly(A)$^+$ RNA from αT3-1 cells and total TNA from control pituitaries were run in parallel on Northern blots, and hybridized to cDNA probes for each of the indicated pituitary hormones. The cDNA probes used were as follows: mouse α subunit (Chin *et al.*, 1981), rat LHβ subunit (Tepper and Roberts, 1984), rat FSHβ subunit (Maurer, 1987), mouse TSHβ subunit (Gurr *et al.*, 1983), rat GH (Seeburg *et al.*, 1977), rat PRL (Cooke *et al.*, 1980), and rat POMC (Eberwine and Roberts, 1984; a genomic fragment containing the third exon).

C. HORMONE RESPONSIVENESS OF VARIOUS α–Tag CELL LINES

Synthesis of α-subunit mRNA is stimulated by gonadotropin-releasing hormone (GnRH) in gonadotropes, while in thyrotropes it is regulated by thyrotropin-releasing hormone (TRH; Labrie *et al.*, 1979). Several cell lines derived from the αT3 and αT4 tumors were therefore tested for response of α-subunit mRNA levels to either nafarelin (a GnRH analog) or to TRH (Fig. 5). The cell lines fall into two classes: those that induce α-subunit mRNA levels in response to GnRH, but not to TRH, and those that fail to respond to either hormone. In cell lines which respond, the endogenous mouse α-subunit RNA and the α-Tag transgene RNA (data not shown) were both induced by three- to eightfold. No cell lines which responded to TRH were obtained. Therefore, at least some of the cell lines from the αT3 and αT4 tumors appear to be of the gonadotrope lineage, since they respond to GnRH, a characteristic limited to gonadotropes in the anterior pituitary.

To further characterize the stimulation of α-subunit mRNA levels by GnRH, a concentration curve and time course were performed using the αT3-1 cell line

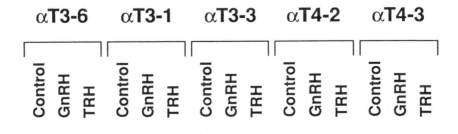

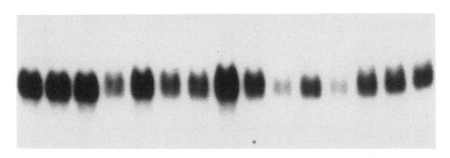

FIG. 5. GnRH and TRH treatment of various α–Tag cell lines demonstrates a range of responsiveness to GnRH and a lack of responsiveness to TRH. Total RNA was prepared from cells incubated for 16 hours without hormone or with 0.1 m*M* nafarelin (a GnRH analog obtained from Syntex) or 0.1 m*M* TRH, run on a Northern blot and hybridized to the α-subunit cDNA probe.

(Fig. 6). A strong response was observed by 8 hours and continued to increase through 48 hours. The induction could be observed with 10^{-9} M nafarelin and increased with greater concentrations of the analog. The increase at 24 hours with 10^{-7} M nafarelin was 8.6-fold. In addition, the response to nafarelin is completely blocked by simultaneous addition of an equal concentration of a GnRH antagonist (0.1 mM; [Ac-D2Nal[1],D4ClPhe[2],D3Pal[3],Arg[5],4-(p-methoxy-benzoyl)-D-2-aminobutyric acid[6],DAla[10]]GnRH; data not shown; Rivier *et al.*, 1986). The finding that the GnRH response is dose dependent and is blocked by a specific antagonist is consistent with GnRH acting through the GnRH receptor. Thus, a number of these cell lines are of the gonadotrope lineage since the presence of GnRH receptors represents a specialized function of pituitary gonadotropes.

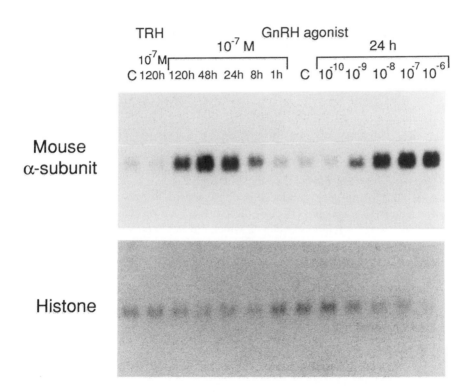

FIG. 6. Time course and concentration curve for response of α–Tag cell line to GnRH are in the physiological range. αT4-1 cells were incubated with the indicated concentration of nafarelin (or TRH, first lane) for the indicated length of time. Total RNA was run on a Northern blot and hybridized to the α-subunit cDNA probe, then the filter was washed and rehybridized to a histone probe.

III.   Hypothalamic GnRH Neurons

The GnRH-producing hypothalamic neurosecretory neurons control reproductive function in mammals by secreting GnRH in a pulsatile pattern to regulate the secretion and synthesis of gonadotropic hormones from the anterior pituitary. Unfortunately, the small number and scattered localization of GnRH neurosecretory neurons in the rostral hypothalamus make study of their cell and molecular biology difficult (Silverman, 1988). The finding that the mouse GnRH gene was expressed appropriately when introduced into transgenic mice (Mason *et al.*, 1986b) and corrected the sterile phenotype of hypogonadal mutant mice [which carry a deletion of the GnRH gene (Mason *et al.*, 1986a)], supported the feasibility of using transgenic technology to target oncogene expression to the GnRH neurons.

We have produced specific tumors of GnRH-secreting neurons by introduction of a hybrid gene composed of the GnRH promoter coupled to the coding region for Tag into transgenic mice (Mellon *et al.*, 1990; Fig. 1). By targeting expression of the oncogene using the regulatory regions of the GnRH gene, we have produced tumors in transgenic mice from which it was possible to derive clonal GnRH-secreting cell lines which represent immortal, differentiated hypothalamic neurons.

A.   INTRODUCTION OF A GnRH–T ANTIGEN TRANSGENE RESULTS IN BRAIN TUMORS

Expression of Tag directed by the 5′ flanking DNA of the rat GnRH gene (GnRH–Tag hybrid gene) produced uniform sterility in nine transgenic founder mice, indicating a specific effect of the transgene on reproductive function. In four animals, brain tumors developed which did not express GnRH mRNA and may have been derived from choroid plexus, a tumor type observed commonly in transgenic mice carrying the Tag gene (Brinster *et al.*, 1984). However, in two other transgenic mice, anterior hypothalamic tumors were observed. Northern blot analysis demonstrated that the tumor from the mouse GT-1 expressed high levels of GnRH and Tag mRNAs and that this expression was specific to the tumor tissue (Fig. 7).

B.   DERIVATION OF IMMORTALIZED NEURONAL CELL LINES

The dispersed GT-1 tumor cells were heterogeneous and included cells with neural and glial phenotypes. Repeated passage using differential plating to separate glia and other cell types generated characteristic cultures of a pure cell population (GT-1) which were cloned by serial dilution (GT1-1, GT1-3, GT1-7).

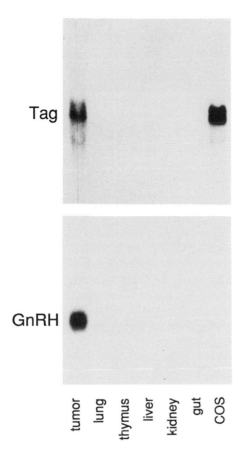

FIG. 7. GnRH and Tag mRNAs are specifically expressed in the GnRH–Tag transgenic mouse hypothalamic tumor. Northern blot of several tissues of the GT-1 mouse hybridized to the SV40 Tag gene or rat GnRH cDNA shows specific expression of both mRNAs in tumor tissue.

These cell lines have been propagated continuously for 2 years with no apparent change in phenotype and can be stored in liquid nitrogen.

Northern blotting (Fig. 8) demonstrated the presence of GnRH and Tag mRNA in the pure culture and in the individual cell lines. Growth rates for these cell lines vary from doubling every 36 hours (GT1-3 and GT1-7) to every 3–4 days (GT1-1). GT cells attached to plastic culture dishes exhibit a distinct neuronal phenotype, including the extension of multiple lengthy neurites which often contact distant cells or end in apparent growth cones (Fig. 9). This neuronal phenotype is more profound when cells are cultured in the absence of serum to inhibit cell division.

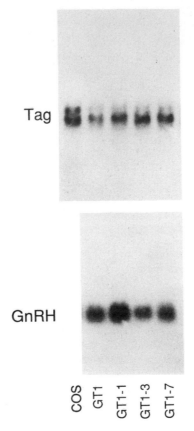

Tag

GnRH

COS GT1 GT1-1 GT1-3 GT1-7

FIG. 8.   GT-1 cell population and clonal cell lines show consistent expression of both Tag and GnRH mRNAs. Northern blot of GT-1 cell population and clonal cell lines (GT1-1, GT1-3, GT1-7) shows consistent expression of both Tag and GnRH and mRNAs after subcloning and culturing for extended time. COS cell mRNA was included as a positive control for Tag mRNA (Mellon *et al.*, 1981).

C.   GT CELLS EXPRESS NEURONAL BUT NOT GLIAL MARKERS

The expression of several well-characterized neuronal- and glial-specific mRNAs was investigated by Northern blotting using poly(A)$^+$ RNA from GT1-7 cells compared to poly(A)$^+$ RNA from mouse brain. The glycolytic enzyme enolase occurs in three forms: nonneuronal enolase, expressed in glial cells; a muscle-specific form; and neuron-specific enolase (NSE; Forss-Petter *et al.*, 1986), strictly localized to central and peripheral neurons. The mRNA for

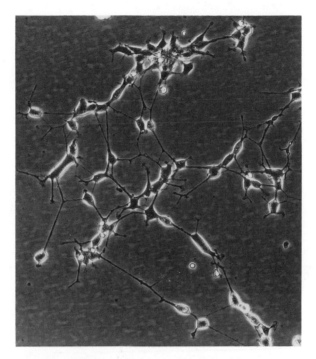

FIG. 9. GT cells develop neurites in culture. A phase-contrast micrograph demonstrates the neuronal phenotype of GT1-3 cells, including neurite formation, growth cones, and cell–cell contacts; ×175. Cells were cultured on plastic tissue culture plates in OptiMEM (GIBCO Laboratories, Grand Island, NY) without serum for 3 days.

the neuronal form was present in higher concentration in GT cells than in total brain (Fig. 10). Neurofilament proteins (NF) are neuron-specific intermediate filament proteins which serve as a structural matrix in axons, dendrites, and perikarya (Lewis and Cowan, 1985). The two mRNAs for the 80-kDa NF protein are also present in GT cells (Fig. 10). In contrast, GT cells fail to express RNAs characteristic of glial cells in the CNS (Lemke, 1988) such as glial fibrillary acidic protein (GFAP; Lewis *et al.*, 1984), myelin basic protein (MBP; Roach *et al.*, 1983), or myelin proteolipid protein (PLP; Milner *et al.*, 1985) (Fig. 10).

The neuroendocrine function of GT cells appears to be limited to GnRH expression in that somatostatin (Montminy *et al.*, 1984), proopiomelanocortin (POMC; Eberwine and Roberts, 1984), corticotropin-releasing hormone (CRF; Imaki *et al.*, 1989), and growth hormone-releasing hormone (GRF; Mayo *et al.*, 1985) mRNAs were not detected.

Several genes which encode proteins specific to the presynaptic membrane have been identified (Trimble and Scheller, 1988). Two related proteins, VAMP-1

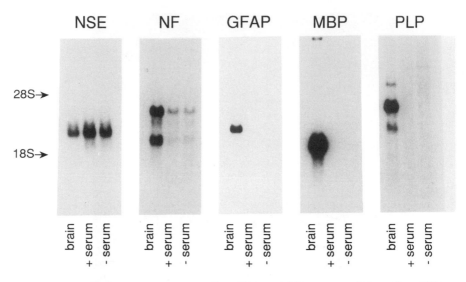

FIG. 10. GT cells express neuron-specific mRNAs and fail to express glial-specific mRNAs. Northern blot of GT1-7 poly(A)+ RNA demonstrates expression of neuron-specific enolase (NSE), the 68-kDa neurofilament protein mRNA (NF), and lack of detectable expression of glial fibrillary acidic protein (GFAP), myelin basic protein (MBP), and myelin proteolipid protein (PLP) mRNAs. RNA from GT1-7 cells cultured in DME with (+) or OptiMEM without (−) serum is compared to poly(A)+ RNA from mouse brain. Expression of neuronal markers does not differ with and without serum, although the cell morphology becomes more neuronal in the absence of mitosis (without serum).

and VAMP-2, are associated with synaptic vesicle membranes and are differentially expressed in the CNS (Elferink *et al.*, 1989). The GT cells express the mRNA for VAMP-2 but not for VAMP-1 (Fig. 11), consistent with the known expression of VAMP-2 in the hypothalamus (Trimble *et al.*, 1990). A 25-kDa synaptosomal protein, SNAP-25, is known to be associated with the presynaptic membrane and is specific to neuron (Oyler *et al.*, 1989). The mRNA for SNAP-25 is present at high levels in GT cells (Fig. 11). Chromogranin B (ChrB, also termed secretogranin I; Forss-Petter, *et al.*, 1989) is a neuroendocrine secretory vesicle protein found in the brain, adrenal medulla, and anterior pituitary. It is particularly abundant in the hypothalamus and is present in brain throughout development (Forss-Petter *et al.*, 1989). GT cells contain ChrB mRNA at significant levels (Fig. 11) consistent with secretory function. The presence of these markers of neuronal, neurosecretory, and synaptic membrane function in GT cells further supports their identity as neurons and documents their differentiated state.

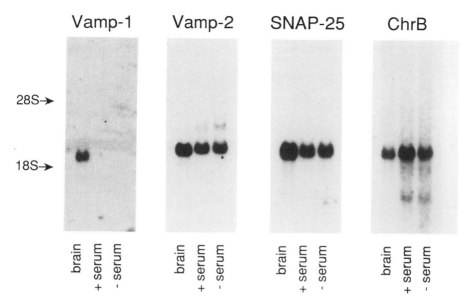

FIG. 11.   GT cells express synapse-specific and neurosecretory-specific mRNAs. Northern blot of GT1-7 poly(A)+ RNA demonstrates expression of VAMP-2 and SNAP-25 mRNAs, lack of expression of VAMP-1 mRNA and high-level expression of Chromogranin B mRNA (ChrB). RNA from GT1-7 cells cultured in DME with (+) or OptiMEM without (−) serum is compared to poly(A)+ RNA from mouse brain.

D.   IMMUNOHISTOCHEMICAL AND ULTRASTRUCTURAL CHARACTERIZATION OF GT CELLS

GT1-3 was characterized extensively by immunocytochemistry and electron microscopy. The cells immunostain for GnRH, Tag, NSE (Fig. 12), and GnRH-associated peptide (GAP; data not shown) (Nikolics *et al.*, 1985) and do not stain for GFAP (Fig. 12) or tyrosine hydroxylase (tyrosine monooxygenase) (data not shown). GnRH is localized in the cytoplasm of cells as well as in neurites (Fig. 12A). Tag immunostaining is nuclear except in dividing cells, where staining is most intense at the periphery. NSE is stained in cell bodies and neurites, while no staining was seen for GFAP. In addition, GnRH can be detected in the media of all three GT cell lines by radioimmunoassay (350 pg/24 hours/10⁶ cells for GT1-3).

The cellular morphology changes with cell cycle. Rounded, dividing cells sit above flattened, extended cells attached firmly to the dish. This is easily seen in Fig. 12B, where the rounded, dividing cells stain peripherally for Tag. Both dividing and nondividing cell stain intensely for GnRH (Fig. 13).

A

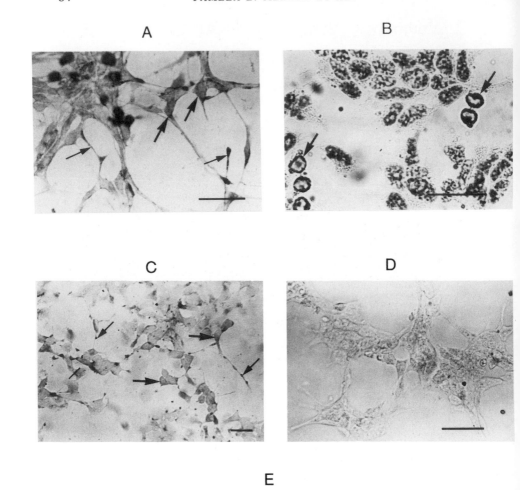

B

C

D

E

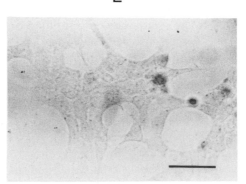

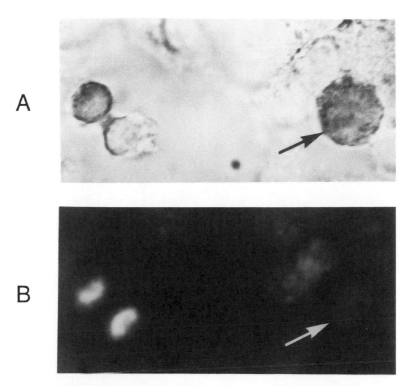

FIG. 13.   GT cells contain GnRH during cell division. GT1-3 cells were immunostained for GnRH, and then fluorescent stained for DNA using Hoechst dye. (A) Transmission light microscopy reveals DAB-labeled GnRH in the peripheral cytoplasm of two apposed, postmitotic cells, and one enlarged nondividing cell (large arrow) (×3150). (B) Epifluorescence microscopy of the same field with wide-band UV excitation reveals the chromatin configuration in the dividing cells. The GnRH-immu-noreactive, postmitotic cells appear to be in telophase. However, in the enlarged cell, the DNA is not condensed and the fluorescence is diffuse (large arrow) (×3150).

←——

FIG. 12.   GT cells immunostain for GnRH, Tag, and neuron-specific enolase but not for GFAP. (A) The cytoplasm within the cell bodies (large arrows) and neuronal processes (small arrows) of many GT1-3 cells stained intensely for GnRH. Dark staining in rounded cells displaced on top of the monolayer was also present in controls, suggesting these cells were no longer viable; (bar = 25 mm). (B) Nuclei of most GT1-3 cells stained positively for Tag, with the exception of dividing cells (arrows) in which the cytoplasm was intensely stained. (C) Cytoplasm of cell bodies (large arrows) and veracosities of neuritic processes (small arrows) of most cells stained strongly for NSE. (D) No specific staining was observed for GFAP. (E) Little background staining was seen in controls in which the primary antibody was substituted with normal rabbit serum.

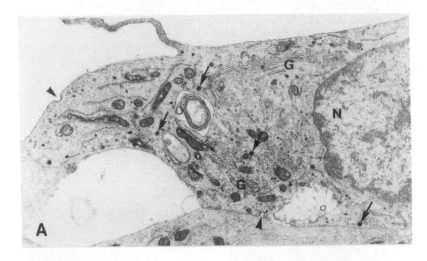

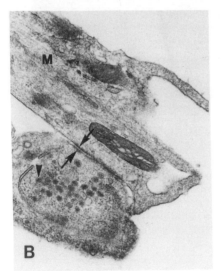

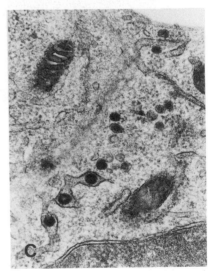

FIG. 14. Cultured GT cells display characteristics of neurosecretory neurons by transmission electron microscopy. (A) A large nucleus (N) and prominent Golgi apparatus (G) occupy this GT1-7 cell body cut parallel to the culture surface. Secretory granules (arrows), cisternae of endoplasmic reticulum, and mitochondria continue into a major dendritic process. Clathrin-coated invaginations (arrowheads) sometimes appear along the plasmalemma (×19,200). (B) Evidence of synaptic-like morphogenesis at points of contact between microtubule-containing processes (M) include aggregations of dense-cored granules, symmetrical thickening of apposed, parallel membranes bordering a cleft (arrows), and occasional coated vesicles (arrowhead) (×43,030). (C) A region of perikaryal cytoplasm shows typical neurosecretory granules condensing in a cistern of smooth endoplasmic reticulum (×62,120).

At the ultrastructural level, GT cell bodies contain both forming and mature neurosecretory granules, Golgi apparatus, and rough endoplasmic reticulum (Fig. 14). Neuritic processes frequently extend from perikarya which contain varying numbers of neurosecretory granules (Fig. 14B). At points of neurite contact, fine structural specializations are suggestive of synaptic-like morphogenesis (Fig. 14B). The condensation of neurosecretory material within smooth endoplasmic cisternae is often indicative of rapid peptide processing (Fig. 14C). Thus, immunohistochemically, as well as morphologically, GT cells appear neuronal in origin and have distinct characteristics of neurosecretory neurons (Mellon *et al.*, 1990).

E. DEPOLARIZATION INDUCES SECRETION OF GnRH

Depolarization of GnRH neurosecretory neurons in brain slice causes rapid release of the neuropeptide (Drouva *et al.*, 1981). The release of GnRH from GT-1 cells was stimulated fourfold ($p < 0.001$) in 15 minutes by treatment with 56 mM K$^+$ (Fig. 15). The GnRH response to K$^+$ was rapid and reached 72% of the maximum in 1 minute. Veratridine depolarizes neurons via the opening of fast, tetrodotoxin-sensitive Na$^+$ channels (Ohta *et al.*, 1973). Depolarization of GT1-7 cells with 50 mM veratridine caused a sixfold ($p < 0.01$) increase in GnRH release. The response to veratridine was blocked by a 30-minute pretreatment of the cells with 50 mM tetrodotoxin. This blockade demonstrates the presence of fast Na$^+$ channels which are necessary for the occurrence of propagated action potentials. In contrast, the action of K$^+$ was not affected by

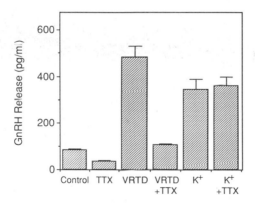

FIG. 15   Depolarization of GT cells with K$^+$ or veratridine (VRDT) stimulates release of GnRH. Treatment of GT1-7 cells with 56 mM K$^+$ or 50 mM veratridine for 15 minutes dramatically increased radioimmunoassayable GnRH release into the medium. The veratridine but not the K$^+$ stimulation was blocked by pretreatment with tetrodotoxin (TTX) for 30 minutues. The Na$^+$ concentration was decreased to maintain isotonicity of the buffer.

tetrodotoxin, indicating that K^+ stimulates GnRH secretion via a different mechanism from veratridine. Studies with hypothalamic slices showing that high K^+ releases GnRH via voltage-dependent Ca^{2+} channels are consistent with this result (Drouva *et al.*, 1981). Tetrodotoxin alone also significantly decreased basal GnRH release, linking basal secretion to spontaneous Na^+ channel activity. These findings demonstrate that hormone release can be regulated in these cultured neurosecretory neurons by depolarization.

IV. Future Prospects

Targeted oncogenesis in transgenic mice provides a means for immortalizing individual cell types. These immortalized cells may be cultured to derive clonal cell lines which maintain many of the differentiated characteristics of the cognate cell type *in vivo*. The generality of the approach is yet to be firmly established by future endeavor. However, in those cases in which the method has proved successful, cell lines which accurately reflect the *in vivo* phenotypes can prove to be invaluable for the study of the cellular and molecular biology of the specific system.

We have successfully used genetic targeting of tumorigenesis to transform and immortalize anterior pituitary cells of the gonadotrope lineage and to derive cell lines which maintain both synthesis and secretion of the α-subunit protein and responsiveness to GnRH (Windle *et al.*, 1990). Study of the molecular and cellular biology of the gonadotrope will be greatly facilitated by the availability of stably transformed gonadotrope cell lines. These cells are proving useful for studies of the mechanisms involved in α-subunit gene expression, synthesis, and secretion, as well as the second messengers and cis-acting sequences in the α-subunit gene which are involved in the response to GnRH. In addition, since these cells display a strong response to GnRH, they provide the first cell model system for the study of the GnRH receptor.

One explanation for the lack of β-subunit expression in these tumors and cell lines is that β-subunit expression correlates with the fully differentiated state of gonadotropes and thyrotropes, and transformation by Tag results in partial dedifferentiation and loss of this function. It is unlikely that dedifferentiation has occurred as a result of culturing, since the pattern of α, Tag, and β expression in the cultured cells duplicates that of the original tumors. We are currently investigating the possibility that dedifferentiation occurs *in vivo* following initiation of Tag expression using lines of α-Tag transgenic mice which develop pituitary tumors at a fairly consistent age.

A more likely explanation for the absence of β-subunit expression is that precursor cells were transformed and arrested in an earlier developmental stage at which they have not yet fully differentiated into gonadotropes or thyrotropes. This is consistent with observations that α-subunit expression appears in ontogeny substantially before LHβ (Simmons *et al.*, 1990). In support of this idea, pituitary tumors from transgenic mice carrying the human or rat LHβ promoter linked to T antigen (J. J. Windle and P. L. Mellon, manuscript in preparation) produce both the α-subunit mRNA and those for the LH and FSH β-subunits. These tumors are apparently derived from more fully differentiated gonadotropes and thus may have been arrested in a later developmental stage by coupling T antigen expression to the LHβ promoter. In this case, it appears that targeted oncogenesis can immortalize cells at specific developmental stages correlated with the onset of expression of the promoter which is used to drive expression of the oncogene. Thus, this method provides a potential tool for the study of cell lineage in mammalian development, an important question which has been difficult to study in higher eukaryotes.

The molecular and cellular events involved in the specific differentiated functions of neurons of the central nervous system have been quite inaccessible to study by molecular and biochemical techniques. The particular difficulty that has long hampered study of the central nervous system has been the inability to transform CNS neurons while maintaining their differentiated state. Immortalized CNS neurons would be invaluable models for the multitude of highly differentiated individual neuronal cell types which compose the brain. The neuroblastoma cell lines which currently exist are derived from peripheral nerve cells. Some cell types of the CNS are represented as cell lines in culture but these are of glial or neuronal/glial precursor origin.

The strategy of targeted oncogenesis has allowed the derivation of clonal cell lines of the hypothalamic neurosecretory neurons which produce GnRH (Mellon *et al.*, 1990). The presence of a GnRH–Tag transgene in the mouse genome throughout development appears to have directed expression of the oncogene during early stages of cellular differentiation. Early in fetal development, GnRH neurons become immunopositive for GnRH while they still remain capable of dividing (Schwanzel-Fukuda and Pfaff, 1989). Thus, targeted oncogene expression in these neurons throughout differentiation may have provided the potential for transformation within the appropriate developmental window. In addition, cells which dedifferentiate and lose expression of GnRH will also lose the expression of the GnRH–Tag transgene. Accordingly, coupling of oncogene expression to the control of a specific marker of differentiation for this population of neurons may select specifically for differentiated cells. Thus, by targeting oncogenesis to a specific, small population of neurons using the regulatory region of a gene which is expressed late in the differentiation of that cell lineage, it was possible to immortalize hypothalamic GnRH neurons which maintain many differentiated functions in culture.

The possibility exists that GnRH neurons represent a unique situation for transformation of CNS neurons. Schwanzel-Fukuda and Pfaff (1989) have shown that GnRH neurons migrate from the olfactory placode into the brain, and continue to divide after expressing GnRH (see also Wray *et al.*, 1989). The developmental origin of olfactory placode is unclear and the possibility exists that it is derived from neural crest rather than neural tube. Tumors of tissues derived from neural crest occur more frequently and such cells may be immortalized more easily.

Nevertheless, these studies demonstrate the feasibility of using targeted oncogenesis to transform specific differentiated neurons which can be established in culture with maintenance of highly differentiated phenotypes. As the field of molecular neurobiology advances, more genes are identified which are expressed in the brain. With further cloning and characterization of promoters from genes expressed specifically in particular neurons or neuronal layers (Travis and Sutcliffe, 1988), the potential exists for obtaining cultured cell models for additional neuronal cell types of the CNS. A promoter specific to the cerebellar Purkinje cells and retinal bipolar neurons has been demonstrated to target these cells in transgenic mice (Oberdick *et al.*, 1990).

In summary, targeted oncogenesis to neuroendocrine cells has demonstrated the potential for immortalization of specific, physiologically important cell types. It has provided a gonadotrope precursor cell which expresses the α-subunit gene and responds to GnRH and which represents a crucial cultured model system for the study of gonadotropin biology. Moreover, it has provided a clonal differentiated neuronal cell line in which to investigate the regulation of expression of the hypothalamic releasing hormone genes and to elucidate the mechanisms involved in specifying gene expression to individual unique populations of neurons in the CNS.

ACKNOWLEDGMENTS

We thank C. Pedula and P. Goldsmith for contributions to the analysis of the GT cell lines. This work was previously presented in Mellon *et al.* (1990) and Windle *et al* (1990). It was supported by NIH Grant HD 20377 and NCI Core Grant CA 14195 (P.L.M.) and NIH Grant HD 08924 (R.I.W.). J.J.W. was a fellow of the American Cancer Society, California division (#S-60-89); current address is Division of Cellular Biology, Cancer Therapy and Research Center, 4450 Medical Dr., San Antonio, TX 78229.

REFERENCES

Baetge, E. E., Behringer, R. R., Messing, A., Brinster, R. L., and Palmiter, R. D. (1988). *Proc. Natl. Acad. Sci. U.S.A.* **85**, 3648–3652.
Botteri, F. M. van der Putten, H., Wong, D. F., Sauvage, C. A., and Evans, R. M. (1987). *Mol. Cell. Biol.* **7**, 3178–3184.
Brinster, R. L., Chen, H. Y., Messing, A., van Dyke, T., Levine, A. J., and Palmiter, R. D. (1984). *Cell (Cambridge, Mass.)***37**, 367–379.
Carr, F. E., and Chin, W. W. (1985). *Endocrinology (Baltimore)* **116**, 1151–1157.
Cepko, C. L. (1988). *Neuron* **1**, 345–353.

Cepko, C. L. (1989). *Annu. Rev. Neurosci.* **12,** 47–65.

Chin, W. W., Kronenberg, H. M., Dee, P. C., Maloof, F., and Habener, J. F. (1981). *Proc. Natl. Acad. Sci. U.S.A.* **78,** 5329–5333.

Cooke, N. E., Coit, D., Weiner, R. I., Baxter, J., and Martial, J. A. (1980). *J. Biol. Chem.* **225,** 6502–6510.

de Vitry, F. (1978). *In* "Cell Biology of Hypothalamic Neurosecretion" (J.-D. Vincent and C. Kordon, eds.), pp. 803–818. Colloq. Int. CNRS, Paris.

de Vitry, F., Camier, M., Czernichow, P., Benda, P., Cohen, P., and Tixier-Vidal, A. (1974). *Proc. Natl. Acad. Sci. U.S.A.* **71,** 3575–3579.

Dichter, M. A., Tischler, A. S., and Greene, L. A. (1977). *Nature (London)* **268,** 501–504.

Drouin, J., Charron, J., Gagner, J.-P., Jeannotte, L., Nemer, M., Plante, R. K., and Wrange, O. (1987). *J. Cell. Biochem.* **35,** 293–304.

Drouva, S. V., Epelbaum, J., Hery, M., Tapia-Arancibia, L., Laplante, E., and Kordon, C. (1981). *Neuroendocrinology* **32,** 155–162.

Eberwine, J. H., and Roberts, J. L. (1984). *J. Biol. Chem.* **259,** 2166–2170.

Efrat, S., Linde, S., Kofod, H., Spector, D., Delannoy, M., Grant, S., Hanahan, D., and Baekkeskov, S. (1988a). *Proc. Natl. Acad. Sci. U.S.A.* **85,** 9037–9041.

Efrat, S., Teitelman, G., Anwar, M., Ruggiero, D., and Hanahan, D. (1988b). *Neuron* **1,** 605–613.

Elferink, L. A., Trimble, W. S., and Scheller, R. H. (1989). *J. Biol. Chem.* **264,** 11061–11064.

Farquhar, M. G., Skutelsky, E. H., and Hopkins, C. R. (1975). *In* "The Anterior Pituitary" (A. Tixier-Vidal and M. G. Farquhar, eds.), pp. 83–135. Academic Press, New York.

Fiddes, J. C., and Goodman, H. M. (1981). *J. Mol. Appl. Genet.* **1,** 3–18.

Forss-Petter, S., Danielson, P., and Sutcliffe, J. G. (1986). *J. Neurosci. Res.* **16,** 141–156.

Forss-Petter, S., Danielson, P., Battenberg, E., Bloom, F., and Sutcliffe, J. G. (1989). *Neuron* **1,** 63–75.

Fox, N., and Solter, D. (1988). *Mol. Cell. Biol.* **8,** 5470–5476.

Frederiksen, K., Jat, P. S., Valtz, N., Levy, D., and McKay, R. (1988). *Neuron* **1,** 439–448.

Giotta, G. J., and Cohn, M. (1981). *J. Cell. Physiol.* **197,** 219–230.

Giotta, G. J., Heitzmann, J., and Cohn, M. (1980). *Brain Res.* **202,** 445–458.

Gurr, J. A., Catterall, J. F., and Kourides, I. A. (1983). *Proc. Natl. Acad. Sci. U.S.A.* **80,** 2122–2126.

Hammang, J., Baetge, E., Behringer, R., Brinster, R., Palmiter, R., and Messing, A. (1990). *Neuron* **4,** 775–782.

Hanahan, D. (1985). *Nature (London)* **315,** 115–122.

Hanahan, D. (1989). *Science* **246,** 1265–75.

Imada, M., and Sueoka, N. (1978). *Dev. Biol.* **66,** 97–108.

Imaki, T., Nahon, J. L., Sawchenko, P. E., and Vale, W. (1989). *Brain Res.* **496,** 35–44.

Jenkins, N. A., and Copeland, N. G. (1989). *In* "Important Advances in Oncology" (V. T. DeVita, S. Hellm, and S. A. Rosenberg, eds.), pp. 61–77. Lippincott, Philadelphia, Pennsylvania.

Labrie, F., Borgeat, P., Drouin, J., Beaulieu, M., Lagace, L., Ferland, L., and Raymond, V. (1979). *Annu. Rev. Physiol.* **41,** 555–569.

Laerum, O. D., Mork, S. J., and De Ridder, L. (1984). *Prog. Exp. Tumor Res.* **27,** 17–31.

Lemke, G. (1988). *Neuron* **1,** 535–543.

Lewis, S. A., and Cowan, N. J. (1985). *J. Cell Biol.* **100,** 843–850.

Lewis, S. A., Balcarek, J. M., Krek, V., Shelanski, M., and Cowan, N. J. (1984). *Proc. Natl. Acad. Sci. U.S.A.* **81,** 2743–2746.

Mason, A. J., Hayflick, J. S., Zoeller, R. T., Young, W. S., Phillips, H. S., Nikolics, K., and Seeburg, P. H. (1986a). *Science* **234,** 1366–1371.

Mason, A. J., Pitts, S. L., Nikolics, K., Szonyi, E., Wilcox, J. N., Seeburg, P. H., and Stewart, T. A. (1986b). *Science* **234,** 1371–1378.

Maurer, R. A. (1987). *Mol. Endocrinol.* **1**, 717–723.

Mayo, K. E., Cerrelli, G. M., Rosenfeld, M. G., and Evans, R. M. (1985). *Nature (London)* **314**, 464–467.

McKay, R. D. G. (1989). *Cell (Cambridge, Mass.)* **58**, 815–821.

Mellon, P. L., Parker, V., Gluzman, Y., and Maniatis, T. (1981). *Cell (Cambridge, Mass.)* **27**, 279–288.

Mellon, P. L., Windle, J. J., Goldsmith, P., Pedula, C., Roberts, J., and Weiner, R. I. (1990). *Neuron* **5**, 1–10.

Milner, R. J., Lai, C., Nave, K.-A., Lenoir, C., Ogata, J., and Sutcliffe, J. G. (1985). *Cell (Cambridge, Mass.)* **42**, 931–939.

Montminy, M. R., Goodman, R. H., Horovitch, S. J., and Habener, J. F. (1984). *Proc. Natl. Acad. Sci. U.S.A.* **81**, 3337–3340.

Nakamura, M., Nakanishi, S., Sueoka, S., Imura, H., and Numa, S. (1978). *Eur. J. Biochem.* **86**, 61–66.

Nelson, C., Albert, V. R., Elsholtz, H. P., Lu, L.-W., and Rosenfeld, M. G. (1988). *Science* **239**, 1400–1405.

Nikolics, K., Mason, A. J., Szonyi, E., Ramachandran, J., and Seeburg, P. H. (1985). *Nature (London)* **316**, 511–517.

Oberdick, J., Smeyne, R., Mann, J., Zackson, S., and Morgan J. (1990). *Science* **248**, 223–226.

Ohta, M. Narohashi, T., and Keller, R. F. (1973). *J. Pharmacol. Exp. Ther.* **184**, 143–154.

Ornitz, D. M., Hammer, R. E., Messing, A., Palmiter, R. D., and Brinster, R. L. (1987). *Science* **238**, 188–193.

Oyler, G. A., Higgins, G. A., Hart, R. A., Battenberg, E., Billingsley, M., Bloom, F. E., and Wilson, M. C. (1989). *J. Cell Biol.* **109**, 3039–3052.

Pierce, J., and Parsons, T. (1981). *Annu. Rev. Biochem.* **50**, 465–495.

Rivier, J. E., Porter, J., Rivier, C. L., Perrin, M., Corrigan, A., Hook, W. A., Siraganian, R. P., and Vale, W. W. (1986). *J. Med. Chem.* **29**, 1846–1851.

Roach, A., Boylan, K., Horvath, S., Prusiner, S. B., and Hood, L. E. (1983). *Cell (Cambridge, Mass.)* **34**, 799–806.

Ronnett, G., Hester, L., Nye, J., Connors, K., and Snyder, S. (1990). *Science* **248**, 603–605.

Rosenfeld, M. G., Crenshaw, E., 3rd, Lira, S. A., Swanson, L., Borrelli, E., Heyman, R., and Evans, R. M. (1988). *Annu. Rev. Neurosci.* **11**, 353–372.

Rubenstein, L. J., Herman, M. M., and VandenBerg, S. R. (1984). *Prog. Exp. Tumor Res.* **27**, 32–48.

Schwanzel-Fukuda, M., and Pfaff, D. W. (1989). *Nature (London)* **338**, 161–164.

Seeburg, P. H., Shine, J., Martial, J. A., Baxter, J. D., and Goodman, H. M. (1977). *Nature (London)* **270**, 486–494.

Silverman, A. (1988). *In* "The Physiology of Reproduction" (E. Knobil and J. D. Neill, eds.), pp. 1283–1304. Raven Press, New York.

Simmons, D. M., Voss, J. W., Ingraham, H. A., Holloway, J. M., Broide, R. S., Rosenfeld, M. G., and Swanson, L. W. (1990). *Genes Dev.* **4**, 695–711.

Stewart, T. A., Pattengale, P. K., and Leder, P. (1984). *Cell (Cambridge, Mass.)* **38**, 627–637.

Tashjian, A. H. J., Yasumura, Y., Levine, L., Sato, G. H., and Parker, M. L. (1968). *Endocrinology (Baltimore)* **82**, 342–352.

Tepper, M. A., and Roberts, J. L. (1984). *Endocrinology (Baltimore)* **115**, 385–391.

Tougard, C., and Tixier-Vidal, A. (1988). *In* "The Physiology of Reproduction" (E. Knobil and J. D. Neill, eds.), pp. 1305–1333. Raven Press, New York.

Travis, G. H., and Sutcliffe, J. G. (1988). *Proc. Natl. Acad. Sci. U.S.A.* **85**, 1696–1700.

Trimble, W. S. and Scheller, R. H. (1988). *Trends Neurosci.* **11**, 241–242.

Trimble, W. S., Gray, T. S., Elferink, L. A., Wilson, M. C., and Scheller, R. H. (1990). *J. Neurosci.* **10**, 1380–1387.

Windle, J., Weiner, R., and Mellon, P. (1990). *Mol. Endocrinol.* **4**, 597–603.
Wray, S., Grant, P., and Gainer, H. (1989). *Proc. Natl. Acad. Sci. U.S.A.* **86**, 8132–8136.

DISCUSSION

J. Levine.   I would like to ask you about the pulsatility of LHRH release from these cells. It is an absolutely remarkable finding that these neurons make contact with one another, and apparently release LHRH in a synchronized, pulsatile pattern. This would not only be important in terms of LHRH regulation, but really would serve as a prototype for the study of pulsatile hormone release in any axis. This has been totally lacking in the field. So one really has to be very cautious in interpreting the data. Do you have any biological marker for these increments which would confirm that they are truly pulses that are relevant to the *in vivo* situation? Obviously you cannot correlate LHRH pulses with LH secretion, but, for example, do you see any correlation between secretion and electrophysiological activity? Or, do you at least see any phasic electrophysiological activity in the time domain of pulses, that is, once every 30 minutes?

P. Mellon.   Such experiments are being done by Martha Bosma and Bertil Hille (University of Washington, Seattle). It is really quite difficult to attempt to hold a patch for over half an hour. To believe that you see a pulse, you need to hold the patch for about an hour and a half. They are trying to do this, but it is extremely difficult technically. We did see spontaneous firing.

The other method we are going to attempt is calcium imaging. We believe it is a calcium-dependent event, and Roger Tsien's laboratory at UC, San Diego has agreed to try to do these experiments. He has really marvelous dyes. We are going to try to determine whether an individual cell can pulse or whether cells require cell–cell contact with one another. These are very difficult experiments. We are just in the initial stages and the pulsatility result is actually very preliminary.

W. F. Crowley, Jr.   I was going to say the same thing. That observation does not surprise me given what we know about the networking of GnRH neurons from what Joan King has shown immunocytochemically with the expansion and contraction of the network size across the cycle as well as from the synchrony that is known to occur in neural tissue. Thus, it would be interesting to see if these cells acquire their synchrony of secretion and, if so, over what time course. One theory I have about the onset of puberty is that there is a sort of growing resonance and harmony among widely separated GnRH neurons and as they develop and mature at puberty they begin to coordinate their pulses. Have you studied their early versus late secretory patterns in culture?

P. Mellon.   I am not sure we have cells frozen from many stages through that period, but we could test earlier cultures. However, we break their contacts periodically. In order to subculture these cells we treat them with trypsin, isolate individual cells, and plate them in new dishes. They have to recreate these contacts every time we subculture them, and in the particular situation we discussed, the cells were being grown on Cytodex beads. The beads are packed to a small syringe, so they can set up contacts even between beads. In order to see synchrony, virtually every cell in the syringe will have to be simultaneously pulsing. The cells on the beads in the syringe form gap junctions within 10 to 20 minutes, then they synchronize. When you remove them and analyze cell–cell contacts by electron microscopy, you see that they have all formed contacts even from bead to bead.

W. F. Crowley, Jr.   A similar synchronization of firing of pacemaker tissue is known to occur in heart. Two separate frog hearts placed in the same physiological milieu *in vitro* soon wind up contracting in harmony without any neuronal connection. Is your GSEB responsive to GnRH?

P. Mellon.   When we map GnRH response it does not map to the GSEB binding site. Instead, it seems to map to the cyclic AMP response elements, which is quite confusing since cyclic AMP does not change in response to GnRH.

W. F. Crowley, Jr.   Perhaps part of the problem may well be that you cut your constructs off at −223, and yet you have had a big loss of responsivity with deletions between −1.8 kb and −223.

P. Mellon. We do not have a loss of GnRH responsivity, we only have a loss of basal level.

W. F. Crowley, Jr. Perhaps some of these response elements may be acting by some of the mechanisms that Bert O'Malley was discussing earlier, such as by cooperativity among the CRE elements. Possibly you're picking up only the latter half of these cooperating transcriptional regulators.

Do the α and LH β tumors secrete their products into the circulation? You did not mention anything about the hormonal profile of the animals?

P. Mellon. Do they secrete protein? I know the αT cells make easily measurable alpha. We have not tested the βT cells yet.

W. F. Crowley, Jr. One of the other big mysteries your models may help solve is what the biologic role of free α subunit is, and if you can make it in abundance, you may be able to see changes in other putative target organs. The specific one I would be interested in would be the adrenal cortex. Have you studied this in your animals?

P. Mellon. No. The best way to make α would be with recombinant DNA: overexpress it in COS cells, rather than use this cell. I am not sure it makes as much as an overexpressing cell would.

W. F. Crowley, Jr. How do you account for the fact that they are infertile? Is it just the tumor size and disruption for the hypophyseal–portal blood supply or is it due to their secretory products?

P. Mellon. I would think that the process of cycling is probably severely interrupted by the tumor because by the time we actually excise a tumor from these animals it has invaded the hypothalamus. I believe that the females do not cycle and that the males are fertile only for a short period of time before the tumor gets too large.

N. Schwartz. What would concern me about the pulsatility data is that the pump may introduce pulsatility. A few papers have been published showing apparent basal pulsatility of hypothalamic explants in culture. For some of these the pump itself may have an irregularity in pumping medium past the tissue.

P. Mellon. It has been done in two laboratories using different apparatus. Dick Weiner (UCSF) sees it when he cultures the cells on coverslips and perfuses them in a dish, while the Negro-Vilar group (NIEHS) sees it on Cytodex beads in a syringe. There are many differences between those two methodologies. Secretion is also calcium dependent.

N. Schwartz. The calcium dependence may be the best argument. Before you tried to immortalize them, in the tumor cells in which you were getting some FSH β regulation, could you suppress with inhibin or enhance with activin?

P. Mellon. No, actually we saw no change in α mRNA or secretion in response to inhibin or activin in the αT cells. These cells are at an early developmental step, so perhaps they do not have inhibin or activin receptors yet, and perhaps the LH-secreting cells may provide a better opportunity to study this.

B. O'Malley. Is there any possibility GSEB represents a small family of proteins?

P. Mellon. The Southwestern blot shows a single band.

B. O'Malley. It is a thick band. You could easily have two or three proteins in that band.

P. Mellon. When we purify it, it really looks like a single band.

B. O'Malley. Do you think the β LH gene may have regulation not because something else is there but because there is no CRE element?

P. Mellon. The CRE is crucial for GnRH regulation, but there are proteins that bind CREs that heterodimerize with jun and react to kinase C. We are investigating whether this is the mechanism.

B. O'Malley. Does depolarization lead to an increase in GnRH mRNA?

P. Mellon. We have not been able to show any effects, in the short term, by depolarization which induce or repress mRNA. I think what we need to do is put a marker with a shorter half-life onto a promoter and do it by transfection. This is what we are planning to do now. GnRH mRNA may have a half-life that is a little too long to allow detection of an effect that is not quite as dramatic as TPA.

J. Marshall. Are some of your clonal cells expressing FSH β message? It could be a function of what can be detected. There is data from Dr. Wierman's group and others that FSH β in RNA appears to be stabilized by testosterone, certainly in rat gonadotrope cells. Have you had the opportunity to culture the clonal cells in the presence of testosterone? If so, did they respond in any way? Are there any data on change in steady-state message levels after culture with other steroids?

P. Mellon. The α cells do not seem to. They have been shown by Penti Siiteri (UCSF) to have estrogen receptor, but the α RNA does not respond. We are just starting studies with the βT cells, which are not yet clonal. We intend to start with estrogen and progesterone, but testosterone is also very important. We do not yet know whether they bind steroids or whether they regulate the mRNAs.

G. Teitelman. I am fascinated by the fact that you can get neuronal cell lines from the CNS. I wonder what the origin of the GnRH neurons in the normal animal is. Do they come from the neural tube or do they migrate from outside?

P. Mellon. They migrate from outside. So in terms of whether this methodology will work for immortalizing other kinds of CNS neurons, we do not yet know, but we are going to try to find out.

G. Teitelman. In my experience neurons of the CNS do not proliferate after they withdraw from the cell cycle, even when they contain an oncogene such as Tag. Dr. Hanahan and I have been studying the brains of transgenic mice harboring a hybrid gene composed of the rat insulin II promoter linked to *Tag*, and we found, for some reason, TAG antigen expressed in the nucleus of the tractus solitarius (r/TS) of the brainstem. We examined transgenic mice containing different hybrid genes and, for unknown reasons, neurons of the r/TS express *Tag*. However, these neurons do not divide.

P. Mellon. Do you have evidence with the promoters that you have used whether these genes are turned on prior to cells becoming postmitotic?

G. Teitelman. Not in the r/TS. However, in the hypothalamus we have clear evidence that *Tag* is present in neuroblasts from day 10 of development.

P. Mellon. Doesn't insulin expression in the neural tube turn off pretty early?

G. Teitelman. In the hypothalamus *Tag* persists throughout life. In early embryo it is expressed by proliferating cells located around the neural tube. In older embryos it is present in glial cells where it persists. However, they do not form. So it might be that *Tag* is unable to induce proliferation of all CNS cells, both neuron and glial cells. The hypothalamic neurons of your transgenic line may be able to divide and become neoplastic cells because they originate from an ectodermal placade, not the neural tube.

P. Mellon. Yes, that is possible. I think that if you find a marker that is turned on prior to postmitosis and is maintained throughout development and past that, you might have a better chance. I do not know which markers you have used, but that would be a prerequisite.

G. Teitelman. In transgenic mice harboring a hybrid glucagon–*Tag* gene we found that *Tag* was expressed in proliferating precursor cells and persisted in differentiated neurons. However, these cells were never found in brain tumors. It should be noted that the oncogene was expressed by α cells of the pancreas and these cells proliferated and became tumor cells. [Efrat *et al.* (1986b).]

P. Mellon. Yes, that was a very nice paper. I hope that that is not going to be the story for every CNS neuron marker gene.

G. Teitelman. With all the constructs you were studying in the pituitary, do you have any idea when *Tag* appears in the pituitary precursors? I know it is difficult because your animals die and you cannot generate transgenic lines.

P. Mellon. We have been trying to do this *in situ* with Larry Swanson's group (Salk Institute). They have not yet seen the signals. We think they might be a little too low so we do not yet know.

G. Teitelman. Have you tried histochemistry with *in situ*?

P. Mellon. Yes, but now we are going back to immunohistochemistry.

S. McKnight. I want to refer back to the studies in which you were studying GnRH regulation of the α-subunit promoter. I find it very interesting that it maps to the CRE element, but a little confusing. It might be worth looking at the mouse α promoter instead of the human promoter, because the mouse promoter, as I recall, actually has mutations in the CRE elements which probably render them inactive. Using the human promoter may actually be complicating your experiments.

P. Mellon. That may be very true. We studied the mouse promoter and have shown that it does induce in transfections. We have not yet mapped the activation region. We know that it does not bind CREB at the site that is homologous to the CRE, and so we are quite interested in where GnRH activation will map in the mouse promoter.

M. Wierman. My question concerns the GnRH-expressing cell line. Have you had a chance to study whether gonadal steroids directly affect either the pulsatile release of GnRH or the GnRH message in the cell lines?

P. Mellon. They do not affect GnRH message so far as we can tell. We have not done very extensive time courses and concentration curves, but within the normal paradigms we have tried, we do not see changes in the messenger RNA. We have not yet examined pulsatility changes.

M. Wierman. Do you know if these cell lines express steroid receptor messages or protein?

P. Mellon. We have not looked for message. The binding studies done by Siiteri (UCSF) showed that these cells had a marginal amount or no estrogen at all. I do not think he measured progesterone receptor.

J. Baxter. Does the GSEB require the CREB element to work?

P. Mellon. Yes, it does.

J. Baxter. Can you replace the CREB element with sites that bind other transcription factors?

P. Mellon. We have put an AP-1 site in that position in the human α gene and tested it in placental cells. There we can replace the CRE. It shows about 30% of the activity of the cyclic AMP response elements, so we think that TSEB in the placental cells can interact with other leucine zipper proteins. We are not yet sure in the case of the pituitary.

J. Baxter. Can you place either of the tissue-specific elements right next to the TATA box and obtain CREB-independent stimulation?

P. Mellon. We have not tried to do this. They work fine if we move them as far as 3 kb downstream.

J. Baxter. I was trying to get at the issue of whether the tissue-specific elements really do need the intermediate transcriptional control element, or whether the CREB element is required only because the tissue-specific elements are located at some distance away from the TATA box.

P. Mellon. If we multimerize the TSEB element upstream of the CAAT box of the TK promoter, five copies will give us a small amount of activity which is tissue-specific in the absence of cyclic AMP response elements. Thus, if we really push the system we can see some activity.

J. Baxter. Have you mapped the phorbol ester response element on the GnRH promoter?

P. Mellon. It is quite complicated. There is actually a complete AP-1 site but we can delete it. It has no effect on the down regulation. It seems to be elsewhere, so it may not be through an AP-1 site.

W. Vale. Do the αT3 cells express Pit1?

P. Mellon. No, there is no Pit1 RNA or protein in αT3 cells.

P. Epstein. Many of your constructs produced tumors in the pancreas; do you know if those were exocrine or endocrine?

P. Mellon. No.

P. Epstein. Do you always find a correlation between the time of onset of the tumor and the rate of growth of the tumor cells in culture?

P. Mellon. Every time we tried to culture a tumor that has arisen beyond about 200 days of age, we failed. The αT3 cells come from one of the earliest tumors and they are one of the fastest growing cell lines. I would say yes with our limited experience. Also the level of T antigen RNA is higher in tumors that develop more quickly.

Neuroendocrine Regulation of the Luteinizing Hormone-Releasing Hormone Pulse Generator in the Rat

JON E. LEVINE,* ANGELA C. BAUER-DANTOIN,* LESLIE M.
BESECKE,* LISA A. CONAGHAN,* SANDRA J. LEGAN,† JOHN M.
MEREDITH,* FRANK J. STROBL,*,[1] JANICE H. URBAN,* KIRSTEN M.
VOGELSONG,* AND ANDREW M. WOLFE*

*Department of Neurobiology and Physiology, Northwestern University, Evanston, Illinois
60208 and †Department of Physiology and Biophysics, University of Kentucky, Lexington,
Kentucky 40536

I. Introduction

Reproduction in male and female mammals is critically dependent on the appropriate neurosecretion of luteinizing hormone-releasing hormone (LHRH). This unassailable fact has been best illustrated by experiments in which the actions of the decapeptide have been blocked by immunoneutralization (Fraser *et al.*, 1975; Ellis *et al.*, 1983) or receptor antagonism (Ellis *et al.*, 1983; Grady *et al.*, 1985; Kartun and Schwartz, 1987), treatments which invariably lead to cessation or reduction of gonadotropin secretion, disruption of gonadal function, and infertility. Sterility in mutant, non-LHRH-producing mice (Silverman *et al.*, 1985), and human infertility associated with LHRH insufficiency (Crowley *et al.*, 1985) have also provided clear demonstrations of the disastrous reproductive consequences of inappropriate or deficient LHRH neurosecretion.

While the biological necessity of LHRH neurosecretion can be easily demonstrated, direct analysis of the secretory activity of LHRH neurons has proved to be profoundly difficult; the small dimensions and physical inaccessibility of the hypothalamo-hypophyseal portal plexus, susceptibility of the neurosecretory process to alterations by anesthetics and surgical trauma, and the scattered topography of LHRH neurons have all posed extraordinary technical problems in the study of LHRH neurosecretion. Nevertheless, the past decade of neuroendocrine research has witnessed development of improved approaches to monitor and analyze patterns of LHRH and LH secretion, and a clearer picture of the mechanisms regulating LHRH release has inexorably emerged. Unquestionably, major

---

[1] Present address: Department of Internal Medicine, University of Virginia Medical Center, Charlottesville, Virginia 22908.

physiological issues remain completely unresolved, not the least of these being the physiological and cellular basis of pulsatile LHRH release. That there has been progress, however, is perhaps best evidenced by the increased number of studies which have attempted to examine not only *if*, but additionally *how* LHRH neurosecretion is regulated under a given set of neuroendocrine circumstances. In this article we have attempted to investigate methodically mechanisms by which known regulators of *in vivo* LHRH release may exert their influences. Our quest to identify specific regulators of the "LHRH pulse generator," as described below, has served as a primary impetus for these studies.

II. The LHRH Neurosecretory System

Although the distribution and overall density of LHRH perikarya can vary considerably among species, a common feature of all mammalian LHRH neuronal groups is the existence of at least one major tuberal tract with apparent secretory function (Silverman *et al.*, 1987). In the rat, most of these fibers arise from preoptic, periventricular, and septal neurons and terminate on fenestrated capillaries in the zona externa of the median eminence. Within these neurons the LHRH decapeptide, as well as gonadotropin releasing hormone (GnRH)-associated peptide (GAP) are derived from posttranslational proteolytic cleavage of a 92-amino acid precursor molecule (Seeburg and Adelman, 1984; Nikolics *et al.*, 1985). After axoplasmic transport of LHRH- and/or GAP-containing granules to the secretory terminal, LHRH and/or GAP are released via a Ca^{2+}-dependent, exocytotic mechanism that can be influenced by a variety of synaptic and endocrine influences (Ramirez *et al.*, 1985). The LHRH decapeptide is released from neurovascular terminals in rhythmic bursts (Carmel *et al.*, 1976; see The LHRH Pulse Generator, below) and conveyed to the anterior pituitary via the hypothalamo-hypophyseal portal vessel system. After diffusion from the secondary portal plexus into the extracellular spaces of the anterior pituitary, LHRH molecules bind to specific membrane receptors on gonadotropes to activate one or more second messenger systems (e.g., calcium/calmodulin; see Hawes and Conn, 1989, for review), thereby regulating the secretion and synthesis of the gonadotropins.

Where and how can one best study this seemingly unapproachable physiological system? Given the foregoing cascade of neuroendocrine events, it follows logically that physiological changes in LHRH neurosecretory rates may be directly reflected in the pattern of LHRH levels in (1) the neurovascular spaces of the median eminence, (2) portal vessel plasma, and (3) the extracellular fluid of the adenohypophysis. Under circumstances where direct feedback regulation of gonadotropin secretion is known to be absent, e.g., in gonadectomized animals, the pattern of LH in the peripheral circulation is also generally held to be a reasonably reliable reflection of the underlying pattern of LHRH neurosecretion.

Not surprisingly, a variety of methods have been devised for sampling of LHRH and/or LH at each loci, including portal cannula/catheter systems (Fink *et al.*, 1967; Fink and Jamieson, 1976; Porter and Smith, 1967), focal perfusion techniques (Levine and Ramirez, 1980), and local microdialysis approaches (Levine and Powell, 1989). Some of the same sampling systems have also been used to monitor LHRH levels in the cerebrospinal fluid (Van Vugt *et al.*, 1985), pituitary venous drainage (Irvine and Alexander, 1987), and peripheral blood (Elkind-Hirsch *et al.*, 1982).

At the present time, there exists no "best" way to study LHRH neurosecretion in the rat. Portal vessel LHRH measurements in anesthetized rats best allow for tracking of gross changes in total LHRH release (Sarkar *et al.*, 1976), but are less useful in characterizing physiological changes in the amplitude or frequency of pulsatile LHRH release. Conversely, local perfusion approaches allow for analysis of LHRH pulse parameters, and additionally obviate limitations imposed by pituitary removal and anesthesia (Levine and Ramirez, 1986); absolute recovery rates using these procedures, however, are relatively low and therefore comparisons of mean LHRH levels are more difficult. Measurements of LH secretion in peripheral blood samples can be carried out with the greatest precision and sensitivity, yet LHRH pulse amplitude and mean LHRH level are not always faithfully mirrored by LH release. Do the respective limitations on all of these approaches mean that we will never be able to interpret available (and often, ostensibly opposing) data until better approaches are born (Kalra and Kalra, 1989)? In this article we champion a more optimistic viewpoint. An overlying theme of the work presented in this article holds that we can indeed come closer to solving several long-standing neuroendocrine conundrums, such as the negative feedback actions of gonadal hormones, provided that (1) complementary monitoring approaches are used, (2) adjunct experiments are devised in which one can control and manipulate LHRH stimuli, and thereby analyze the physiological significance of observed LHRH release profiles, and (3) analysis of data is founded on a logical set of premises regarding the cellular mechanisms governing LHRH release.

III. The LHRH Pulse Generator

It is now well established that LHRH is released in rhythmic secretory bursts (Carmel *et al.*, 1976; Levine and Ramirez, 1980, 1982; Levine *et al.*, 1982; Clarke and Cummins, 1982), and that this pulsatile pattern of secretion functions as the primary neural determinant of pulsatile LH secretion (Levine *et al.*, 1982; Clarke and Cummins, 1982). Temporal associations between LHRH and LH pulses have been confirmed in monkeys (Levine *et al.*, 1985; Van Vugt *et al.*, 1985; Pau *et al.*, 1989), sheep (Levine *et al.*, 1982; Clarke and Cummins, 1982; Karsch *et al.*, 1987; Caraty and Locatelli, 1988), rabbits (Pau *et al.*, 1986), and rats (Levine and Duffy, 1988; Urbanski *et al.*, 1988). Release of LHRH may also support FSH secretion, although in rats the evidence is less compelling that pulses of LHRH directly stimulate pulses of FSH (Levine and Duffy, 1988; Culler and Negro-Vilar, 1986). The LHRH decapeptide may instead primarily support other processes leading to secretion of follicle-stimulating hormone (FSH). Sluggish and/or restrained FSH secretory dynamics may also obscure obvious LHRH/FSH relationships (Grady and Schwartz, 1981). The two gonadotropins, in turn, promote gametogenesis, stimulate gonadal hormone production, and trigger ovulation. Thus, since a pulsatile pattern of LHRH release has been demonstrated to be indispensible for sustained secretion of the gonadotropins (Knobil, 1980, 1981; Valk *et al.*, 1980), one can, by extension, assert that virtually all major reproductive processes are ultimately dependent on the proper functioning of the neural mechanisms governing pulsatile LHRH release.

What is the LHRH pulse generator? At best, we can give the vague description of the LHRH pulse generator as a set of neurons that periodically fire a high-frequency volley of action potentials which culminate in the neurosecretion of a pulse of LHRH (Kawakami *et al.*, 1982; Kaufman *et al.*, 1985). The neuronal groups and/or synaptic circuits composing the LHRH pulse-generating mechanism are still poorly understood. While LHRH neurosecretory cells compose at least the effector portion of the pulse-generating mechanism, it is not clear whether they are intrinsically capable of pulse generation, or if they are primarily driven by pulsatile synaptic signals from an extrinsic pulse-generating source. Restoration of LH pulsatility in hypogonadal mice following icv transplantation of preoptic tissue (Kokoris *et al.*, 1988) suggests that LHRH neurons are either intrinsically capable of pulsatile release, or can derive a pulsatile drive from afferent synaptic contacts within the median eminence. Pulsatile LHRH release from superfused hypothalamic fragments (Dluzen and Ramirez, 1986a) or, more recently, from immortalized LHRH-producing neurons (Mellon *et al.*, 1990; Martinez de la Escalera *et al.*, 1990) also suggests that the LHRH neuron is itself capable of releasing LHRH in a pulsatile fashion. A relationship between these *in vitro* release episodes, and either pulsatile electrophysiological activity or the *in vivo* pattern of pulsatile LHRH and LH secretion awaits confirmation. Others have ascribed the pulsatile rhythm to one or more rhythmic synaptic inputs, such as those which may be conveyed from noradrenergic neurons (Teresawa *et al.*, 1988).

Given the uncertainty as to the neurochemical content of LHRH pulse-generating cells, it comes as no surprise that the anatomical location of the pulse generator in the rat is also not well defined. In the monkey, there is general agreement that the integrity of the arcuate nucleus is both necessary and sufficient for the maintainence of pulsatile LH secretion (Krey *et al.*, 1975). That the bulk of LHRH perikarya are located within this area is consistent with the idea that

LHRH neurons or, at the very least, closely situated afferents to LHRH neurons, are themselves responsible for generating pulsatile LHRH neurosecretory patterns. Although early deafferentation (Blake and Sawyer, 1974) and lesion (Soper and Weick, 1980) experiments suggested that a similar situation may prevail in rats, immunocytochemical and *in situ* hybridization studies have not allowed for a similar consensus to be reached. While some pro-LHRH immunopositive and pro-LHRH mRNA-expressing neurons can be found in the mediobasal hypothalamus of the rat, the preponderance of these neurons is located much more rostrally in septal, diagonal band, OVLT, and medial preoptic areas (Ronnelkleiv *et al.*, 1989). Pharmacological experiments, namely, local application of an α-adrenergic receptor blocker (Jarry *et al.*, 1990), have also provided evidence that the latter area is of particular importance in the regulation of pulsatility. How, then, can pulsatile LH secretion proceed after anterior hypothalamic deafferentation (Blake and Sawyer, 1974)? One possibility is that some descending LHRH-containing tracts may escape section in the latter type of experiment. It may also be possible that following removal of a large number of cells from the pool of LHRH pulse-generating neurons, a remaining few from an original functional syncytium can assume pacemaking activity, and thereby maintain continued pulse generation.

Apart from the location and identity of the pacemaker cell(s) which drive pulsatile LHRH release, other important issues also remain completely unresolved. How are the *in vivo*, electrophysiological activities of widely dispersed LHRH perikarya synchronized? What are the cellular mechanisms which underlie bursting activity in neurosecretory cells, particularly at such long interburst intervals? Through which intercellular routes, and by which intracellular and molecular mechanisms, is the activity of the LHRH pulse generator subject to neural and endocrine regulation? Under which physiological circumstances do these regulatory mechanisms operate? Described below are experiments that were designed as initial steps in a long-term plan to address these questions. An immediate objective of these experiments was to identify specific neurochemical and endocrine inputs to the pulse-generating mechanism; our longer term goal is to ultimately use these as pharmacological, biochemical, and molecular tools to probe the LHRH pulse-generating process, and to characterize the mechanisms by which this activity is regulated. A corollary objective in these experiments was, of course, to identify neuroendocrine regulators which may influence LHRH neurosecretion by other means.

IV. A Working Premise

While countless neural and endocrine factors likely regulate the secretion of LHRH, in most cases it is not known whether a particular regulatory input is manifest at the level of the LHRH pulse-generating mechanism, or if it instead (or additionally) regulates some other cellular process leading to secretion of LHRH. The latter route could include direct synaptic excitation (parallel to pulsatile stimulation), or regulation of LHRH synthesis, posttranslational processing, or stimulus–secretion coupling (Fig. 1). Hypothalamic regulatory mechanisms may also have counterpart actions at the level of the adenohypophysis; these have been particularly troublesome to identify, especially in the analysis of negative feedback actions of gonadal hormones.

In attempts to distinguish regulatory actions on the pulse-generating mechanism, we have relied on the following premise: LHRH pulse amplitude can depend on any one of a number of variables, including (1) characteristics of the underlying action potential volley, (2) size of the readily releasable LHRH pool, (3) stimulus–secretion coupling dynamics, (4) degree of recruitment of cells into the pulsing neuronal pool, and (5) degree of synchronization of activity among a population of pulsing neurons. The frequency of LHRH pulses, however, can really depend only on the frequency at which action potential volleys are generated; this, by definition, is a function of the LHRH pulse generator. Thus, specific inputs to the pulse generator should be identifiable if they suppress or enhance the frequency, but not the amplitude, of pulsatile LHRH release.

The foregoing premise is applied with two important caveats. The first of these is that our assertion "frequency modulation indicates regulation of the pulse generator" is not accompanied by the proposition "amplitude modulation indicates regulation of some other mechanism(s)." Indeed, we have been careful to note that amplitude modulation could conceivably arise through alterations in the characteristics of pulse-generating action potential volleys. Such changes would almost certainly involve regulation of some aspect of the pulse-generating process. The purpose of the working premise is to provide a means to identify neural and endocrine factors which we can be reasonably sure regulate *only* the pulse-generating mechanism; while amplitude-modulating factors may indeed regulate the pulse generator, we simply are not able to isolate experimentally these actions from those which may be exerted elsewhere, for example, at the level of the stimulus–secretion coupling mechanism. A second amendment to the working premise allows for the possibility, even the likelihood, that LHRH pulse amplitude may be altered secondarily to changes in LHRH pulse frequency. These amplitude changes, however, would be expected to be inversely related to the frequency change, since they could occur as a result of the differing degree of recovery of the readily releasable peptide pool during interpulse intervals of different lengths (O'Dea *et al.*, 1989).

A limitation of our approach—the analysis of LHRH or LH data to assess pulse generator activity—is that we do not directly monitor the electrophysiological activity of the pulse generator. We must, therefore, infer that secretory changes observed under particular experimental circumstances occur as a result of alterations in this activity. The relative strength of our approach, however, is that we directly monitor the output of the pulse generator. Thus,

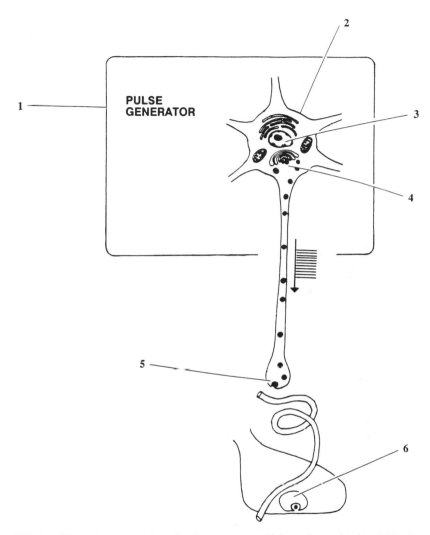

FIG. 1.   Schematic representation of various processes which may be regulated to yield a change in luteinizing hormone-releasing hormone (LHRH) release or actions. These include (1) LHRH pulse generation, (2) direct synaptic activation or inhibition of LHRH neurons, (3) proLHRH biosynthesis, (4) proLHRH posttranslational processing, (5) stimulus–secretion coupling, and (6) LHRH-stimulated LH secretion.

changes in both frequency *and* amplitude of LHRH release can be identified. This advantage becomes particularly important in the identification of factors which not only regulate pulse generator activity, but may also, or instead, exert effects on LHRH release downstream from the pulse generator, e.g., at the secretory terminal. An example of the latter almost certainly includes the endogenous opioid peptides, which may suppress both pulse generator activity (Williams *et al.*, 1990) and stimulus–secretion coupling (Drouva *et al.*, 1981; Lincoln *et al.*, 1985).

V. Experimental Methods

A. MONITORING LHRH RELEASE PATTERNS

In many of the experiments described within this article, LHRH release patterns were monitored either by push–pull perfusion (Fig. 2) of the mediobasal

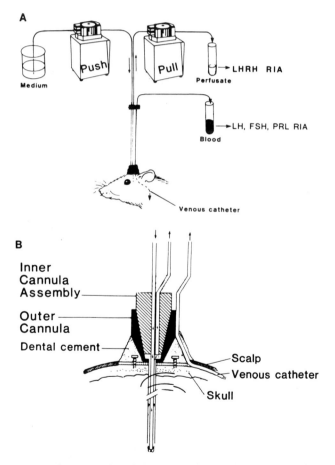

FIG. 2.   Diagram of the push–pull perfusion system (A) and the cannnulas and catheter assemblies (B). For details, see text. LH, Luteinizing hormone; FSH, follicle stimulating hormone; PRL, prolactin; RIA, radioimmunoassay. (From Levine and Ramirez, 1986.)

hypothalamus (Levine and Ramirez, 1980, 1986) or intrahypophyseal micro-dialysis, (Fig. 3) (Levine and Powell, 1989). In some experiments in gonadec-tomized animals, LHRH patterns were instead estimated by means of high-frequency (2.5 or 5 minute) sampling of peripheral LH levels.

Push–pull perfusion was first introduced by Levine and Ramirez (1980) as a method for measurement of LHRH release in conscious, freely moving animals (Levine and Ramirez, 1986). It has since been used to examine LHRH release from mediobasal hypothalamic/median eminence or adenohypophyseal loci in rats (Levine and Duffy, 1988; Dluzen and Ramirez, 1987), rabbits (Ramirez *et al.*, 1986; Pau and Spies, 1986a), sheep (Levine *et al.*, 1982; Schillo *et al.*, 1985), and monkeys (Levine, 1986; Pau *et al.*, 1989; Teresawa *et al.*, 1988). The approach makes use of an arrangement of concentric cannulas which allows for continual diffusion exchange between substances in extracellular fluid and perfu-sate. Perfusion fluid is simultaneously delivered to, and withdrawn from, a locus in the brain such that a small sphere of tissue is continually washed by the perfusion medium. A fraction of the total LHRH released within the perfusion sphere diffuses into the perfusate at a constant rate (Levine and Ramirez, 1986) and can be measured by radioimmunoassay (RIA). Fluctuations of LHRH in successive perfusate fractions have been shown by various validative measures (Levine and Ramirez, 1986) to reflect physiological release patterns. Push–pull perfusion has been used to identify temporal relationships between LHRH re-lease and gonadotropin pulses (Levine *et al.*, 1982; Levine and Duffy, 1988; Urbanski *et al.*, 1988), and to characterize LHRH release profiles during the

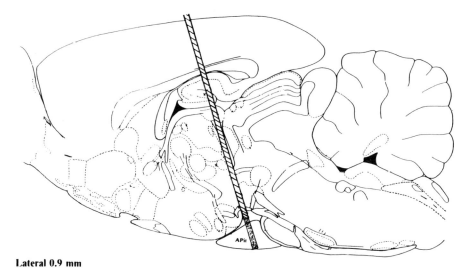

Lateral 0.9 mm

FIG. 3.   Schematic representation of the placement of the intrahypophyseal microdialysis probe. (From Levine and Powell, 1989.)

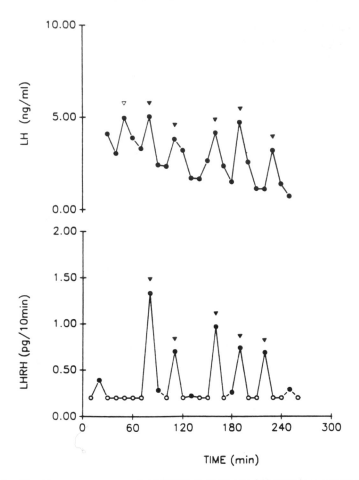

FIG. 4. Simultaneous measurement of LHRH (bottom) and LH (top) in a conscious, freely moving castrate rat. Release rate of LHRH was determined by RIA of mediobasal hypothalamic/median eminence perfusates collected in continuous 10-min fractions. The LH levels were assessed by LH RIA of blood samples obtained at the start of each perfusate collection interval. Open circles denote hormone levels that fell below detectable limits in the particular RIA. Inverted triangles denote hormone pulses as identified by PULSAR analysis (Merriam and Wachter, 1982). Open triangles denote silent LHRH pulses in this graph, as well as the few LH pulses that were not associated with LHRH pulses in subsequent figures. Abscissa label refers to time relative to the beginning of push–pull perfusion procedures. (From Levine and Duffy, 1988.)

estrous cycle (Levine and Ramirez, 1982; Park and Ramirez, 1989), puberty (Watanabe and Teresawa, 1989), hyperprolactinemia (Voogt *et al.*, 1987), following gonadectomy (Dluzen and Ramirez, 1985, 1987; Levine and Duffy, 1988), in aging (Rubin and Bridges, 1989), and after various systemic treatments known to affect secretion of LH (Levine and Ramirez, 1980; Levine, 1986;

Dluzen and Ramirez, 1987); it has also been used to examine the effects of cupric acetate (Pau and Spies, 1986a), neuropeptide Y (Khorram et al., 1987), norepinephrine (Pau and Spies, 1986b; Ramirez et al., 1986), adrenergic agonists and antagonists (Teresawa et al., 1988; Pau et al., 1989), naloxone (Orstead and Spies, 1987; Karahalios and Levine, 1988), corticotropin releasing factor (CRF; Nikolarakis et al., 1988), local depolarization with supraphysiological K^+ (Levine and Ramirez, 1980), and local application of progesterone (Ramirez et al., 1985). The advantages and limitations of the push–pull perfusion approach have been discussed previously (Levine and Ramirez, 1986). Figure 4 depicts simultaneous LHRH and LH profiles as measured in a conscious, freely moving castrate male rat, using hypothalamic push–pull perfusion in combination with peripheral blood sampling (Levine and Duffy, 1988).

Microdialysis is a relatively new monitoring approach that we have adapted to measure LHRH levels in the extracellular fluid of the anterior pituitary gland (Fig. 3). The basic principle of sampling by microdialysis is similar to that which has been described for push–pull perfusion techniques: a moving pool of fluid (infusate) is brought into contact with a relatively stationary pool of fluid (extracellular fluid) and exchange of solute from the latter to the former pool occurs by diffusion. The major difference between the approaches is that microdialysis makes use of a semipermeable membrane as a physical boundary between infusate and extracellular fluid. With the exception of the addition of a cylinder of dialysis membrane to the tip of the probe, the arrangement of microdialysis cannulas is essentially the same as the concentric push–pull cannula design. The potential advantages and disadvantages of the use of microdialysis for monitoring neurohumoral factors have also been discussed in two reviews (Kendricks, 1989; Levine and Powell, 1989).

High-frequency measurements of pulsatile LH patterns were used in some experiments as a less direct estimate of LHRH release patterns. The application of this approach, however, was limited to acute pharmacological experiments in gonadectomized animals. In these situations, we and others have previously confirmed that pulsatile LH secretion is indeed a reliable endocrine marker for pulsatile LHRH release (Levine et al., 1982, 1985; Levine and Duffy, 1988).

B. A HYPOTHALAMIC CLAMP APPROACH IN THE RAT

The ability to manipulate and deliver LHRH stimuli to a surgically isolated pituitary gland has proved to be of enormous benefit in the study of the hypothalamo-hypophyseal unit of the monkey (Knobil, 1980; Norman et al., 1982). We have developed an isolated pituitary paradigm in rats (Strobl and Levine, 1988, 1990; Strobl et al., 1989) which operates on the same principle as that for experiments conducted in monkeys or sheep with stalk transections (Clarke and Cummins, 1984) or lesions of the arcuate nucleus (Knobil, 1980), or in human patients with hypogonadotropic hypogonadism (Crowley et al., 1985). In this

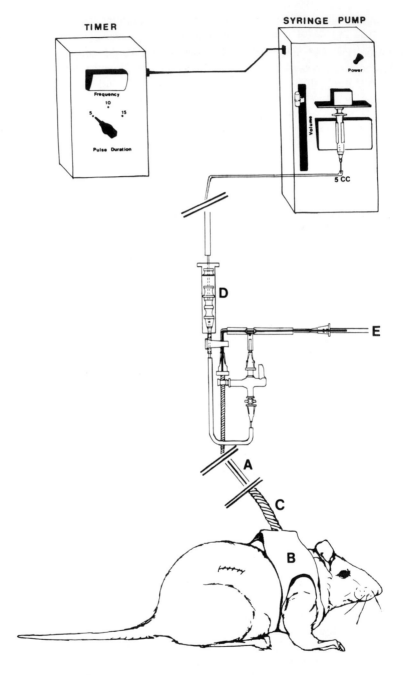

FIG. 5.   Schematic diagram of the *in vivo* isolated pituitary paradigm. See text for details. (From Strobl and Levine, 1990.)

type of approach, pulsatile LHRH stimuli are presented to a functioning pituitary gland that is devoid of endogenous hypothalamic drive. Our version of this "hypothalamic clamp" technique in the rat is the hypophysectomized, pituitary-grafted animal (Fig. 5). In this animal model, hypophysectomized rats receive anterior pituitary transplants under the kidney capsule, and are fitted with a concentric atrial catheter device which allows for infusions through a center catheter and blood sampling from a side port. A timer directs an infusion pump to deliver pulsatile LHRH infusions. Since the system is protected by a jacket, tether, and swivel, the animal can recover, move freely, and receive the infusions for an unlimited period of time. Use of the technique has been described in detail (Strobl and Levine, 1990); below we review its use in experiments which complement our analysis of gonadal hormone feedback mechanisms at the hypothalamic level.

A similar "withdrawal and replacement" strategy was used in proestrus rats to examine the neurohumoral regulation of LH surges (see Section VI,D). In these experiments, however, only acute blockade of hypothalamic drive was required to examine the effects of immediate replacement with exogenous LHRH and/or neuropeptide Y. The well-established pentobarbital-blocked, proestrous rat (Everett and Sawyer, 1950) was utilized for this purpose.

VI.   Neural and Endocrine Regulation of LHRH Release

Of literally dozens of neuroendocrine factors which likely regulate the output of LHRH neurons, those that have received the most attention include (1) gonadal hormones (operating as negative feedback regulators), (2) endogenous opioid peptides, (3) catecholamines, (4) estrogen and progesterone (operating as positive feedback regulators in females), and (5) neuropeptide Y (NPY). While the involvement of each of these factors in the normal operation of the hypothalamic–pituitary–gonadal axis is now well recognized, the loci and mechanism(s) of their physiological actions have been the subject of much investigation and, often, considerable confusion and debate (Kalra and Kalra, 1989). Below we consider the potential role(s) of these neuroendocrine factors in the physiological regulation of pulsatile LHRH/LH release. For each, we have attempted to better understand both "how" and "where" the regulatory input is manifest. Which specifically regulate the pulse generator, according to our working premise? Which may have effects downstream from the pulse-generating mechanism, at the secretory terminal, or at the level of the anterior pituitary?

A.   NEGATIVE FEEDBACK ACTIONS OF GONADAL HORMONES

1.   Effects of Castration on LHRH Release

In virtually all male mammals, LH secretion is predominantly regulated by two influences: pulsatile LHRH stimulation, and the negative feedback actions of gonadal hormones. In testes-intact male rats, LH pulses are of low and irregular amplitude and frequency (Ellis and Desjardins, 1982), and elimination of circulating gonadal hormones by castration results in rapid and continued increases in mean LH levels, LH pulse frequency, and LH pulse amplitude (Ellis and Desjardins, 1984). Indeed, even by 24 hours following removal of the testes, both LH pulse frequency and amplitude exhibit robust increases in this species. Treatment of castrates with physiological testosterone replacement therapy has been shown to return LH pulse amplitude and frequency to levels characteristic of the gonadally intact male (Summerville and Schwartz, 1981; Steiner *et al.*, 1982).

The site(s) and mechanism(s) by which testosterone may exert its negative feedback actions have remained unclear, given previous unexpected findings that LHRH levels in anesthetized, stalk-transected rats are not changed in castrates (Brar *et al.*, 1985), and that LHRH content (Kalra and Kalra, 1982) and *in vitro*

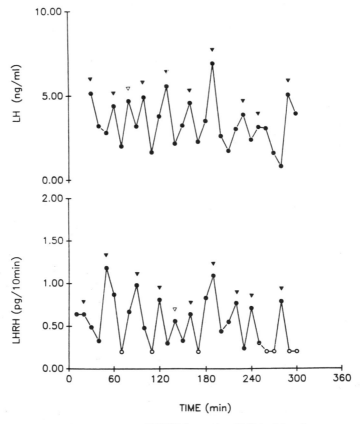

FIG. 6.   Simultaneous measurement of LHRH (bottom) and LH (top) in a short-term castrate rat. See Fig. 4 for explanation of symbols. (From Levine and Duffy, 1988.)

release (Kalra *et al.*, 1987), are decreased with time following orchidectomy. One hypothesis, proposed on the basis of experiments with arcuate nucleus-lesioned monkeys (Plant and Dubey, 1984), has held that testosterone may retard the activity of the LHRH pulse generator, and thereby exert its feedback suppression of LH secretion. According to this view, removal of testosterone by castration would result in an acceleration of LHRH pulse generator activity, namely, an increase in the frequency of pulsatile LHRH release. In another study (Levine and Duffy, 1988), we tested this hypothesis by simultaneously monitoring *in vivo* LHRH and LH release at 24–32 hours following castration or sham surgery. Push–pull perfusion of the mediobasal hypothalamus was used in combination with peripheral blood sampling procedures to measure LHRH and LH in perfusates and plasma samples, respectively. In the castrate rats, LHRH release was of relatively high frequency (Fig. 6), and nearly all identified LHRH pulses were temporally associated with LH pulses. In testes-intact rats, by contrast, LHRH

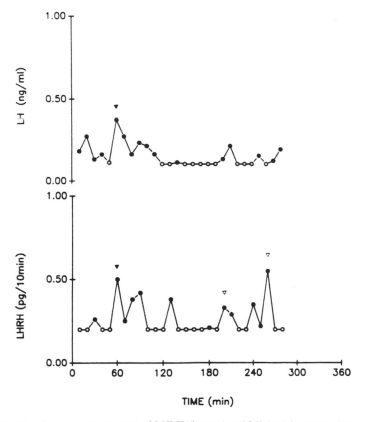

FIG 7.   Simultaneous measurement of LHRH (bottom) and LH (top) in a testes-intact rat. See Fig. 4 for explanation of symbols. (From Levine and Duffy, 1988.)

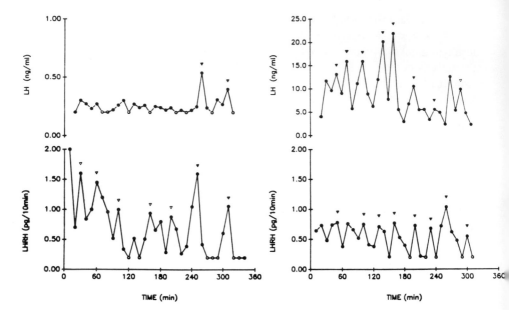

FIG. 8.   Simultaneous measurement of LHRH and LH in a testes-intact rat (left-hand side) and a short-term castrate rat (right-hand side). Note the abundance of silent LHRH pulses in the testes-intact rat, but not in the castrate rat. See Fig. 4 for explanation of symbols. (From Levine and Duffy, 1988.)

pulses were less frequent (Fig. 7), and many "silent" LHRH pulses (those not associated with corresponding LH pulses) were observed (Fig. 8). A comparison of the summary data from the two groups is given in Table I, which reveals that LHRH pulse frequency, but not LHRH pulse amplitude or mean level, was significantly increased in the castrate animals. While LHRH pulse amplitude was

TABLE I

Characteristics of Pulsatile LHRH Release in Intact and Short-Term Castrate Rats as Determined by Push–Pull Perfusion[a]

	LHRH levels	
Characteristic	Intact rat	Castrate rat
Number (*n*)	7	7
Pulse frequency (pulses/hour)	0.83 ± 0.13	1.30 ± 0.16
Pulse amplitude (pg, trough to peak)	0.83 ± 0.24	0.58 ± 0.06
Hormone level (pg/10 min)	0.46 ± 0.11	0.47 ± 0.04

[a]Values are mean ± SEM.

not changed significantly in the castrates, it did decline to the extent that it offset the increase in LHRH pulse frequency to yield an unchanged mean level.

2. A Model for the Hypothalamic–Pituitary–Testicular Axis of the Rat

The two major findings of the hypothalamic push–pull perfusion study—increased LHRH pulse frequency in castrates, and "silent" LHRH pulses in intacts—indicate that gonadal hormones normally exert two different feedback actions on the hypothalamic-hypophyseal unit, these being (1) retardation of the activity of the LHRH pulse generator, and (2) suppression of pituitary responsiveness to the decapeptide. On the basis of these observations and the earlier LH and testosterone measurements of Ellis and Desjardins (1982), we proposed a working model (Fig. 9) for the normal functioning of the rat hypothalamic–pituitary–testicular axis (Levine and Duffy, 1988). The model holds that there is a continual cycling of the following neuroendocrine events within the male rat reproductive axis: an LHRH pulse initiates an LH pulse, one or more LH pulses evoke an episode of heightened testosterone secretion (Ellis and Desjardins, 1982), and testosterone suppresses responsiveness of the pituitary to further stimulation by the LHRH decapeptide (Nansel et al., 1979). The pituitary suppression is self-limiting, as it temporarily eliminates LH support of testosterone secretion. The model incorporates the idea that there is little, if any, moment-to-moment endocrine regulation of LHRH release, provided that testosterone levels remain within physiological limits; if levels of testosterone drop below these limits for a prolonged period (e.g., following castration), then pulse generation would gradually increase in frequency, in response to the removal of negative feedback suppression of the pulse generator. According to our model, the

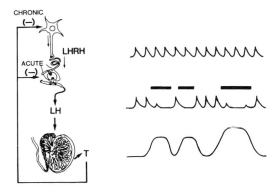

FIG. 9. Schematic representation of a working model for the operation of the hypothalamic–hypophyseal–gonadal axis of the adult male rat. From top to bottom on right-hand side, lines represent pulsatile LHRH release, episodes of enhanced direct negative feedback (dark bars), LH secretion, and testosterone secretion. See text for details of the proposed model. (From Levine et al., 1990.)

"housekeeping" homeostasis within the male reproductive axis is maintained by the ebb and flow of feedforward and feedback regulation between pituitary and gonad. Longer term shifts in the activity of the axis, or compensatory responses to unusual perturbations such as castration, are manifest via changes in the frequency of LHRH pulse generation.

3. Testing the Model: Is LHRH Pulse Frequency Really Increased following Castration?

We have conducted a series of experimental tests of the foregoing model. In one test, we have attempted to confirm that the LHRH release profiles observed in our hypothalamic push–pull perfusion study are indicative of the pattern of LHRH stimulation reaching gonadotropes. Using intrahypophyseal micro-

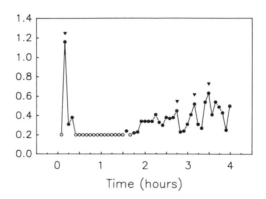

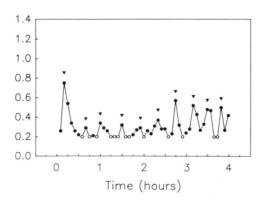

FIG. 10.   Patterns of LHRH levels in intrahypophyseal microdialysates in a testes-intact rat (top) and a short-term castrate (bottom) male rat. Inverted triangles in this and subsequent figures denote significant hormone pulses as identified by ULTRA pulse analysis (Van Cauter *et al.*, 1989). (From Levine and Meredith, 1990.)

dialysis, we monitored LHRH levels in the extracellular, hypophyseal fluid of testes-intact and castrate male rats (Levine and Meredith, 1990). Commercially available probes (Carnegie-Medicin, Stockholm, Sweden; Bioanalytical Systems, West Lafayette, IN) were used to dialyze the extracellular fluid of the adenohypophysis. The probes were used at a flow rate of 2.5 μl/min, which was found in *in vitro* recovery experiments to yield a 4% exchange rate (Levine and Meredith, 1990). We have determined that the polycarbonate/polyether membrane at the tip of these CMA/10 probes allows consistent, rapid exchange of the decapeptide, at a level which is sufficient to permit *in vivo* experimentation in male rats (Levine and Meredith, 1990), and in steroid-treated, ovariectomized female rats (unpublished observations). The overall mean level in dialysates from all animals in these experiments was 0.44 pg/10 min (Levine and Meredith, 1990). Using the 4% relative recovery rate, it is estimated that the average extracellular concentration of LHRH is 0.37 nM, which compares well with EC50 values for LHRH-stimulated LH secretion (0.8–20 nM).

In these experiments, the overall LHRH mean level was found to be unchanged following castration (Levine and Meredith, 1990). A significant effect on LHRH pulse frequency, however, emerged when we analyzed the pattern of LHRH levels in intrahypophyseal microdialysates obtained over several hours of experimentation in short-term castrate versus testes-intact rats (Levine and Meredith, 1990). Figure 10 depicts the LHRH data obtained in two rats, one following castration and the other following sham surgery. The results from these experiments are in direct agreement with data obtained in our earlier push–pull perfusion study; there is a moderate, but significant, increase in the frequency of LHRH pulses in microdialysates obtained from castrate animals (Levine and Meredith, 1990). Moreover, like the earlier push–pull perfusion data, the LHRH pulse amplitude was not increased in the castrates compared to the intacts; if anything, there was again a slight (nonsignificant) decrease in LHRH pulse amplitude in the castrates which offset the increase in LHRH pulse frequency, yielding no overall net change in the mean LHRH level. Interestingly, mean LHRH pulse amplitudes in both groups were smaller than those observed in their counterpart groups studied by hypothalamic push–pull perfusion, despite the fact that mean levels observed in both studies were remarkably similar. This difference may be explainable on the basis of the possibility that dilution of LHRH pulses occurs during transit through the portal vasculature. This difference notwithstanding, the intrahypophyseal microdialysis data provide confirmation from the "other end" of the portal vessels that gonadal hormones normally suppress the activity of the LHRH pulse generator.

4. Testing the Model: Do Gonadal Hormones Really Suppress Pituitary Responsiveness?

To test the pituitary component of the model directly, we used the *in vivo*, isolated pituitary paradigm to examine the responses of ectopic pituitaries to castration during unvarying pulsatile LHRH stimulation (Strobl *et al.*, 1989). Hypophysectomized male rats received anterior pituitary transplants under the kidney capsule, and were fitted with the infusion/sampling system described in Fig. 5. The animals received hourly pulsatile LHRH stimulation for 5 days, and were then castrated or underwent sham surgeries. Blood samples were obtained 2 hours prior to surgeries, and at 2-hour intervals for 24 hours thereafter, and analyzed by LH RIA. It was reasoned that if castration were to induce LH secretion from the ectopic pituitaries under conditions of fixed LHRH drive, then conclusive evidence would be provided for the existence of a major hypophyseal site of feedback regulation. As demonstrated in Fig. 11, support was indeed gained from these experiments for direct feedback suppression of LHRH-stimulated LH secretion (Strobl *et al.*, 1989). In the hypophysectomized, pituitary-grafted rats that received sham surgeries, LH levels remained constant throughout the 24-hour postsurgery period. By contrast, plasma LH concentrations in the hypophysectomized, pituitary-grafted castrates increased steadily for approximately 18 hours, when they reached a plateau at levels approximately threefold higher than initial values. Since the pulsatile LHRH drive was not changed during the course of these experiments, it must be concluded that the rise resulted from the removal of direct pituitary feedback suppression by the testes. That this reflects the response of the normal pituitary is suggested by the fact that the absolute amounts of immunoreactive LH, and the trajectory of the LH rise in the hypophysectomized, pituitary-grafted castrates, closely re-

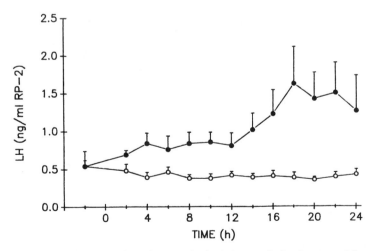

FIG. 11. Plasma LH levels in hypophysectomized, pituitary-grafted male rats receiving pulsatile LHRH infusions. Animals were either castrated (filled circles) or underwent sham surgery (open circles) at time 0, as indicated on the abscissa. All symbols represent the mean ± SEM for $n = 5$ animals/group. (From Strobl *et al.*, 1989.)

sembled those in normal castrate counterparts during the initial 20 hours following removal of the testes. Interestingly, these similarities were not apparent at longer time points following castration, as LH levels continue to rise in the normal castrates, while they plateau in the hypophysectomized, pituitary-grafted castrates. Thus, release from direct pituitary suppression can account for most of the acute response to castration, while the continued postcastration rise in LH secretion may increasingly depend on additional hypothalamic input; according to the proposed model, this input consists of a gradual increase in the frequency of LHRH pulse generation (Levine and Duffy, 1988).

5. Testing the Model: How Is the Readily Releasable LHRH Pool Changed following Castration?

A third test of the proposed model was based on the following rationale: if LHRH pulse frequency is increased following castration, while LHRH biosynthesis is either unchanged (Wiemann et al., 1990) or even decreased (Rothfeld et al., 1987; Park et al., 1988), then one might predict that the amount of LHRH available for release following a given depolarizing stimulus would actually decrease with time following castration. Indeed, this has already been demonstrated to be the case for in vitro LHRH release, either in response to depolarizing extracellular potassium concentrations (Kalra et al., 1987) or electrical stimulation (Dyer et al., 1980) and for in vivo LHRH release following electrical stimulation of the median eminence in anesthetized, portal vessel-cannulated castrate rats (Brar et al., 1985). In analogous experiments in unanesthetized rats, the LHRH secretagogue N-methyl-DL-aspartate (NMA; Price et al., 1978) was used to compare LH secretion to a standard excitatory stimulus in testes-intact and castrate rats. Peripheral administration of the aspartate analog, NMA, has been shown to evoke an acute increase in the secretion of LH in rats (Price et al., 1978), sheep (Estienne et al., 1990), and monkeys (Wilson and Knobil, 1982). The mechanism by which NMA induces this LH secretion almost certainly involves an acute increase in LHRH neurosecretion, since it has been demonstrated that NMA-induced LH secretion can be fully antagonized by prior administration of an LHRH antagonist (Gay and Plant, 1987). Furthermore, it has also been shown that NMA directly stimulates LHRH release from hypothalamic tissues in vitro (Bourguignon et al., 1989). It has been proposed that peripherally administered NMA primarily exerts effects at locations where the blood–brain barrier is diminished, such as the arcuate nucleus (Gay and Plant, 1987). We therefore used peripheral NMA injections to examine and compare secretory responses of sham-operated and castrate (6 day) rats (Strobl et al., 1990); LH responses to a standard dose of excitatory amino acid analog were considered to be indirect estimates of the readily releasable LHRH pool under the different gonadal conditions. Figure 12 depicts the relative LH responses of castrate and testes-intact male rats to challenges with 5 mg NMA, iv. Significant

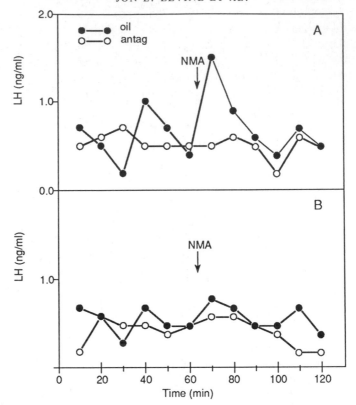

FIG. 12.   Plasma LH levels in intact (A) or castrate (B) rats receiving 5 mg N-methyl-DL-aspartate (NMA) at the time indicated on each graph. Open circles denote data obtained from groups pretreated with LHRH antagonist, and closed symbols represent data from animals pretreated with oil vehicle. (From Strobl et al., 1990.)

increases in LH secretion following NMA were recorded in the testes-intact rats, and these responses were fully blocked by pretreatment with LHRH antagonist; by contrast, LH responses in the castrates were diminished (Fig. 12). Our LH measurements, therefore, indicate that LHRH secretory responses to a standard excitatory stimulus are attenuated following castration. Given that we have shown that LHRH pulse frequency is increased in castrates, and others have shown a similar inverse relationship between hormone pulse frequency and amplitude (O'Dea et al., 1989), our NMA data are consistent with the hypothesis that diminishment of the LHRH-releasable pool accompanies, and may even be secondary to, increased LHRH pulse generator activity.

6.   Out of Neuroendocrine Chaos: Is Any of This Beginning to Make Sense?

The effects of castration on LHRH release have been difficult to comprehend, especially when gross correlates of LHRH neuronal function (tissue content, mean release) have been considered without respect to the pulse-generating mechanisms which govern the LHRH release process. Clearly, the frequency of LHRH pulses is an extremely important determinant of the ongoing level of LH secretion (Knobil, 1980) and, as we suggest below, the mean LHRH level (or LHRH pulse amplitude) may be dissociable from the frequency of LHRH release, or even inversely related to it. A second confounding variable in many studies has been the inability to determine the extent to which direct suppression of LHRH-stimulated LH secretion may contribute to the entirety of gonadal negative feedback. Below we attempt to construct a postcastration neuroendocrine scenario that incorporates the findings of several groups that have examined different parameters of LHRH neuronal function following castration.

We view the following consequences of castration as being relatively undisputed: (1) the LHRH content of the median eminence is gradually decreased (Kalra and Kalra, 1982); (2) mean LHRH release from *in vitro* hypothalamic fragments is decreased (Kalra *et al.*, 1987); (3) LHRH mRNA levels are either unchanged (Wiemann *et al.*, 1990) or decreased (Rothfeld *et al.*, 1987; Park *et al.*, 1988); (4) LHRH immunocytochemical signal is generally decreased (Gross, 1980); (5) proLHRH (proGnRH) tissue content is increased (Roselli *et al.*, 1990); and, most importantly, (6) LH pulse amplitude *and* frequency increase rapidly over 24 hours and continue to increase thereafter for prolonged, if not indefinite, periods (Ellis and Desjardins, 1984). How are these postcastration events interrelated? We have proposed that the negative feedback actions of gonadal hormones are principally manifest through two integrated mechanisms, these being suppression of pituitary responsiveness to LHRH, and retardation of the LHRH pulse generator. The former mechanism may be important in the acute homeostatic regulation of LH secretion, while the latter may mediate longer term shifts in the activity of the axis. Our model (Fig. 13) holds that castration results in a relatively rapid escape from direct suppression of pituitary responsiveness (< 24 hours), and a more gradual increase in LHRH pulse frequency. All of the consequences of castration that are enumerated above are consistent with this hypothesis, namely, (1) LHRH content is decreased with time following castration due to a gradual diminishment of the releasable LHRH pool, (2) mean LHRH release *in vitro* is diminished for the same reason, (3) LHRH mRNA levels are not inversely related to circulating steroid levels because gonadal feedback actions are manifest via the pulse generator and not through the LHRH biosynthetic mechanism; (4) the LHRH immunocytochemical signal within the median eminence is decreased as a result of increased LHRH pulsatility without a compensatory increase in LHRH biosynthesis; (5) proLHRH levels may increase as a result of decreased posttranslational processing (an action unrelated to feedback suppression; see Roselli *et al.*, 1990); and (6) the frequency of LH is increased as a result of a continued increase in LHRH pulse frequency. The LH pulse amplitude is likely increased following castration as a result of the removal of direct feedback inhibition of LHRH-stimulated LH secretion (Nansel *et al.*,

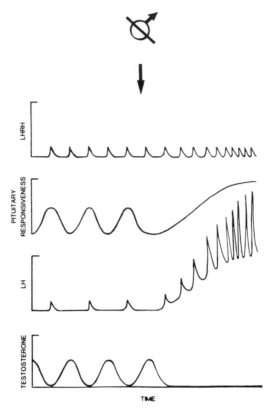

FIG. 13. Proposed model for neuroendocrine responses to castration in male rats. See text for details. (From Levine *et al.*, 1990.)

1979; Strobl *et al.*, 1989), and mean LH levels probably increase after castration as a function of the combined release from both hypothalamic and pituitary mechanisms.

Do other *in vivo* experimental data support this view? Portal vessel LHRH measurements in anesthetized, stalk-transected male rats have indicated that mean LHRH levels in portal vessel blood are not changed following castration (Brar *et al.*, 1985). This is certainly compatible with our model, if indeed LHRH pulse frequency is increased after castration at the expense of LHRH pulse amplitude. Even more compelling is the observation from the same study (Brar *et al.*, 1985) that while mean LHRH levels are unchanged in castrates, the electrically stimulated LHRH release is actually decreased in these animals versus testes-intact rats. This is also compatible with our suggestion that the LHRH-releasable pool is decreased in castrates as a secondary consequence of increased

LHRH pulse frequency. A study of the effects of castration on portal blood LHRH patterns in conscious rams also provided clear evidence that LHRH pulse frequency is increased with time following orchidectomy, in agreement with our findings in male rats. In an original study by Dluzen and Ramirez (1985), LHRH patterns were monitored in hypothalamic push–pull perfusates and found to be decreased in amplitude, and possibly increased in frequency, following castration. While this result is consistent with our findings, it was later found by the same group that LHRH patterns in push–pull perfusates of the adenohypophysis were increased in both amplitude and frequency following castration (Dluzen and Ramirez, 1987). The authors proposed that LHRH neurons may become increasingly synchronized with time following castration, and this mechanism may explain why pituitary, but not hypothalamic, perfusate LHRH levels are increased in castrates. While this is certainly an intriguing possibility, our data (both hypothalamic and pituitary) appear to be at odds with two aspects of this hypothesis: first, we have found no increase in mean LHRH levels at either the hypothalamic or pituitary site and, second, high-resolution measurement of LH pulses (Ellis and Desjardins, 1984) shows no indication that the profiles become any less irregular with time following castration; indeed, we have found that LH pulses in short-term ($<$ 32 hours) castrates are every bit as uniform as those measured at longer times following castration. Hopefully, time will tell whether the discordant results obtained in the latter study compared to our own can be explained on technical grounds, or perhaps on the basis of experimental factors (e.g., length of postcastration interval). The latter disagreement notwithstanding, it is clear from virtually all examinations of LHRH and/or LH pulsatility in male rats and other species that one response to castration must be an increase in the activity of the LHRH pulse generator. In future experiments in hypophysectomized, pituitary-grafted rats, we hope to assess the importance of this feedback mechanism in determining the trajectory of the postcastration LH rise. The relative importance of the hypothalamic versus pituitary mechanism at increasing times following castration will also require extensive analysis.

While the negative feedback actions of gonadal hormones (principally estrogen and progesterone) are beyond the intended scope of this article, brief note can be made of some of the similarities in their apparent actions when compared to the testicular hormones. Ovariectomy results in reduced LHRH tissue content (Kalra, 1976), reduced or unchanged *in vitro* LHRH release (Dluzen and Ramirez, 1986b), reduced LHRH mRNA levels (Park *et al.*, 1990; Roberts *et al.*, 1989), and reduced immunocytochemical signal (Kobayashi *et al.*, 1978). At the same time (albeit with a longer time course than in male rats) LH pulse amplitude and frequency are increased with time following gonadectomy (Leipheimer and Gallo, 1983). We have demonstrated that acute estrogen treatment can suppress pituitary responsiveness by a rapid action on the isolated pituitary gland (Figs. 14 and 15), indicating that a pituitary feedback mechanism

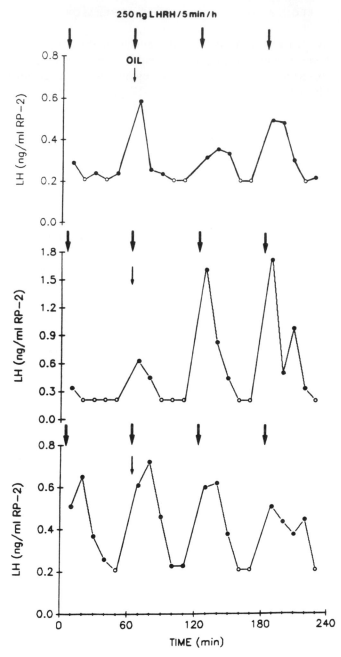

FIG. 14. Plasma LH levels in hypophysectomized, pituitary-grafted female rats receiving pulsatile infusions of LHRH and single oil injections. Larger arrows denote times of pulsatile LHRH infusions. Smaller arrows indicate times at which oil vehicle was administered. Sampling sessions were carried out 5 days after transplantation and initiation of pulsatile LHRH infusions. (From Strobl and Levine, 1988.)

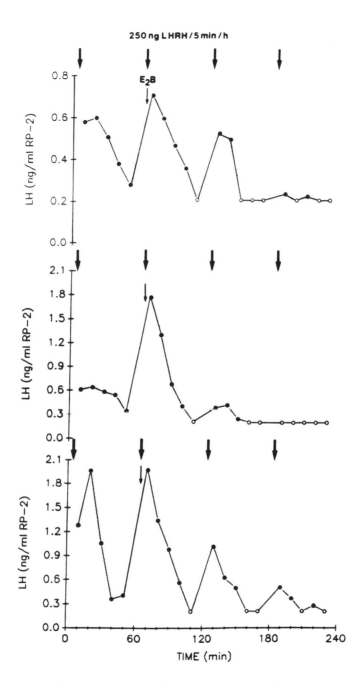

FIG. 15. Plasma LH levels in hypophysectomized, pituitary-grafted rats receiving pulsatile LHRH infusions and a single injection of estradiol benzoate. Smaller arrows indicate time of injection of 2 μg estradiol benzoate. Note the near-total suppression of LH secretion within 2–3 hours following estradiol benzoate administration. (From Strobl and Levine, 1988.)

can operate in females, as it appears to in males (Strobl et al., 1989). The nature and degree of the hypothalamic component of gonadal feedback in females have, unfortunately, proved to be as difficult to characterize as the counterpart mechanism in males. The mean LHRH level in portal vessel blood of anesthetized, stalk-transected rats has been found to increase following ovariectomy (Sarkar and Fink, 1980), although other studies have failed to document such an increase in mean level (Eskay et al., 1977). Rigorous analysis of the LHRH pulse frequency following ovariectomy in unanesthetized rats has not been possible, given limitations of current techniques; in view of the progressive increase in LH pulse frequency following gonadectomy, however, it is difficult to understand how LHRH pulse frequency could *not* be increased. Pending analysis of LHRH pulse frequency following ovariectomy, and the effects of ramp increases in LHRH pulse frequency on the isolated pituitary gland, we would predict that the general mechanisms of negative feedback in females resemble those in males, and would include retardation of the LHRH pulse generator and suppression of pituitary responsiveness to the LHRH decapeptide (Strobl and Levine, 1988). Increases or decreases in mean LHRH levels in portal blood or tissue may likewise be secondary consequences of changes in the level of pulse generator activity.

In summary, the LHRH pulse generator appears to operate as the primary neural target of gonadal hormone negative feedback. Below, we consider the possibility that endogenous opioid peptides (EOPs) suppress release of LHRH through a similar mechanism, especially since it has been proposed that gonadal negative feedback may be mediated by EOP neurons (Cicero et al., 1979; Van Vugt et al., 1982; Bhanot and Wilkinson, 1983).

B.   ENDOGENOUS OPIOID PEPTIDES

It has been demonstrated in several studies that EOP neurons tonically inhibit secretion of LH in the rat (Cicero et al., 1976; Bruni et al., 1977; Pang et al., 1977; Kalra and Kalra, 1983) through the intrahypothalamic inhibition of LHRH release (Drouva et al., 1981; Wilkes and Yen, 1981; Ching, 1983; Leadem et al., 1985; Nikolarakis et al., 1986). Administration of opiate receptor antagonists, such as naloxone, stimulates release of LHRH *in vitro* (Wilkes and Yen, 1981; Leadem et al., 1985; Nikolarakis et al., 1986; Kalra et al., 1987) and *in vivo* (Ching, 1983; Blank et al., 1985; Sarkar and Yen, 1985; Orstead and Spies, 1987; Karahalios and Levine, 1988), and can reverse exogenous opiate suppression of LHRH secretion (Ching, 1983). Furthermore, intracerebral implantation or infusion of naloxone (Kalra, 1981; Orstead and Spies, 1987) or β-endorphin antiserum (Schulz et al., 1981) evokes LH secretion, and local administration of β-endorphin exerts the opposite effect (Wiesner *et al.*, 1985). The physiological significance of this EOP inhibitory mechanism remains obscure.

A prevailing theory holds that the tonic EOP inhibition of LHRH release may be important in the transmission of the negative feedback actions of gonadal steroids (Cicero *et al.*, 1979; Van Vugt *et al.*, 1982; Bhanot and Wilkinson, 1983). This hypothesis is largely based on the observations that opiate receptor antagonists can reverse steroid-induced suppression of LH secretion (Cicero *et al.*, 1979; Sylvester *et al.*, 1982; Van Vugt *et al.*, 1982), and that opioid inhibitory tone appears to be diminished in long-term gonadectomized animals (Bhanot and Wilkinson, 1983; Petraglia *et al.*, 1984). Other groups, however, have obtained contradictory results (Blank *et al.*, 1985; Kalra *et al.*, 1987; Orstead and Spies, 1987), which suggests that removal of negative feedback may not be directly associated with loss of opioid inhibitory tone. Because of these apparent contradictions, and because the hypothesis had not been addressed in un-anesthetized rats, we attempted to characterize the LHRH responses of intact and gonadectomized rats to opiate receptor antagonism (Karahalios and Levine, 1988). We reasoned that if EOP neurons mediate the negative feedback actions of gonadal steroids, then it would be expected that (1) the effects of opiate receptor antagonism on LHRH release would be similar to those produced by gonadec-tomy, and (2) opioid inhibitory tone would decline during the escape of the hypothalamic–hypophyseal unit from gonadal negative feedback.

Using push–pull perfusion, we examined the effects of naloxone (2.5 mg/kg sc) on pulsatile LHRH release in intact rats during metestrus, and in rats ovariec-tomized 4 or 8 days earlier. Figure 16 depicts LHRH responses to naloxone or saline vehicle in individual animals in each group. In each case it was found that naloxone prompts an obvious increase in LHRH pulse amplitude. Although the responses varied in mean increment, there was consistently a transient, signifi-cant increase in LHRH pulse amplitude (and, hence, in mean level as shown in Fig. 17) following naloxone injections. Interestingly, responses were not discern-ibly diminished at 8 days following ovariectomy, when considered as percentage increases over preinjection levels.

The primary actions of EOPs operating through μ receptors in the rat appear to involve suppression of the amplitude of LHRH release; as such, we conclude that they act in a qualitatively different manner than gonadal hormones. The opioid peptides appear to primarily reduce the amount of LHRH liberated per pulse, while gonadal hormones decelerate the rate of pulse generation. Since EOPs and gonadal hormones have different primary actions, and because opioid inhibitory tone is not diminished after escape of the hypothalamic/pituitary unit from nega-tive feedback, it would appear (contrary to a widely held theory) that the negative feedback actions of gonadal steroids are not serially mediated by EOP neurons. Rather, opioid inhibition of LHRH release and gonadal negative feedback may operate in parallel to control LH secretion in the rat. Given that EOP inhibitory

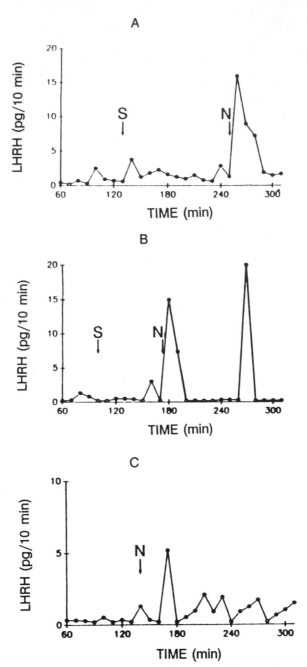

FIG. 16. Individual LHRH release profiles before and after saline or naloxone injections (2.5 mg/kg sc) in a metestrous rat (A) and in two rats at 4 days (B) or 8 days (C) following ovariectomy. Naloxone (N) or saline (S) injections were administered during the beginning of 10-min intervals. Note doubling of the release rate scale for the data depicted in (C). (From Karahalios and Levine, 1988.)

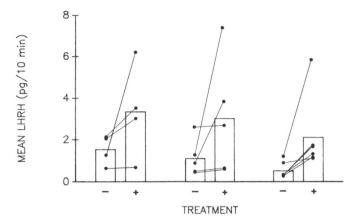

FIG. 17.   Mean LHRH levels in push–pull perfusates of the mediobasal hypothalamus/median eminence in metestrous rats (left) and in rats at 4 days (middle) or 8 days (right) following ovariectomy. Plus and minus symbols denote data collapsed over 1 hour before and 1 hour after naloxone administration. Points connected by lines represent mean levels in individual animals before and after treatment. (From Karahalios and Levine, 1988.)

tone appears to be manifest via suppression of LHRH pulse amplitude, it may instead be the case that this mechanism plays a more important role in the amplitude modulation of LHRH release that results from the positive feedback actions of steroids (Levine and Ramirez, 1980).

How do the EOPs influence LHRH release? The LHRH pulse amplitude could be affected through a number of mechanisms, including direct synaptic inhibition, or presynaptic inhibition. Their influence may also be indirect; it has been proposed, for example, that noradrenergic neurons may mediate the effects of opioid peptides (Kalra and Kalra, 1984), an issue that will be addressed later in this article (see Section VI,C). One likely target for the EOPs, as previously suggested by Lincoln (Lincoln et al., 1985), is the stimulus–secretion coupling mechanism. This scenario puts the opioid input downstream from the pulse generator, and is consistent with the finding that opiates suppress the K^+-stimulated release of LHRH from mediobasal hypothalamic fragments in vitro (Drouva et al., 1981). Thus, while opioid peptides may also exert effects on the pulse generator (Knobil, 1989), we hypothesize that the primary effects of EOPs are exerted through direct modulation of release (Wilkes and Yen, 1981), specifically by suppressing the amount of LHRH released during a given episode of electrophysiological stimulation (Drouva et al., 1981). That the opioid influence appears to be exerted downstream from the α_1-adrenergic regulation of the pulse generator (see Section VI,C) is also in accordance with this view.

C. α_1-ADRENERGIC INPUTS[2]

Do catecholamines regulate the activity of the LHRH pulse generator in the rat? It has been known since the early 1970s that α-adrenergic receptor antagonism in ovariectomized monkeys (Bhattacharya et al., 1972) results in the cessation of pulsatile LH secretion. An analogous α-adrenergic mechanism in rats, however, has been much more difficult to characterize. For example, both stimulatory and inhibitory effects of norepinephrine have been reported in rats, with positive actions most often found in steroid-primed animals, and negative actions dominating in gonadectomized rats (Gallo and Drouva, 1979). Also, various authors have reported that LHRH/LH pulse frequency and amplitude (Estes et al., 1982), or just frequency or amplitude (Kaufman et al., 1985; Teresawa et al., 1988), are regulated by α-adrenergic mechanisms. The apparent multiplicity of noradrenergic and/or adrenergic actions (negative vs positive, amplitude vs frequency) may be explainable on the basis of the differential activation of various receptor subtypes or different anatomical distribution of receptors. They may also be explainable on technical grounds; low sampling frequencies and less specific adrenergic drugs (Estes et al., 1982), for example, may have made it difficult in some studies to discriminate between effects of adrenergic drugs on LH pulse amplitude versus frequency.

We have reexamined the catecholaminergic regulation of pulsatile LH secretion in ovariectomized rats, using a frequent-sampling paradigm and a specific α_1-adrenergic antagonist, prazosin (Vogelsong and Levine, 1990). Our objective in these experiments was to determine if a unique, receptor-mediated mechanism operates in ovariectomized rats, through which catecholamines may regulate only the activity of the LHRH pulse generator, while not affecting other processes leading to LHRH release. Figure 18 depicts the LH profiles of ovariectomized rats that were bled at 5-min intervals and treated with one of several doses of prazosin (0, 0.2, 0.5, 1.0, or 2.0 mg/kg, ip). Administration of prazosin resulted in reductions in LH pulse frequency, without an accompanying decline in LH pulse amplitude. As shown in Fig. 19, this effect was dose dependent and reached significance at the two highest levels of prazosin. Saline vehicle injections were without effect on either LH pulse amplitude or frequency. In these experiments there were two other notable observations: first, that virtually no pulses of intermediate amplitude occurred during the onset of action of prazosin, or recovery from action of the drug; second, when pulses returned after the initial pause following prazosin injection, the LH pulses occurred at a lower frequency which, in many cases, was seen to slowly return to the original interpulse interval

---

[2] The work reviewed in this section focuses only on α_1-adrenergic regulation of pulsatile LHRH/LH release in the ovariectomized rat; for comprehensive treatments of catecholaminergic regulation of the reproductive axis, the reader is directed to several excellent reviews, including Ramirez et al. (1984) and Kalra and Kalra (1983).

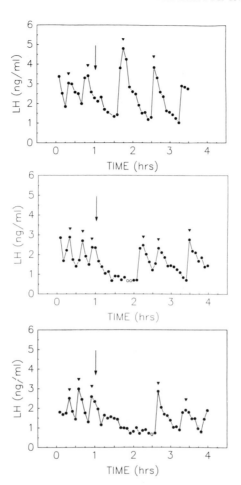

FIG. 18.   Effects of α_1-adrenergic receptor antagonism on pulsatile LH secretion in ovariec-tomized rats. Data are from experiments with three different animals, each receiving either 0.2 mg/kg (top), 0.5 mg/kg (middle), or 2 mg/kg (bottom) of prazosin, ip. Arrows indicate time of administration. (From Vogelsong and Levine, 1990.)

toward the end of the sampling period. Taken together with the observation that prazosin did not suppress pulse amplitude (if anything, first pulses following prazosin tended to rebound with greater amplitude; see Fig. 18) the suppression of frequency by prazosin indicates that α_1-adrenergic inputs meet the criterion for a specific regulator of the pulse generator. We conclude that α_1-adrenergic synaptic inputs act by means of a mechanism similar (albeit opposite in sign) to that by which gonadal hormones regulate LHRH release—they specifically regulate the activity of the LHRH pulse generator.

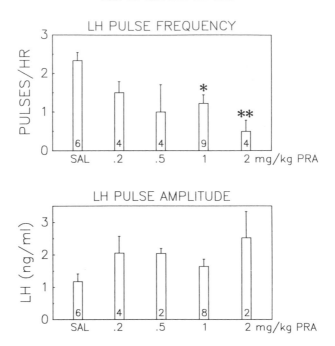

FIG. 19.   Summary data showing effects of α_1-adrenergic antagonism on LH secretion. Prazosin (PRA) treatments produced a dose-dependent suppression of LH pulse frequency, but not amplitude. Bars represent the mean ± SEM of pulse frequency and amplitude values for groups receiving the prazosin doses indicated at the bottom of each panel. Asterisks denote significant suppression versus values from the pretreatment control periods for each group. SAL, Saline; *, $p < .025$; **, $p < .01$.

Facilitation of the LHRH pulse generator through α_1-adrenergic receptor activation is likely mediated by noradrenergic and/or adrenergic neurotransmission. One would predict, therefore, that inhibition of noradrenergic and/or adrenergic turnover would produce the same effect as antagonism of the α_1-adrenergic receptor, namely, reduction of frequency, but not amplitude, of LHRH/LH secretion. We tested this hypothesis by examining the effects of the dopamine β-hydroxylase (dopamine β-monooxygenase) inhibitor, FLA-63, on the frequency and amplitude of pulsatile LH secretion (Legan et al., 1990). Treatment with FLA-63 at 25 mg/kg produces a profound and prolonged reduction in tissue norepinephrine and epinephrine levels, thereby reducing the amounts of these catecholamines available for release into the synaptic cleft. Figure 20 depicts results from a vehicle-treated (top panel) and FLA-treated rat (bottom panel), from which blood samples were obtained at 5-min intervals for 2 hours before, and between 2 and 4 hours after treatments. While LH pulse amplitude and frequency were unchanged following treatment with the propylene

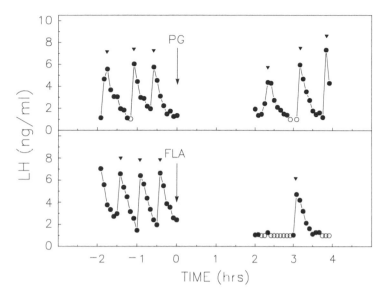

FIG. 20. Pulsatile LH patterns in plasma in an ovariectomized rat receiving propylene glycol (PG) vehicle (top) and one receiving the dopamine β-hydroxylase inhibitor, FLA-63 (bottom). Both rats were sampled at 5-min intervals for 2 hours before, and at 2–4 hours after, injections. LH pulse amplitude and frequency were unchanged following vehicle administration, while FLA-63 treatment resulted in a dramatic reduction in the frequency, but not the amplitude, of pulsatile LH release.

glycol vehicle, the administration of FLA-63 produced a profound depression of LH pulse frequency, without altering LH pulse amplitude. The finding was consistent and significant for the group of animals receiving this treatment (Legan et al., 1990). Thus, presumptive depression of noradrenergic and adrenergic neurotransmission produces exactly the same result as does antagonism of α_1-adrenergic receptors—a reduction of LH pulse frequency, but not amplitude.

How might the α_1-adrenergic-mediated mechanism facilitate the pulse generator? Evidence from studies using local injection of antagonist in hemi-POA lesioned, ovariectomized rats (Jarry et al., 1990) suggests that the locus of this facilitation is at the level of the LHRH cell bodies. Noradrenergic fibers and terminals have been found in close proximity to LHRH cell bodies in the same area (Hoffman et al., 1982; Jennes et al., 1982), these likely originating from A1, A2, and A6 brain stem catecholaminergic cell groups. Application of methoxamine or norepinephrine to hypothalamic slice preparations (Condon et al., 1989) can result in an augmentation of bursting activity in LHRH-producing cells. The authors of that study propose that the α_1-adrenergic input involves reduction in an early outward potassium current, gK_a. Taken together with the demonstration that norepinephrine can stimulate secretion of LHRH from immor-

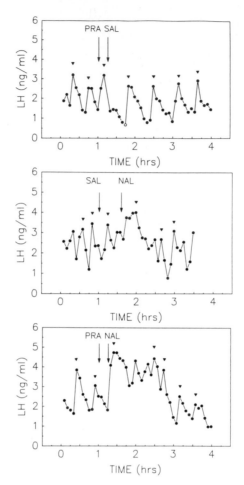

FIG. 21. Pulsatile LH patterns in plasma from ovariectomized rats receiving injections of prazosin (PRA) and saline (SAL) (top), saline and naloxone (NAL) (middle), or prazosin and naloxone (bottom). Prazosin reduced frequency, whereas naloxone increased amplitude of LH secretion. Pretreatment with prazosin did not block the effects of naloxone. (From Vogelsong and Levine, 1990.)

talized, LHRH-producing cells *in vitro* (Mellon *et al.*, 1990), the bulk of the findings to date is consistent with the idea that an α_1-adrenergic receptor may directly mediate noradrenergic facilitation of LHRH pulse generation within LHRH cell bodies. This effect on the pulse generator may be distinct from inhibitory mechanisms exerted by norepinephrine at more caudal sites, or stimulatory mechanisms which may directly influence LHRH release at the level of the median eminence (Negro-Vilar *et al.*, 1979). It is also currently not known if the noradrenergic regulation of the pulse generator that we have described is the key component of the noradrenergic facilitation of the LH surge (Barraclough *et al.*, 1984), or if other mechanisms, such as direct stimulation of release from LHRH axon terminals (Negro-Vilar *et al.*, 1979), figure more importantly in this physiological function.

Apart from its potential involvement in generating the preovulatory LH surge, the significance of α_1-adrenergic regulation of the pulse generator in ovariec-tomized, untreated rats is not clear. This adrenergic input could represent just one of a cadre of afferents whose additive activity determines the basal activity of the LHRH pulse generator. A noradrenergic influence may also serve to mediate the effects of several other regulators; other neural or endocrine factors may influence noradrenergic neurons which may, in turn, integrate these incoming regulatory signals and provide a final pathway for regulating LHRH neurons. Endogenous opioid peptides, γ-aminobutyric acid (GABA), and gonadal steroid (negative feedback as well as positive feedback) influences, for example, have been proposed to be mediated by noradrenergic neurons (Kalra and Kalra, 1984; Lincoln *et al.*, 1985). Evidence from our laboratory suggests that this may not be the case for the EOPs in regulating pulsatile release, however, since the primary effects of α_1-adrenergic inputs and the EOPs appear to be dissimilar (amplitude versus frequency regulation), and because we have found that the stimulatory effects of naloxone can still be observed in animals pretreated with an α_1-adrenergic antagonist (Fig. 21). Nevertheless, it still may be true that the α_1-adrenergic mechanism may mediate the effects of other neural and hormonal modulators. We have found, for example, that prazosin treatments are without effect in gonadally intact female rats, even at doses which profoundly reduce pulse generator activity in ovariectomized rats (Vogelsong and Levine, 1990). One explanation for this is that ovarian hormones suppress the activity of the LHRH pulse generator by inhibiting noradrenergic synaptic transmission.

D. POSITIVE FEEDBACK ACTIONS OF GONADAL HORMONES

Secretion of LH during the ovulatory cycle of the rat is pulsatile (Gallo, 1981; Fox and Smith, 1985), and is maintained at low levels on estrus, metestrus, diestrus, and the morning of proestrus through gonadal negative feedback inhibition (Leipheimer *et al.*, 1984). Under the influence of a rising tide of circulating estrogen, which is secreted by the ripening follicles, the hypothalamo-hypophyseal unit releases an abrupt surge of LH on the afternoon of proestrus which triggers ovulation on the ensuing morning of estrus. It has been known for many years that the preovulatory release of estrogen, together with a daily signal generated by a 24-hour neuronal clock (Everett and Sawyer, 1950), are the primary determinants of the LH surge. Progesterone released just prior to or during the surge (Kalra and Kalra, 1983) appears to both amplify the surge and to prevent its recurrence during the same "critical period" on the following day (Freeman *et al.*, 1976). In ovariectomized animals, LH surges of physiological magnitude and timing can be produced by administration of proper sequential administration of estrogen and progesterone (Caligaris *et al.*, 1971). Prolonged administration of estrogen alone via silastic capsule implants has been shown to evoke daily LH surges (Legan and Karsch, 1975), although these are diminished in amplitude and more variable in timing without accompanying progesterone treatment. Estrogen therefore appears to activate a circuitry within the brain which generates a signal for the release of the LH surge, and as directed by the circadian clock, the conveyance of this signal to the LH-releasing apparatus is limited to the afternoon hours. An important locus for the integration of these signals appears to be the preoptic/anterior hypothalamic area (Kalra and McCann, 1975). The mechanisms by which estrogen and progesterone exert these feedback effects have been the subject of much study, and many hypotheses have been advanced to explain the interneuronal events which may intervene between steroid binding and the generation of the surge. A strong case has been made, for example, that noradrenergic neurons may mediate the positive feedback actions of estrogen and progesterone (Barraclough *et al.*, 1984). It is likely, in any case, that steroids do not act directly upon the LHRH neuron (Shivers *et al.*, 1983) to evoke an LHRH surge. Thus, the interneuronal transmission of their positive feedback actions remains to be defined. In the discussion below, we consider how these feedback actions are ultimately exerted on the LHRH secretory process. For detailed treatments of the possible involvement of aminergic or other peptidergic neurons in transmitting positive feedback influences to the LHRH neuron, the reader is directed to Kalra and Kalra (1983), Ramirez *et al.* (1984), and Freeman (1988). In the discussion below, we consider only how these feedback actions are ultimately exerted on the LHRH secretory process, and focus on the potential role of the LHRH pulse generator in initiating the LH surge.

1. Release of LHRH during LH Surges: Role of the LHRH Pulse Generator

It has been known since the work of Sarkar *et al.* (1976) that the culmination of the neural events leading to the LH surge is an acute increase in the amount of LHRH released into the portal vessels on the afternoon of proestrus. What form does this proestrus LHRH surge take? Is it an acute increase in the frequency or amplitude of LHRH pulse generation, or possibly an increase in the basal LHRH release rate? One important clue comes from the analysis of pulsatile LH patterns in sequential blood samples drawn from proestrous rats during initiation of the surge (Fig. 22) (Fox and Smith, 1985). In that study it was found that the frequency of LH pulses was not discernibly increased during the dramatic rise in LH levels that constitutes the ascending limb of the LH surge.

In original push–pull perfusion studies by Levine and Ramirez (1982), it was

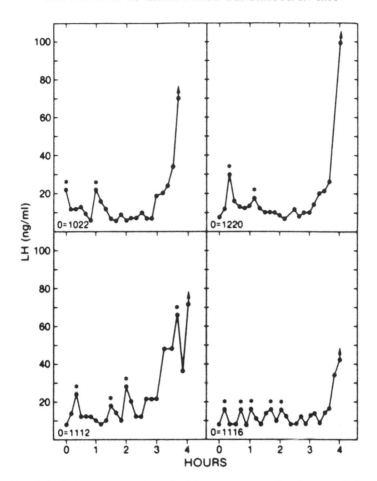

FIG. 22. Pulsatile LH patterns as determined in proestrous rats. Samples were obtained during the hours preceding the initiation of the LH surge, as indicated. The LH pulse frequency was not increased during the ascending limb of the LH surge. (From Fox and Smith, 1985.)

documented that the underlying LHRH release profile in proestrous rats (Fig. 23) or in ovariectomized rats receiving estrogen and progesterone (Levine and Ramirez, 1980) also shows no indication of a rapid increase in LHRH pulse frequency during initiation of the LH surge. Rather, the increase in LHRH release appears to take the form of pulses of increased amplitude (Levine and Ramirez 1980, 1982; Park and Ramirez, 1989) during the rising phase of the LH surge. The effects of higher doses of estrogen in the absence of progesterone treatment have proved to be more difficult to document; portal vessel studies in

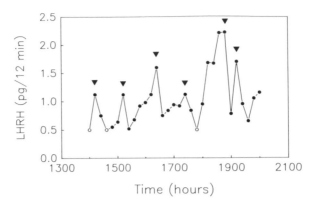

FIG. 23.   Pulsatile LHRH release profile in a proestrous rat. Pulse amplitude, but not frequency, was increased during the time of the initiation of the LH surge. (From Levine and Ramirez, 1982.)

anesthetized, stalk-transected animals have shown that estrogen induces a significant increase in mean LHRH levels during the afternoon hours (Sarkar and Fink, 1979). In push–pull perfusion studies, effects of single injections of higher doses of estrogen have not produced such increases (Dluzen and Ramirez, 1986b), although implantation of silastic capsules containing 17β-estradiol was found to be accompanied by the appearance of increased afternoon LHRH release (Levine, 1986). It is likely that the mixed results using estrogen alone are attributable to the fact that LH surges produced by these treatments are reduced in amplitude, and more variable in timing, than those generated in proestrous rats, or in estrogen- and progesterone-treated animals. Nevertheless, the data from proestrous and estrogen/progesterone-treated rats suggest that increased LHRH pulse amplitude is the change in LHRH release that may be most important in eliciting LH surges. Although subtle changes in pulse frequency may occur, it would appear that they cannot be called on to account for the rapid ascent of LH or LHRH levels during the generation of the surge.

   If an acceleration of LHRH pulse generation does not seem to mediate the initiation of the surge in the rat, then what role, if any, might the pulse generator play in the surge-generating process? Clearly, the circadian clock generates a daily signal for the release of the LHRH/LH surge, and ovarian steroid exposure opens a neural gate which allows transmission of this signal to the LHRH release process. The signal primarily prompts an increase in LHRH pulse amplitude. We speculate that more than one synaptic pathway may convey this signal to release pulses of increased amplitude in proestrous rats, with some exerting their effects at the level of the pulse generator, and others modifying LHRH downstream from the pulse-generating mechanism. Those acting on the pulse generator, however, would appear to exert their effects through a mechanism which differs from the straightforward frequency regulation seen following gonadectomy or α_1-adrenergic activation in ovariectomized rats. Two ways in which the pulse generator could be involved in amplitude modulation include the following: steroids could increase the strength of synaptic transmission within the circuitry composing the pulse generator (e.g., by increasing transmitter release or postsynaptic receptor densities), or they could activate a surge-generating circuitry which superimposes a surge-release signal on the pulse-generating electrophysiological rhythm. The latter hypothesis is perhaps most compatible with the two beliefs that steroids do not act directly upon LHRH neurons (Shivers et al., 1983) and that the LHRH neuron is intrinsically capable of LHRH pulse generation (Martinez de la Escalerez et al., 1990).

2. Does an Increase in LHRH Biosynthesis Contribute to the Surge?

The activation of LHRH release which prompts the LH surge takes place rapidly, suggesting that a synaptic signal or, more likely, a stereotyped set of synaptic signals, is the proximal cause of the LHRH-induced LH surge. It has also been proposed, however, that an accumulation of intraneuronal LHRH may occur in anticipation of the surge. Such an increase would most likely come about by a steroid-induced increase in LHRH biosynthesis. LHRH pulse amplitude, then, could be increased at least in part by the increased releasable LHRH pool that would be present within the nerve terminal at the time of the initiation of the surge. Measurements of LHRH concentrations in hypothalamic tissues at time points before, during, and after the surge have provided some support for this view, as at least one study has documented that LHRH concentrations increase prior to the surge and decline as the surge proceeds (Wise et al., 1981). Others, however, have noted only the decline in LHRH levels during the elaboration of the surge (Kalra and Kalra, 1977). We tested the hypothesis that the releasable LHRH pool is increased in advance of the LH surge, in experiments utilizing the LHRH secretagogue, NMA (Luderer et al., 1990). Metestrous and proestrous rats were challenged with a standard dose of NMA (5 mg, iv) and the LH responses were recorded. Figure 24 demonstrates that the LH responses to NMA were actually somewhat diminished in proestrous rats at 3 hours prior to the surge, compared to responses in the metestrous animals. This unexpected result suggests that the releasable LHRH pool is not enhanced at this time prior to the surge. Based on this finding, along with the observations by Park et al. (1990) and Zoeller and Young (1988) that LHRH mRNA levels are increased following the initiation of the LH surge, and not during or prior to this event (Fig. 25), it is hypothesized that changes in LHRH biosynthesis do not play an important role in the increase in LHRH release that occurs on the afternoon of proestrus (Levine and Ramirez, 1982). We concur with the proposal that LHRH biosynthesis is not increased in preparation for the surge, but instead may be stimulated as a secondary consequence of this event, so as to replenish depleted stores of the decapeptide at the nerve terminal.

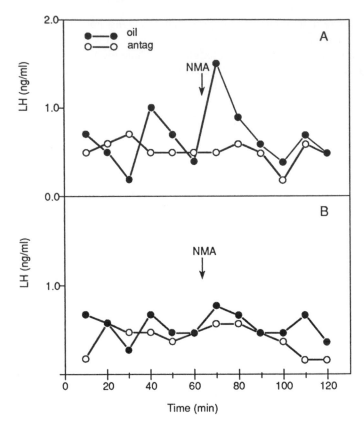

FIG. 24. Plasma LH responses to NMA in metestrous (A) and proestrous (B) rats. Animals received 5 mg NMA, iv, at the times indicated. (From Luderer *et al.*, 1990.)

3. *Direct Feedback Regulation of the Pituitary*

The occurrence of the LH surge in the rat is determined by the timing of the antecedent LHRH surge. The magnitude of the LH surge depends not only on the LHRH trigger, but also on the degree to which the anterior pituitary is sensitized to the actions of the decapeptide. It is well known that the rising titers of estrogen secreted by the developing follicles not only activate LHRH release, but also serve to sensitize the pituitary gland to stimulation by LHRH (Aiyer and Fink, 1974). Both basal and LHRH-stimulated LH secretion from pituitary fragments (Fallest *et al.*, 1989) or cultures (Drouin and Labrie, 1976) are increased when tissues are obtained from proestrous rats. *In vivo* experiments are in agreement with these results, inasmuch as LHRH-stimulated LH secretion is gradually

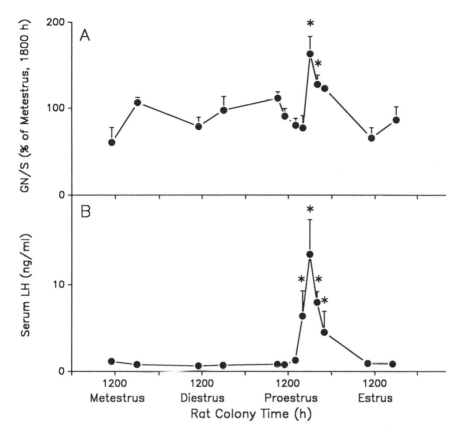

FIG. 25. Relative LHRH mRNA levels as assessed by quantitative *in situ* hybridization (A), and LH levels during the rat estrous cycle. (From Park *et al.*, 1990.)

increased from late diestrus through early proestrus, paralleling the increasing estradiol concentrations (Aiyer *et al.*, 1974). Pituitary sensitivity is then sharply increased on the afternoon of proestrus, a phenomenon that has been attributed to the self-priming actions of LHRH (Aiycr *et al.*, 1974) and the direct actions of progesterone secreted during the surge (Aiyer *et al.*, 1974). In total, the pituitary has been shown to become nearly 50-fold more sensitive to the stimulatory effects of LHRH in proestrous rats (Aiyer *et al.*, 1974).

We used the *in vivo* isolated pituitary paradigm to estimate the degree to which physiological levels of estrogen and progesterone can modify the LHRH-stimulated LH secretion by the ectopic pituitary gland (Strobl and Levine, 1990). Hypophysectomized, ovariectomized rats bearing pituitary transplants were

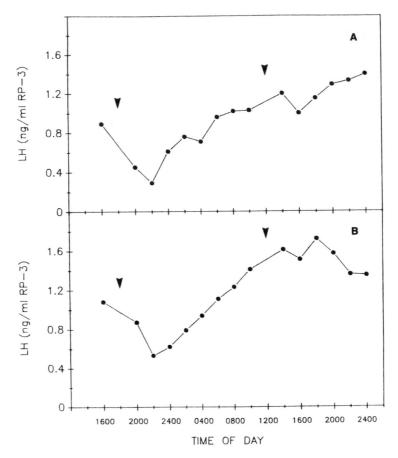

FIG. 26.   Plasma LH levels in individual hypophysectomized, pituitary-grafted rats receiving estrogen and oil (A) or estrogen and progesterone (B). Arrowheads indicate the time of capsule implantation. Hourly LHRH infusions were continued throughout sampling periods. LH levels were first suppressed and then rose above initial levels to the same degree after either treatment regimen, namely, progesterone did not augment either the negative or positive effects of estrogen. (From Strobl and Levine, 1990.)

maintained on pulsatile LHRH infusions for 4 days and then treated with estrogen, or estrogen and progesterone (Fig. 26), in regimens known to elicit LH surges in normal animals (Fig. 26) (Strobl and Levine, 1990). Treatment with estrogen, or estrogen plus progesterone, initially produced a suppression of LH secretion comparable to that produced in normal, pituitary-intact animals receiving similar treatments. This response was then followed by a steady rise in LH secretion in response to the unvarying pulsatile LHRH infusions, yielding levels threefold greater than suppressed values. Added treatments with progesterone neither suppressed nor enhanced to any greater degree than estrogen alone. These results are in agreement with the hypothesis that the initial phase of priming of the pituitary to LHRH is dependent on the direct actions of estrogen; they fail to confirm the idea that progesterone may additionally augment pituitary responsiveness. The estrogen priming action may involve an augmentation of pituitary receptor numbers (Clayton and Catt, 1981) or facilitation at a postreceptor site (Pieper *et al.*, 1984).

4. *Neurohormonal Priming Mechanisms*

The rapid increase in pituitary responsiveness to LHRH that occurs on the afternoon of proestrus has often been attributed to the acute self-priming actions of LHRH, in combination with a "background" of estrogen priming. It has been demonstrated in both *in vivo* and *in vitro* experiments that LHRH self-priming occurs in response to sequential administration of exogenous LHRH. It appears, moreover, that the cellular events which underlie LHRH-stimulated release and LHRH self-priming may differ (Pickering and Fink, 1979). Apart from the demonstration that a small increase in LHRH pulse amplitude does occur prior to the surge (Levine and Ramirez, 1982), however, there is unfortunately little direct

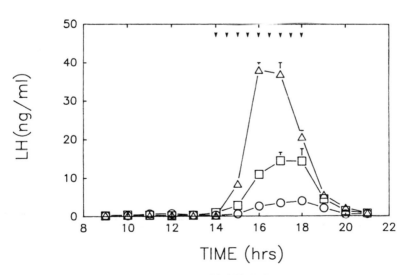

FIG. 27.   Simulation of LH surges in pentobarbital-blocked, proestrous rats by pulsatile administration of LHRH. Treatment with intermittent LHRH infusions at 15 ng (○), 150 ng (□), or 1500 ng/pulse (△) produced LH surges of subphysiological, physiological, and supraphysiological magnitude, respectively. Times of infusions are indicated by arrowheads at the top of this and subsequent graphs. (From Bauer-Dantoin *et al.*, 1991.)

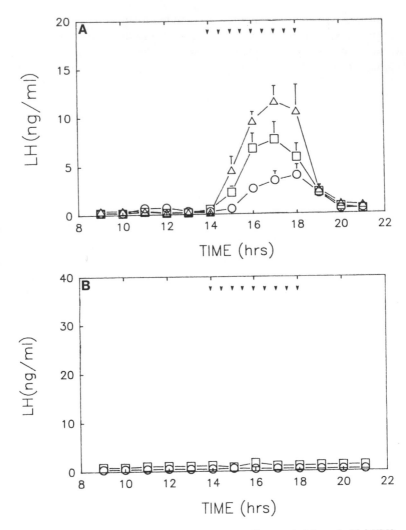

FIG. 28. Effects of pulsatile neuropeptide Y (NPY) or saline, coadministered with LHRH on LH secretion in pentobarbital-blocked, proestrous rats. (A) Coadministration of 0 μg NPY (○), 1 μg NPY (□), and 10 μg NPY (△) with LHRH produced a dose-dependent enhancement of LHRH-stimulated LH surges. (B) Infusions of 1 μg (○) and 10 μg (□) NPY alone were without effect. (From Bauer-Dantoin *et al.*, 1991.)

physiological evidence that an endogenous LHRH self-priming mechanism operates as a component of the surge-generating process.

It may instead be the case that one or more non-LHRH hypothalamic factors are also secreted into the hypophyseal portal vessel system, and that these may modulate the responsiveness of gonadotropes to stimulation by the decapeptide.

Evidence suggests that the 36-amino acid neuropeptide Y (NPY) may function in this neuromodulatory capacity. Portal vessel concentrations of the peptide are increased in parallel with LHRH in proestrous rats (Sutton *et al.*, 1988), and immunoneutralization of the peptide attenuates steroid-induced LH surges. It has also been shown that NPY can potentiate LHRH-stimulated LH release *in vitro* (Crowley *et al.*, 1987). We tested the ability of NPY infusions to potentiate LHRH-stimulated LH surges in pentobarbital-blocked, proestrous rats (Bauer-Dantoin *et al.*, 1991). In proestrous rats treated with the anesthetic, which blocks endogenous LHRH secretion, pulsatile infusions of LHRH were administered to simulate subphysiological, physiological, and supraphysiological LH surges (Fig. 27). Superimposition of NPY infusions on the LHRH infusions caused clear-cut, dose-dependent enhancement of the LHRH-induced surges (Fig. 28A). Infusions of NPY alone were without effect (Fig. 28B), confirming that the actions of NPY are modulatory in nature. The results of the latter study demonstrate that NPY can potentiate LHRH-stimulated LH surges, even to the extent that the LHRH at a lower dose (15 ng/pulse) combined with NPY treatments produced LH surges which were equivalent to those produced by LHRH pulses alone at a 10-fold greater dose. These observations are entirely consistent with the idea that NPY operates as a neuroendocrine modulator during generation of the preovulatory LH surge; in addition to modifying LHRH secretion directly (Crowley and Kalra, 1987; Sabatino *et al.*, 1989), NPY appears to be secreted into the hypophyseal portal vessels on the afternoon of proestrus to prime the pituitary gland to the actions of the decapeptide. This mechanism may explain why the pituitary gland becomes greatly sensitized to the releasing actions of exogenous LHRH during the afternoon of proestrus, and may at least partially account for the fact that a relatively small neurosecretory LHRH trigger can stimulate the secretion of such a massive surge of gonadotropin. This neuromodulatory action of NPY appears to be distinct from its direct effects on LHRH release via intrahypothalamic mechanisms, the latter having been demonstrated recently both *in vivo* (Khorram *et al.*, 1987) and *in vitro* (Crowley and Kalra, 1987; Sabatino *et al.*, 1989). It is possible that both mechanisms are physiologically coordinated on the afternoon of proestrus, so as to ensure successful elaboration of the LH surge.

VII. Summary

We have analyzed the mechanisms by which several known regulators of the LHRH release process may exert their effects. For each, we have attempted to determine how and where the regulatory input is manifest and, according to our working premise, we have attempted to identify factors which specifically regulate the LHRH pulse generator. Of the five regulatory factors examined, we have identified two inputs whose primary locus of action is on the pulse-generating

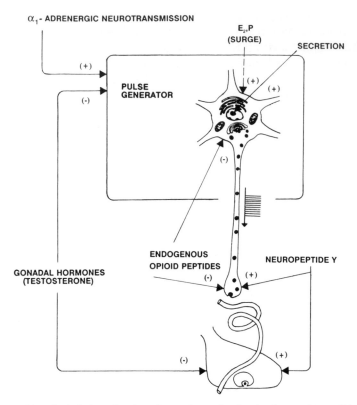

FIG. 29. Hypothetical sites of action of several neural and endocrine regulators. It is proposed that (clockwise) testosterone and α_1-adrenergic afferents regulate the LHRH pulse generator without affecting other processes leading to LHRH secretion. Testosterone also suppresses pituitary responsiveness to stimulation by LHRH. Estrogen (E_2) and progesterone (P) may exert some of their positive feedback actions through the delivery of a unique, punctuated synaptic signal, and this action is likely mediated by afferents to the LHRH neuron. The hypersecretion of LHRH that occurs during the surge may secondarily evoke a compensatory increase in LHRH biosynthesis. The endogenous opioid peptides primarily regulate LHRH pulse amplitude, and may exert their actions through direct synaptic or presynaptic inhibition. Neuropeptide Y exerts modulatory actions at the LHRH nerve terminal (Sabatino *et al.*, 1989) and at the level of the pituitary gland. Note that the model makes no assumption as to the cellular identity of the LHRH pulse generator, or the number of synapses which may intervene between actions on receptors and regulation of LHRH secretion. See text for discussion of the potential physiological roles played by these regulatory mechanisms.

mechanism—one endocrine (gonadal negative feedback), and one synaptic (α_1-adrenergic inputs) (see Fig. 29). Other factors which regulate LHRH and LH release appear to do so in different ways. The endogenous opioid peptides, for example, primarily regulate LHRH pulse amplitude (Karahalios and Levine, 1988), a finding that is consistent with the idea that these peptides exert direct postsynaptic or presynaptic inhibition (Drouva *et al.*, 1981). Gonadal steroids exert positive feedback actions which also result in an increase in the amplitude of LHRH release, and this action may be exerted through a combination of cellular mechanisms which culminate in the production of a unique, punctuated set of synaptic signals. Gonadal hormones and neurohormones such as NPY also exert complementary actions at the level of the pituitary gland, by modifying the responsiveness of the pituitary to the stimulatory actions of LHRH.

The LHRH neurosecretory system thus appears to be regulated at many levels, and by a variety of neural and endocrine factors. We have found examples of (1) neural regulation of the pulse generator, (2) hormonal regulation of the pulse generator, (3) hormonal regulation of a neural circuit which produces a unique, punctuated synaptic signal, (4) hormonal regulation of pituitary responsiveness to LHRH, and (5) neuropeptidergic regulation of pituitary responsiveness to LHRH. While an attempt has been made to place some of these regulatory inputs into a physiological context, it is certainly recognized that the physiological significance of these mechanisms remains to be clarified. We also stress that these represent only a small subset of the neural and endocrine factors which regulate the secretion or actions of LHRH. A more comprehensive list would also include CRF, GABA, serotonin, and a variety of other important regulators. Through a combination of design and chance, however, we have been able to identify at least one major example of each type of regulatory mechanism.

VIII. Conclusions

Why would nature design a neurosecretory system with these types of regulatory controls? Perhaps the best answers to this question emerge from a consideration of the variety of biological tasks that the LHRH neurosecretory system is required to perform. First and foremost, the LHRH neuronal population is charged with the responsibility of maintaining basal activity within the hypothalamic–hypophyseal–gonadal axis. The physiological importance of pulsatile neurohormone release in the maintainance of basal activity within the axis has already been discussed (Knobil, 1980; Lincoln *et al.*, 1985). Indeed, the evidence is overwhelming that a frequency modulation (FM) signal has evolved as the primary means by which the CNS provides continuous control over adenohypophyseal secretions (Knobil, 1980).

The LHRH neuron is also called on to execute major shifts in the activity of the axis—from dormancy to activity, activity to dormancy, and perhaps in some cases from one of these states to some physiological shade of gray in between. Puberty and reproductive seasonality are, of course, two examples of situations in which such shifts occur. The LHRH neuron may mediate these shifts in several possible ways. One important way is through frequency modulation, namely, by regulation of cellular mechanisms controlling frequency within pacemaker cells.

A second important way may be to synchronize the unsynchronized activity of different cells, perhaps by increasing their electrophysiological coupling. Yet another way is by recruiting increasing numbers of cells into the actively pulsing neuronal pool. With any one of these possibilities, however, there likely exists a very efficient regulatory process: by affecting the activity of a very few pacemaker cells, one can dramatically affect the overall activity of the axis.

What else must the LHRH neurons be required to do? The LHRH neurosecretory system is additionally required to do something quite different—to produce an abrupt signal for the initiation of the LH surge. In this case, amplitude modulation is clearly important in the rat. A unique synaptic signal is apparently the mediator of this event, prompting a relatively rapid, transient increase in LHRH pulse amplitude. In this task, the LHRH neuron appears to require the additional coordinated efforts of other regulators. Indeed, it may be too much to ask of a neuronal system to be responsible for continual intermittent signaling *and* for abruptly increasing secretion by over 1000% to initiate the LH surge. Several mechanisms appear to have evolved to compensate for this, including endocrine and neurohormonal modulation of pituitary responsiveness.

The LHRH neurosecretory system must also be able to respond to acute and chronic perturbations. It may respond to an acute perturbation, such as an acute stress, by rapid neuromodulatory changes, such as an acute increase in opioid inhibitory tone. Being at least partially downstream from pulse generation, the opioid input would be in an optimal position to do this as an ultimate or penultimate gate controlling LHRH release. Responses to chronic perturbations, such as prolonged reduction in circulating gonadal hormones, appear to involve longer term resetting of the activity of the axis, and this likely involves resetting of the rate of LHRH pulse generation. The best example of this, of course, is the enhanced frequency of LHRH pulse generation that develops following castration.

What is the LHRH pulse generator? At present, the anatomical, cellular, and molecular bases of LHRH pulse generation remain largely unknown. Likewise, little is known of the inter- and intracellular routes through which the activity of the LHRH pulse generator is regulated *in vivo*. We have identified two regulatory inputs—one endocrine and one synaptic—which appear to regulate the pulse generator without influencing other cellular processes leading to LHRH neurosecretion. We are hopeful that these two types of "smart bait" will enable us to probe the microanatomy and cellular regulation of the pulse generator in future studies.

ACKNOWLEDGMENTS

This research was supported by NIH Grants RO1-HD20677, PO1-HD21921, KO4-00879 (RCDA to J.E.L.), and F33-HD07304 (S.J.L./J.E.L.). The authors wish to thank the NIADDK, Dr. Gordon

Niswender, and Dr. Leo Reichert, Jr. for supplying RIA materials, and Drs. William Ellinwood and Martin J. Kelly for supplying the EL-14 LHRH antiserum. The authors also wish to acknowledge the work of our collaborators, Dr. Ulrike Luderer and Dr. Neena B. Schwartz, on the two NMA-stimulation studies. We also wish to thank Ms. Claudia L. Cryer for her valuable technical advice and assistance in preparation of this manuscript.

REFERENCES

Aiyer, M. S., and Fink G. (1974). *J. Endocrinol.* **62,** 553.

Aiyer, M. S., Fink, G., and Greig, F. (1974). *J. Endocrinol.* **60,** 47.

Barraclough, C. A., Wise, P. M., and Selmanoff, M. K. (1984). *Rec. Prog. Horm. Res.* **60,** 487.

Bauer-Dantoin, A., McDonald, J. K., and Levine, J. E. (1991). *Endocrinology,* in press.

Bhanot, R., and Wilkinson, M. (1983). *Endocrinology (Baltimore)* **112,** 399.

Bhattacharya, A. N., Dierschke, D. J., Yamaji, T., and Knobil, E. (1972). *Endocrinology (Baltimore)* **90,** 778.

Blake, C. A., and Sawyer, C. H. (1974). *Endocrinology (Baltimore)* **94,** 730.

Blank, M. S., Ching, M., Catt, K. J., and Dufau, M. L. (1985). *Endocrinology (Baltimore)* **116,** 1778.

Bourguignon, J.-P., Gerared, A., and Franchimont, P. (1989). *Neuroendocrinology* **49,** 402.

Brar, A. K., McNeilly, A. S., and Fink, G. (1985). *J. Endocrinol.* **104,** 35.

Bruni, J. F., Van Vugt, S., Marshall, S., and Meites, J. (1977). *Life Sci.* **21,** 461.

Caligaris, L., Astrada, J. J., and Taleisnik, S. (1971). *Endocrinology (Baltimore)* **89,** 331.

Caraty, A., and Locatelli, A. (1988). *J. Reprod. Fertil.* **82,** 263.

Carmel, P. W., Araki, S., and Ferin, M. (1976). *Endocrinology (Baltimore)* **99,** 243.

Ching, M. (1983). *Endocrinology (Baltimore)* **112,** 2209.

Cicero, T. J., Meyer, E. R., Bell, R. D., and Koch, G. A. (1976). *Endocrinology (Baltimore)* **98,** 367.

Cicero, T. J., Schainker, B. A., and Meyer, E. R. (1979). *Endocrinology (Baltimore)* **104,** 1286.

Clarke, I. J., and Cummins, J. T. (1982). *Endocrinology (Baltimore)* **111,** 1737.

Clarke, I. J., and Cummins, J. T. (1984). *Neuroendocrinology* **39,** 267.

Clayton, R. N., and Catt, K. (1981). *Endocrinol. Rev.* **2,** 186.

Condon, T. P., Ronnekleiv, O. K., and Kelly, M. J. (1989). *Neuroendocrinology* **50,** 51.

Crowley, W. R., and Kalra, S. P. (1987). *Neuroendocrinology* **46,** 97.

Crowley W. F., Filicori, M., Spratt, D. I., and Santoro, N. F. (1985). *Recent Prog. Horm. Res.* **41,** 473.

Crowley, W. R., Hassid, A., and Kalra, S. P. (1987). *Endocrinology (Baltimore)* **120,** 941.

Culler, M. D., and Negro-Vilar, A. (1986). *Endocrinology (Baltimore)* **118,** 609.

Dluzen, D. E., and Ramirez, V. D. (1985). *J. Endocrinol.* **107,** 331.

Dluzen, D. E., and Ramirez, V. D. (1986a). *Endocrinology (Baltimore)* **118,** 1110.

Dluzen, D. E., and Ramirez, V. D. (1986b). *Neuroendocrinology* **43,** 459.

Dluzen, D. E., and Ramirez, V. D. (1987). *Neuroendocrinology* **45,** 328.

Drouin, J., and Labrie, F. (1976). *Endocrinology (Baltimore)* **98,** 1528.

Drouva, S., Epelbaum, J., Tapia-Arancibia, L., Laplante, E., and Kordon, C. (1981). *Neuroendocrinology* **32,** 163.

Dyer, R. G., Mansfield, S., and Yates, J. O. (1980). *Exp. Brain Res.* **39,** 453.

Elkind-Hirsch, K., Ravnikar V., Schiff, I., Tulchinsky, D., and Ryan, K. J. (1982). *J. Clin. Endocrinol. Metab.* **54,** 602.

Ellis, G. B., and Desjardins, C. (1982). *Endocrinology (Baltimore)* **110,** 1618.

Ellis, G. B., and Desjardins, C. (1984). *Biol. Reprod.* **30,** 619.

Ellis, G. B., Desjardins, C., and Fraser, H. M. (1983). *Neuroendocrinology* **37,** 177.

Eskay, R. L., Mical, R. S., and Porter, J. C. (1977) *Endocrinology (Baltimore)* **110,** 263.

Estes, K. S., Simpkins, J. W., and Kalra, S. P. (1982). *Neuroendocrinology* **35,** 56.

Estienne, M. J., Schillo, K. K., Hileman, S. M., Green, M. A., and Hayes, S. H. (1990). *Biol. Reprod.* **42,** 126.

Everett, J. W., and Sawyer, C. H. (1950). *Endocrinology (Baltimore)* **47,** 198.

Fallest, P. C., Hiatt, E. S., and Schwartz, N. B. (1989). *Endocrinology (Baltimore)* **124,** 1370.

Fink, G., and Jamieson, M. G. (1976). *J. Endocrinol.* **68,** 71.

Fink, G., Naller, R., and Worthington, W. C. (1967). *J. Physiol. (London)* **191,** 407.

Fox, S. R., and Smith, M. S. (1985). *Endocrinology (Baltimore)* **116,** 1485.

Fraser, H. M., Jeffcoate S. L., Gunn, A., and Holland, D. T. (1975). *J. Endocrinol.* **64,** 191.

Freeman, M. E. (1988). *In* "The Physiology of Reproduction" (E. Knobil and J. D. Neill, eds.), p. 1893. Raven Press, New York.

Freeman, M. E., Dupke, K. C., and Croteau, C. M. (1976). *Endocrinology (Baltimore)* **99,** 223.

Gallo, R. V. (1981). *Biol. Reprod.* **24,** 100.

Gallo, R. V., and Drouva, S. V. (1979). *Neuroendocrinology* **29,** 149.

Gay, V. L., and Plant, T. M. (1987). *Endocrinology* **120,** 2289.

Grady, R. R., and Schwartz, N. B. (1981). *In* "Intragonadal Regulators of Reproduction" (P. Franchimont and C. P. Channing, eds.), p. 377. Academic Press, London.

Grady, R. R., Shin, L., Charlesworth, M. C., Cohen-Becker, I. R., Smith, M., Rivier, C., Rivier, J., Vale, W., and Schwartz, N. B. (1985). *Neuroendocrinology* **40,** 246.

Gross, D. S. (1980). *Endocrinology (Baltimore)* **106,** 1442.

Hawes, B. E., and Conn, P. M. (1989). *In* "Neuroendocrine Regulation of Reproduction" (S. S. C. Yen and W. W. Vale, eds.), p. 219. Serono Symposia, U.S.A.

Hoffman, G. E., Wray, S., and Goldstein, M. (1982). *Brain Res. Bull.* **9,** 417.

Irvine, C. H. G., and Alexander, S. L. (1987). *J. Endocrinol.* **113,** 183.

Jarry, H., Leonhardt, S., and Wutke, W. (1990). *Neuroendocrinology* **51,** 337.

Jennes, L., Beckman, W. C., Stumpf, W. E., and Gramma, R. (1982). *Exp. Brain Res.* **46,** 331.

Kalra, P. S., and Kalra, S. P. (1977). *Acta Endocrinol.* **85,** 449.

Kalra, P. S., and Kalra, S. P. (1982). *Endocrinology (Baltimore)* **111,** 24.

Kalra, P. S., Crowley, W. R., and Kalra, S. P. (1987). *Endocrinology (Baltimore)* **120,** 178.

Kalra, S. P. (1981). *Endocrinology (Baltimore)* **109,** 1805.

Kalra, S. P., and Kalra, P. S. (1983). *Endocrinol. Rev.* **4,** 311.

Kalra, S. P., and Kalra, P. S. (1984). *Neuroendocrinology* **38,** 418.

Kalra, S. P., and Kalra, P. S. (1989). *Biol. Reprod.* **41,** 559.

Kalra, S. P., and McCann, S. M. (1975). *Neuroendocrinology* **19,** 289.

Karahalios, D. G., and Levine, J. E. (1988). *Neuroendocrinology* **47,** 504.

Karsch, F. J., Cummins, J. T., Thomas, G. B., and Clarke, I. J. (1987). *Biol. Reprod.* **36,** 1207.

Kartun K., and Schwartz, N. B. (1987). *Biol. Reprod.* **36,** 103.

Kaufman, J-M., Kesner, J. S., Wilson, R. C., and Knobil, E. (1985). *Endocrinology (Baltimore)* **116,** 1327.

Kawakami, M., Uemura, T., and Hayashi, R. (1982). *Neuroendocrinology* **35,** 63.

Kendricks, K. M. (1989). *In* "Methods in Enzymology" (P. M. Conn, ed.), Vol. 168, p. 182. Academic Press, San Diego, California.

Khorram, O., Pau, K.-Y. F., and Spies, H. G. (1987). *Neuroendocrinology* **45,** 290.

Knobil, E. (1980). *Recent Prog. Horm. Res.* **36,** 53.

Knobil, E. (1981). *Biol. Reprod.* **24,** 44.

Knobil, E. (1989). *In* "Neuroendocrine Regulation of Reproduction" (S. S. C. Yen and W. W. Vale, eds.), p. 3. Serono Symposia, U.S.A.

Kobayashi, R. M., Lu, K. H., Moore, R. Y., and Yen, S. S. C. (1978). *Endocrinology (Baltimore)* **102,** 98.

Kokoris, G. J., Lam, N. Y., Ferin, M., Silverman, A. J., and Gibson, M. J. (1988). *Neuroendocrinology* **48**, 52.

Krey, L. C., Butler, W. R., and Knobil, E. (1975). *Endocrinology (Baltimore)* **96**, 1073.

Leadem, C. A., Crowley, W. R., Simpkins, J. W., and Kalra, S. P. (1985). *Neuroendocrinology* **40**, 487.

Legan, S. J., and Karsch, F. J. (1975). *Endocrinology* **96**, 57.

Legan, S. J., Urban, J. H., Mehta, S. S., Conaghan, L. A., and Levine, J. E. (1990). *Soc. Neurosci. 20th Annu. Meet.* Abstr. No. 495.11, p.1201.

Leipheimer, R. E., and Gallo, R. V. (1983). *Neuroendocrinology* **37**, 421.

Leipheimer, R. E., Bona-Gallo, A., and Gallo, R. V. (1984). *Endocrinology (Baltimore)* **114**, 605.

Levine, J. E. (1986). *Ann. N.Y. Acad. Sci.* **473**, 503.

Levine, J. E., and Duffy, M. T. (1988). *Endocrinology (Baltimore)* **122**, 2211.

Levine, J. E., and Meredith, J. (1990). *Soc. Neurosci., 20th Annu. Meet.* Abstr. No. 168.4, p. 396.

Levine, J. E., and Powell, K. D. (1989). *In* "Methods in Enzymology" 168, (P. M. Conn, ed.), Vol. 168, p. 166. Academic Press, San Diego, California.

Levine, J. E., and Ramirez, V. D. (1980). *Endocrinology (Baltimore)* **107**, 1782.

Levine, J. E., and Ramirez, V. D. (1982). *Endocrinology (Baltimore)* **111**, 1439.

Levine, J. E., and Ramirez, V. D. (1986). *In* "Methods in Enzymology" (P. M. Conn, ed.), Vol. 124, p. 466. Academic Press, Orlando, Florida.

Levine, J. E., Pau, K. Y. F., Ramirez, V. D., and Jackson, G. L. (1982). *Endocrinology (Baltimore)* **111**, 1449.

Levine, J. E., Norman, R. L., Gleissman, P. M., Oyama, T. T., Bangsberg, D. R., and Spies, H. G. (1985). *Endocrinology (Baltimore)* **117**, 711.

Levine, J. E., Stroll, F. J., Meredith, J. M., Vogelsong, K. M., Legan, S. J., and Urban, J. H. (1990). *In* "Neuroendocrine Regulation of Reproduction" (S. S. C. Yen and W. W. Vale, eds.), p. 47. Serono Symposia, Norwell, Massachusetts.

Lincoln, D. W., Fraser, H. M., Lincoln, G. A., Martin G. B., and McNeilly, A. S. (1985). *Recent Prog. Horm. Res.* **41**, 369.

Luderer, U., Strobl, F. J., Levine, J. E., and Schwartz, N. B. (1990). *Soc. Neurosci. 20th Annu. Meet.* p. 1202.

Martinez de la Escalera, G., Choi, A. L. H., and Weiner, R. I. (1990). *Soc. Neurosci., 20th Annu. Meet.* Abstr. No. 126.2, p. 284.

Mellon, P. L., Windle, J. J., Goldsmith, P. C., Padula, C. A., Roberts, J. L., and Weiner, R. I. (1990). *Neuron* **5**, 1.

Merriam, G. M., and Wachter, K. W. (1982). *Am. J. Physiol.* **243**, E310.

Nansel, D. D., Aiyer, M. S., Meinzer, W. H., and Bogdanove, E. M. (1979). *Endocrinology (Baltimore)* **104**, 524.

Negro-Vilar, A., Ojeda, S. R., and McCann, S. M. (1979). *Endocrinology (Baltimore)* **104**, 1749.

Nikolarakis, K. E., Pfeiffer, D. G., Almeida, O. F. X., and Herz, A. (1986). *Neuroendocrinology* **44**, 314.

Nikolarakis, K. E., Almeida, O., Sirinathsinghji, D. J. S., and Herz, A. (1988). *Neuroendocrinology* **47**, 545.

Nikolics, K., Mason, A. J., Szonyi, E., Ramachandran, J., and Seeburg, P. H. (1985). *Nature (London)* **316**, 511.

Norman, R. L., Gleissman, P., Lindstrom, S. A., Hill, J., and Spies, H. G. (1982). *Endocrinology (Baltimore)* **111**, 1874.

O'Dea, L., Finkelstein, J. S., Schoenfeld, D. A., Butler, J. P., and Crowley, W. F. (1989). *Am. J. Physiol.* **256**, E510.

Orstead, K. M., and Spies, H. G. (1987). *Neuroendocrinology* **47**, 14.

Pang, C. N., Zimmerman, E., and Sawyer, C. H. (1977). *Endocrinology (Baltimore)* **101**, 1726.

Park, O. K., and Ramirez, V. D. (1989). *Neuroendocrinology* **50**, 66.

Park, O. K., Gugneja, S., and Mayo, K. E. (1990). *Endocrinology (Baltimore)* **127,** 365.

Park, Y., Park, S. D., Cho, W. K., and Kim, K. (1988). *Brain Res. Bull.* **451,** 255.

Pau, K-Y. F., and Spies, H. G. (1986a). *Neuroendocrinology* **43,** 197.

Pau, K-Y. F., and Spies, H. G. (1986b) *Brain Res.* **399,** 15.

Pau, K.-Y. F., Orstead, K. M., Hess, D. L., and Spies, H. G. (1986). *Biol. Reprod.* **35,** 1009.

Pau, K.-Y. F., Hess, D. L., Kaynard, A. H., Ji, W.-Z., Gliessman, P. M., and Spies, H. G. (1989). *Endocrinology (Baltimore)* **124,** 891.

Petraglia, F., Locatelli, V., Penalva, A., Cocchi, D., Genazzani, A. R., and Muller, E. E. (1984). *J. Endocrinol.* **101,** 33.

Pickering, A. J., and Fink, G. (1979). *J. Endocrinol.* **81,** 223.

Pieper, D. R., Gala, R. R., Schiff, M. A., Regiani, S. R., and Marshall, J. C. (1984). *Endocrinology (Baltimore)* **115,** 1190.

Plant, T. M., and Dubey, A. K. (1984). *Endocrinology (Baltimore)* **115,** 2145.

Porter, J. C., and Smith, K. R. (1967). *Endocrinology (Baltimore)* **81,** 1182.

Price, M. T., Olney, J. W., and Cicero, T. J. (1978). *Neuroendocrinology* **26,** 352.

Ramirez, V. D., Feder, H. H., and Sawyer, C. H. (1984). *In* "Frontiers in Neuroendocrinology" (L. Martini and W. F. Ganong, eds.), Vol. 8, p.27. Raven Press, New York.

Ramirez, V. D., Kim, K., and Dluzen, D. (1985). *Recent Prog. Horm. Res.* **41,** 421.

Ramirez, V. D., Ramirez, V. D., Slamet, W., and Nduka, E. (1986). *Endocrinology (Baltimore)* **118,** 2331.

Roberts, J. L., Dutlow, C. M., Jakubowski, M., Blum, M., and Millar, R. P. (1989). *Mol. Brain Res.* **6,** 126.

Ronnekleiv, O. K., Naylor, B. R., Bond, C. T., and Adelman, J. P. (1989). *Mol. Endocrinol.* **3,** 363.

Roselli, C. E., Kelly, M. J., and Ronnelkleiv, O. K. (1990). *Endocrinology (Baltimore)* **126,** 1080.

Rothfeld, J. M., Hejtmancik, J. F., Conn, P. M., and Pfaff, D. W. (1987). *Exp. Brain Res.* **67,** 113.

Rothfeld, J. M., Hejtmancik, J. F., Conn, P. M., and Pfaff, D. W. (1989). *Mol. Brain Res.* **6,** 121.

Rubin, B. S., and Bridges, R. S. (1989). *Neuroendocrinology* **49,** 225.

Sabatino, F. D., Collins, P., and McDonald, J. K. (1989). *Endocrinology (Baltimore)* **124,** 2089.

Sarkar, D. K., and Fink, G. (1979). *J. Endocrinol.* **80,** 303.

Sarkar, D. K., and Fink, G. (1980). *J. Endocrinol.* **86,** 511.

Sarkar, D. K., and Yen, S. S. C. (1985). *Endocrinology (Baltimore)* **116,** 2080.

Sarkar, D. K., Chiappa, S. A., Fink, G., and Sherwood, N. M. (1976). *Nature (London)* **264,** 461.

Schillo, K. K., Leshin, L. S., Kuehl, D., and Jackson, G. L. (1985). *Biol. Reprod.* **33,** 644.

Schulz, R., Wilhelm, A., Pirke, K. M., Gramsch, C., and Herz, A. (1981). *Nature (London)* **294,** 757.

Seeburg, P. H., and Adelman, J. P. (1984). *Nature (London)* **311,** 666.

Shivers, B., Harlan, R., Morrell, J., and Pfaff, D. W. (1983). *Neuroendocrinology* **36,** 1.

Silverman, A. J., Zimmerman, E. A., Gibson, M. J., Perlow, M. J., Charlton, H. M., Kokoris, G. J., and Krieger, D. T. (1985). *Neuroscience* **16,** 69.

Silverman, A. J., Jhamandas, J., and Renaud, L. P. (1987). *J. Neurosci.* **7,** 2312.

Soper, B. D., and Weick, R. F. (1980). *Endocrinology (Baltimore)* **106,** 348.

Steiner, R. A., Bremner, W. J., and Clifton, D. K. (1982). *Endocrinology* **111,** 2055.

Strobl, F. J., and Levine, J. E. (1988). *Endocrinology (Baltimore)* **123,** 622.

Strobl, F. J., and Levine, J. E. (1990). *Methods Neurosci.* **2,** 316.

Strobl, F. J., Gilmore, C., and Levine, J. E. (1989). *Endocrinology (Baltimore)* **124,** 1140.

Strobl, F. J., Luderer, U., Schwartz, N. B., and Levine, J. E. (1990). *Soc. Neurosci.*, *20th Annu. Meet.* Abstr. No. 495.17, p.1202.

Summerville, J. W., and Schwartz, N. B. (1981). *Endocrinology (Baltimore)* **109,** 1442.

Sutton, S. W., Toyama, T. T., Otto, S., and Plotsky, P. M. (1988). *Endocrinology (Baltimore)* **123,** 1208.

Sylvester, P. W., Van Vugt, D. A., Aylsworth, C. A., Hanson, E. A., and Meites, J. (1982). *Neuroendocrinology* **34**, 269.

Teresawa, E., Krook, C., Hei, D. L., Gearing, M., Schultz, N. J., and Davis, G. A. (1988). *Endocrinology (Baltimore)* **123**, 1808.

Urbanski, H. F., Pickle, R. L., and Ramirez, V. D. (1988). *Endocrinology (Baltimore)* **123**, 413.

Valk, T. W., Corley, K. P., Kelch, R. P., and Marshall, J. C. (1980). *J. Clin. Endocrinol. Metab.* **51**, 730.

Van Cauter, E., van Coevorden, A., and Blackman, J. D. (1989). *In* "Neuroendocrine Regulation of Reproduction" (S. S. C. Yen and W. W. Vale, eds.), p.113. Serono Symposia, U.S.A.

Van Vugt, D. A., Sylvester, P. W., Aylsworth, C. F., and Meites, J. (1982). *Neuroendocrinology* **34**, 274.

Van Vugt, D. A., Diefenback, W. D., Alston, E., and Ferin, M. (1985). *Endocrinology (Baltimore)* **117**, 1550.

Vogelsong, K. M., and Levine, J. E. (1990). *Soc. Neurosci., 20th Annu. Meet.* Abstr. No. 495.15, p. 1202.

Voogt, J. L., de Greef, W. J., Visser, T. J., de Koning, J., Vreeburg, J. T. M., and Weber, R. F. A. (1987). *Neuroendocrinology* **46**, 110.

Watanabe, G., and Teresawa, E. (1989). *Endocrinology* **125**, 92.

Wiemann, J. N., Clifton, D. K., and Steiner, R. A. (1990). *Endocrinology (Baltimore)* **127**, 523.

Wiesner, J. B., Koenig, J. I., Drulich, L., and Moss, R. L. (1985). *Endocrinology (Baltimore)* **116**, 475.

Wilkes, M. M., and Yen, S. S. C. (1981). *Life Sci.* **28**, 2355.

Williams, C. L., Nishihara, M., Thalabard, J.-C., Grosser, P. M., Hotchkiss, J., and Knobil, E. (1990). *Neuroendocrinology* **52**, 133.

Wilson, R. C., and Knobil, E. (1982). *Brain Res.* **248**, 177.

Wise, P. M., Rance, N., Selmanoff, M. K., and Barraclough, C. A. (1981). *Endocrinology (Baltimore)* **108**, 79.

Zoeller, R. T., and Young, W. S., III (1988). *Endocrinology (Baltimore)* **123**, 1688.

DISCUSSION

R. Hertz.   One of the major functions of this whole system is the integration of these phenomena with the mating behavior of the animal. In some species this is so closely coordinated that ovulation occurs within a few hours of the period of sexual receptivity, for example, in the rat and guinea pig. I would like your opinion on how the phenomena you have observed relate to the behavioral aspects of reproductive function.

J. Levine.   I can think of two ways that there can be coordination of sexual behavior with the neuroendocrine events which underlie the generation of the preovulatory LH surge. One is that steroids prompt activity within LHRH neurons and, as proposed, I believe, by Renaud, there may be collateral axonal projections off the LHRH neurons which are involved in facilitating sexual behavior. From the work of Pfaff and others we certainly know that LHRH neurons do project to the midbrain central gray and that these appear to be involved in facilitating lordosis in female rats. So it would appear that a very efficient neuronal mechanism may exist which can coordinate the timing of sexual behavior with ovulation. A second type of mechanism may operate apart from the activity within a collateral system. Steroids may simply act to increase activity in other descending facilitory neuronal circuitries. The timing of sexual behavior relative to ovulation may simply be a fixed function of the temporal organization of steroid actions within these circuits versus those in LHRH neurons or afferents to LHRH neurons. That is, the intra- and intercellular sequelae of steroid actions in respective cell groups controlling ovulation and lordosis may coordinate the timing of these events.

N. Schwartz. My questions concern the neuropeptide Y experiments. Are the doses you use at all "physiological"? Do you have any feeling about the mechanism by which neuropeptide Y would act on the gonadotropes or on other pituitary cells?

J. Levine. In reference to your question on the dose of neuropeptide Y used in these experiments, we have taken the same blood samples and sent them to John MacDonald at Emory University for assay with a sensitive and specific neuropeptide Y radioimmunoassay. He has found that we achieve levels of neuropeptide Y with these infusions that are probably somewhat higher, but are at least in the same range as those concentrations that have been reported to be present in portal vessel plasma. So I think we are within a high physiological range. Neuropeptide Y circulates at much higher concentrations than most peptides and is present in the brain at relatively high concentrations compared to other peptidergic modulators.

Regarding your question on the mechanism of action of neuropeptide Y, I assume you mean at the pituitary level. My assumption is that neuropeptide Y probably operates through its own receptor and, by an unknown mechanism, probably facilitates LHRH-stimulated LH secretion at some point along the second messenger cascade activated by LHRH, e.g., the calcium/calmodulin pathway. Perhaps it facilitates calcium mobilization within the cell or at any other step beyond. The mechanism is really not known. This is an interesting physiological mechanism that has not been explored to any great degree at the cellular level. It is a prototype, perhaps, of a neuroendocrine modulator, that is, a substance that on its own has no effect, but will potentiate the actions of another neuroendocrine factor. Its actions may be analogous to a neuromodulator within the central nervous system. The other example of this type of modulation within a neuroendocrine system is, of course, the role of vasopressin in potentiating CRF-stimulated ACTH secretion. I think we are dealing with the same type of neuroendocrine mechanism.

R. Hazelwood. Your comments on neuropeptide Y are correct. It is probably the most widely distributed neuropeptide in the brain. It is highly concentrated in the hypothalamus, particularly in the arcuate nucleus and tracts. Can you reconcile your suggestions about its effect on LHRH with those seen in the diabetic rat, where its levels are extremely high and its synthesis is increased over 500% in the hypothalamus, yet these diabetic animals are very poor procreators?

J. Levine. I am not sure why this would be the case. However, the mechanism I have described is specific for a particular physiological situation, that is, in a proestrous rat after exposure to endogenous steroids which elicit the preovulatory LH surge. Inhibitory effects of neuropeptide Y have also been documented, and perhaps within the animal model that you mentioned—the diabetic rat—negative effects of neuropeptide Y may predominate. These may result either from different types of activities with the same neuropeptide Y cells that under other circumstances may facilitate LHRH release or actions, or from augmented activity within other neuropeptide Y-producing neurons which exclusively mediate inhibitory effects.

N. Schwartz. Can you comment on the role of FSH in some of the situations you have been studying?

J. Levine. The relationship between LHRH and FSH secretion, at least on a moment-to-moment basis, is at best ambiguous. We have seen, for example, in our push–pull perfusion experiments, that when we are able to discern pulses of FSH, they do not appear to be temporally associated beyond a chance level with LHRH pulses. As I mentioned earlier, this does not mean that I believe that LHRH is not involved in the maintenance of FSH secretion. In our hypophysectomized, pituitary-grafted animals, for example, we have seen the following. When we put a transplant into a hypophysectomized animal and then draw a blood sample, there are undetectable or nearly undetectable FSH levels. Clearly these levels increase after several days of pulsatile LHRH infusions. At the same time, we do not see any pulselike responses on the part of FSH to the LHRH infusion, and this is why I think that FSH is dependent in part on LHRH stimulation, but not in the acute sense. The acute stimulation of increments of FSH do not appear to be dependent on LHRH. The maintenance of a basal level of FSH, however, if I can use that term, does appear to be at least partially dependent on LHRH. Another experiment in which we have seen results that agree with this is the one Neena

Schwartz and I have collaborated on in which we have given *N*-methyl-D-aspartate (NMA) injections at doses which evoked LH increments that are of physiological amplitude. At the same time we see absolutely no response on the part of FSH secretion. Thus, we have to conclude that physiological increments in LHRH release, which we assume are being stimulated by NMA, are capable of stimulating LH pulses, while at the same time are incapable of stimulating FSH pulses.

W. Vale. You mentioned that Kelly Mayo found that the proestrus rise in LH preceded the increase in hypothalamic mRNA levels and suggested that LHRH expression was caused by the release of LHRH. The problem with this interpretation is that they were not measuring LHRH mRNA synthesis rates. They were measuring concentrations which because of existing pools will only slowly reflect acute changes in synthesis. In another model, Louise Bilezikjian and Marsha Barinaga studied the relationship between the secretion of GH and the expression of its gene measured by run-on assay of GH transcription rates. We concluded that GRF-induced increases in GH mRNA synthesis occurred within minutes and were independent of GH release. When monitoring GH mRNA levels it took much longer to observe any GRF-induced changes.

J. Levine. Park and Mayo have their work cut out for them in manipulating that model and trying to really discern what the time course of those events are. However, giving additional time for translation, posttranslational processing, and axonal transport, it is difficult to see how the acute increase in synthesis could precede secretion.

R. Hazelwood. I am ignorant of the behavioral changes that may occur immediately before, during, and after the LH surge other than the obvious. Are there any feeding behavioral changes during this time?

J. Levine. I would guess that feeding behavior is increased in correlation with the increase in activity that occurs as the lights go off. That this is even greater during the period of behavioral heat that ensues on the evening of proestrus is likely.

R. Hazelwood. I am looking for further evidence that neuropeptide Y might be involved because it is probably the most powerful endogenous orexigenic agent that exists in mammals.

J. Levine. I believe that Harold Spies has some nice data on that point, implicating neuropeptide Y in respect to feeding in the rabbit.

Gonadotropin-Releasing Hormone Pulses: Regulators of Gonadotropin Synthesis and Ovulatory Cycles

JOHN C. MARSHALL,* ALAN C. DALKIN,† DANIEL J. HAISENLEDER,†
SANDER J. PAUL,† GIROLAMO A. ORTOLANO,† AND
ROBERT P. KELCH†

*Department of Medicine, University of Virginia Health Sciences Center, Charlottesville,
Virginia 22908 and †Division of Endocrinology and Metabolism, University of Michigan Medical
Center, Ann Arbor, Michigan 48109

I. Introduction

Reproductive function is predominantly controlled by the pituitary gonadotropin hormones, luteinizing hormone (LH), and follicle-stimulating hormone (FSH). Both hormones are composed of two glycoprotein subunits, a common α subunit and distinct β subunits which confer biologic specificity (1). The genes coding for the three gonadotropin subunits are located on different chromosomes (2). The two pituitary gonadotropins act synergistically on the gonads, with FSH predominantly regulating gametogenesis, and LH production of the steroid hormones.

Both LH and FSH are secreted by the same pituitary gonadotrope cells and their synthesis and secretion is controlled by the hypothalamic decapeptide, gonadotropin-releasing hormone (GnRH). LH and FSH are known to be secreted differentially in some circumstances, but intensive investigation during the past two decades has failed to provide convincing evidence for other gonadotropin-releasing hormones. Thus, the differential regulation of LH and FSH synthesis and secretion appears to be effected by changes in the pattern of GnRH secretion, together with the direct feedback effects of gonadal steroids and peptides on the gonadotrope cell. Gonadal steroids can alter GnRH secretory patterns and also modify gonadotrope responses to GnRH. Inhibin, a polypeptide also consisting of two subunits and secreted by both the testis and the ovary, circulates in plasma and exerts a selective direct action to predominantly inhibit FSH secretion.

Early studies of the patterns of LH secretion in humans revealed that LH is released into the circulation in a series of pulses (3,4), and studies in animals have shown that pulses of LH reflect pulsatile secretion of GnRH by the hypothalamus (5). This pulsatile mode of GnRH secretion is essential for the maintenance of gonadotropin synthesis and release. Administration of GnRH in a continuous manner, or stimulation by long-acting GnRH agonists, results in

155

desensitization of gonadotropin secretion (6,7). In normal physiologic situations, such as during pubertal maturation and during ovulatory cycles in females, the pattern of pulsatile GnRH secretion changes, with alterations occurring in both the frequency and amplitude of GnRH release (8). Gonadal steroids modify pulsatile GnRH secretion and the pattern of GnRH stimulation appears to be important in the differential synthesis and release of LH and FSH by the pituitary gonadotrope.

In this article, we examine the role of the pattern of GnRH stimulation of the gonadotrope in regulating gonadotropin synthesis and secretion. In addition, we review current evidence on the effects of the changes in pulsatile GnRH release during human pubertal maturation and during ovulatory cycles in women, and suggest that the ability to alter the pattern of GnRH stimulation is an essential component of the regulation of reproduction in mammals.

In the first section, the regulation of gonadotropin subunit gene expression will be reviewed. In the subsequent section, the proposed role of altered GnRH pulsatile secretion in normal physiology and as a cause of reproductive failure are considered.

II. Regulation of Gonadotropin Subunit Gene Expression

The cloning of the cDNAs for the α, LH β, and FSH β subunits has allowed direct measurement of steady state mRNA concentrations. The original cDNA clones used in the present studies were generously supplied by Dr. W. W. Chin (rat α and LH β) (9,10) and Dr. R. A. Maurer (FSH β) (11). Steady state subunit mRNA concentrations were measured in cytoplasmic RNA extracts from pituitary tissue. Dot-blot hybridization was performed using saturating amounts of ^{32}P-labeled cDNA probes, and concentrations of mRNAs are expressed as femtomoles cDNA bound/100 μg pituitary DNA (12). The rates of transcription of the subunit genes were measured by Dr. M. A. Shupnik (University of Virginia, Charlottesville, VA) as previously described (13).

A. PHYSIOLOGIC STUDIES

1. Gonadectomy and Gonadal Steroid Replacement

Initial evidence for differential regulation of the gonadotropin subunit genes was found in studies of the effects of gonadectomy in male and female rats (Fig. 1). Serum LH, FSH, and all three subunit mRNAs increase following gonadectomy in both sexes. However, the timing and magnitude of the changes in serum hormones and steady state mRNAs differ between the sexes (12,14–16). In females, serum LH, and α and LH β subunit mRNAs do not increase until 4–7 days after ovariectomy. Thereafter, α mRNA rises to a plateau (5-fold increase)

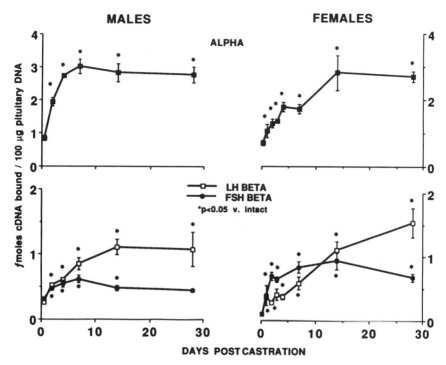

FIG. 1.   Pituitary gonadotropin subunit mRNA concentrations following gonadectomy in male and female rats.

after 14 days, but LH β mRNA continues to increase gradually through 30 days (15-fold increase). In contrast, serum FSH and FSH β mRNA increase more rapidly, reaching a plateau (fourfold rise) after 4 days. In male rats, both α and LH β mRNAs increase within 24 hours after orchidectomy. Thereafter, α mRNA plateaus after 14 days while LH β continues to rise through 30 days. Again, FSH β mRNA increases more rapidly than LH β and peak concentrations occur after 7 days, after which FSH β plateaus or even declines (15).

Replacement of estradiol (E_2) in ovariectomized rats suppresses all three gonadotropin subunit mRNAs (14–16), but if E_2 is replaced in physiologic concentrations, only partial suppression occurs over 7 days (Fig. 2). E_2 replacement at the time of ovariectomy has yielded interesting results and differential effects on subunit gene expression are seen (17). Estradiol prevented the increase in LH β, but α mRNA concentrations continued to increase to a plateau after 10 days. The rise in FSH β mRNA concentrations was only partly suppressed by replacement of E_2 alone, or both E_2 and progesterone (P). Administration of a GnRH antagonist to ovariectomized females, in the presence or absence of E_2

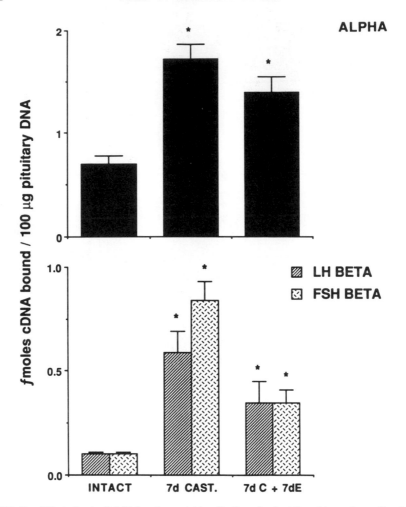

FIG. 2.   Effect of estradiol (E$_2$) replacement by silastic sc implant (to achieve plasma E$_2$ of 40 pg/ml) in 7-day ovariectomized female rats. 7dC + 7dE, 7 days of E$_2$ given to rats castrated 7 days earlier. *$p < 0.05$ vs intact.

and P, abolished the increase in α and LH β mRNAs, but did not prevent the rise in FSH β mRNA. This suggests that the increase in gonadotropin subunit mRNA following ovariectomy, particularly that of α and LH β, is a consequence of increased GnRH secretion. However, the failure of a GnRH antagonist to block the increase in FSH β suggests that additional ovarian hormones are involved in inhibiting FSH β mRNA expression in intact females (17). Definitive evidence *in vivo* is lacking, but *in vitro* data suggest that inhibin is required to maintain FSH β mRNA expression at the level found in intact female rats. The addition of inhibin to pituitary cells in culture resulted in a rapid decline in the concentrations of FSH β mRNA (18–20).

In male rats, replacement of testosterone (T) at the time of castration prevents the increase in all three subunit mRNAs (21). As shown in Fig. 3 replacement of physiologic concentrations of serum T (~2.5 ng/ml) in previously castrated animals suppresses subunit mRNAs to intact values (17). However, when higher

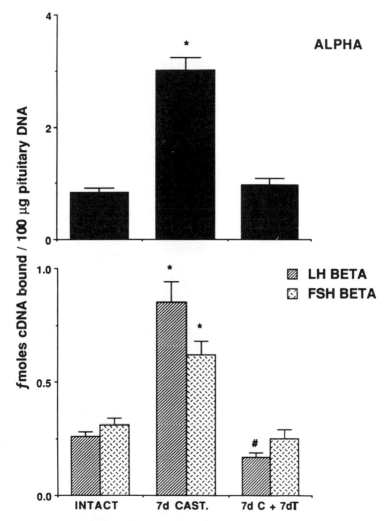

FIG. 3.   Effect of testosterone (T) replacement by silastic sc implant (plasma T was 2.5 ng/ml) in 7-day orchidectomized male rats. *$p < 0.05$ vs intact.

doses of T were given to castrate animals by intermittent injections, FSH β mRNA was not fully suppressed (22,23). These differences were of interest and could reflect the effects of different concentrations of testosterone acting either on FSH β transcription, or on the stability of FSH β mRNA in cytosol. Administration of a GnRH antagonist to castrate male rats reduced FSH β mRNA concentrations with a half-disappearance time of ~20 hours. In the presence of testosterone, however, steady state FSH β mRNA concentrations declined more slowly (~50 hours) after a GnRH antagonist, while the rate of decline of α and LH β mRNAs was unchanged (24). These data suggested that T may prolong FSH β transcription or stabilize FSH β mRNA, and other studies have suggested that stabilization of mRNA may be the mechanism of action. When a GnRH antagonist was administered for 4 days to castrate male rats, replacement of physiologic concentrations of testosterone increased FSH β mRNA twofold, without altering α or LH β concentrations (25,26). Measurement of FSH β mRNA transcription rates in GnRH antagonist-treated rats showed that transcription was not increased in the presence of testosterone, which suggests that the action of testosterone in increasing FSH β mRNA is exerted at a posttranscriptional level (25). Some studies have also suggested that other compounds may regulate FSH β mRNA stability. *In vitro,* addition of activin to pituitary cell cultures rapidly increased FSH β mRNA (19), and preliminary data show that this is due to enhanced stability of the FSH β mRNA (27). Thus, while data are presently incomplete, evidence suggests that testosterone and activin may enhance FSH β mRNA stability. It is also possible that compounds such as inhibin decrease FSH β mRNA by actions which include reducing the stability of FSH mRNA in cytosol.

Overall, gonadal steroid replacement results in suppression of gonadotropin subunit mRNA concentrations, but the degree and mechanisms of this action may depend on plasma steroid concentrations. Testosterone suppresses pulsatile GnRH secretion and can also modify gonadotrope responses to GnRH. Estradiol has variable effects on pulsatile GnRH secretion *in vivo*, but estradiol replacement to castrate animals results in reduced subunit mRNA transcription rates (28). *In vitro* studies showed that estradiol did not reduce α and FSH β transcription rates and actually increased LH beta transcription, suggesting that the predominant actions of estradiol *in vivo* are exerted by inhibiting hypothalamic GnRH secretion (29,30). The positive actions of estradiol on LH secretion and LH β mRNA expression may be exerted directly at the level of the gonadotrope. While not all evidence is in agreement, the latter view is supported by the effect of estradiol in increasing LH β transcription *in vitro*, and the presence of an estrogen response element in the 5' region of the rat LH β gene (31). Progesterone, particularly in the presence of estradiol, reduces pulsatile GnRH secretion *in vivo* (32,33). Thus, progesterone could suppress subunit gene expression via hypothalamic mechanisms *in vivo*, but may also exert direct negative effects on the gonadotrope. *In vitro* studies using pituitaries from sheep showed that progesterone directly suppressed FSH β gene transcription (34), although studies using rat pituitary cells suggest that this action may also require the presence of estradiol (35).

Thus, at the present time data are incomplete, but it appears likely that gonadal steroids and peptides exert actions both at the hypothalamic level to reduce GnRH secretion and directly on the gonadotrope cell to modify expression of subunit mRNAs, hormone synthesis, and secretion. The relative degree and magnitude of actions at these sites may be modified by steroid concentration and duration of exposure to steroid hormones.

2. Gonadotropin Subunit mRNAs during the Rat Estrous Cycle

Serum LH, FSH, α, and LH and FSH β mRNA subunit concentrations in the rat estrous cycle are shown in Fig. 4.

During the 4-day cycle in female rats, serum LH and FSH concentrations remain low, except during the preovulatory surge during the late afternoon and evening of proestrus (36). α, LH β, and FSH β mRNAs show evidence of differential expression during the cycle, with increased expression of both β mRNAs occurring at the time of the gonadotropin surge on proestrus (37,38). On the morning of metestrus, FSH β mRNA was increased twofold, gradually falling to basal levels by the evening. α and LH β mRNAs were stable on metestrus, but both transiently increased twofold during a 12-hour period on diestrus when FSH β was unchanged. These changes in mRNA expression occurred in the absence of increased secretion of LH and FSH. On the afternoon of proestrus, LH β mRNA increased threefold, rising before the preovulatory surge of serum LH. FSH β mRNA increased fourfold, but maximum concentrations occurred 2 hours after the onset of the serum FSH surge. In contrast to these changes in β subunit mRNA concentrations, α mRNA was unchanged during the gonadotropin surge. These data suggest that both coordinate and differential regulation of expression of the three subunit genes occur during the estrous cycle. Coordinate increases in α and LH β mRNA are seen on diestrus, and both β subunit mRNAs increase during the proestrus gonadotropin surges. On metestrus, however, only FSH β mRNA increases when both α and LH β mRNAs are stable.

The mechanisms regulating subunit gene expression and hormone secretion during the cycle remain uncertain, but may include the effects of changes in GnRH secretory pattern and the direct actions of ovarian steroids and peptides on the gonadotrope. Secretion of GnRH changes during the cycle, and both the amplitude and frequency of pulsatile GnRH release, are increased during the proestrus LH surge (39,40). This coincides with the increased LH β mRNA concentrations and a twofold increase in the rate of LH β transcription (29), suggesting that the LH β mRNA changes are dependent on GnRH secretion.

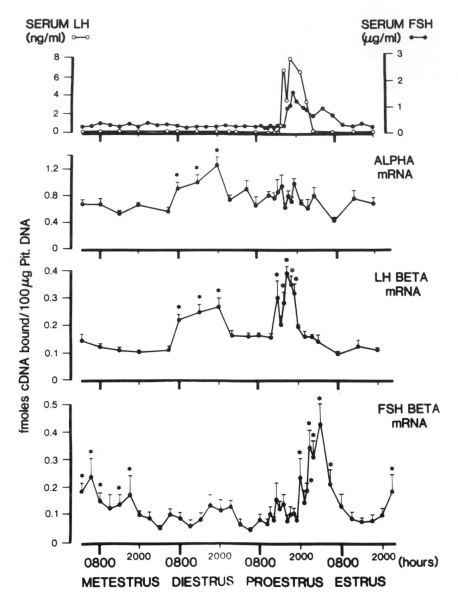

FIG. 4.   Serum LH, FSH, and gonadotropin subunit mRNA concentrations during the 4-day estrous cycle in rats. *$p < 0.05$ vs basal values. (Reproduced from Refs. 37 and 38 with permission.)

While it is likely that changes in the pattern of GnRH stimulation regulate gene expression, the direct actions of ovarian steroids and peptides on the gonadotrope probably act to differentially regulate mRNA expression. Ovarian inhibin may be important in this regard, as inhibin can selectively inhibit FSH release and reduce FSH β mRNA (18,19). We have measured serum inhibin and FSH, and FSH β subunit mRNA, during the 4-day cycle. The results showed that serum inhibin fell during the late morning and early afternoon of proestrus, prior to the evening increase in FSH β mRNA during the FSH surge. Inhibin transiently rose at the time of the gonadotropin surge, a few hours prior to the acute decline in FSH β mRNA during the night of proestrus–estrus. Serum inhibin was stable on estrus and early metestrus, but rose during the latter half of metestrus when FSH β mRNA concentrations were declining. These data show a general inverse relationship between serum inhibin and FSH β mRNA concentrations in the pituitary, and suggest that inhibin may play an important role in regulating FSH β mRNA expression during the 4-day estrous cycle (41).

In efforts to elucidate the relationship of subunit mRNA expression and the gonadotropin surges, we have also examined α and LH β mRNAs in ovariectomized estradiol-replaced rats. This model exhibits daily surges of LH, which are similar in timing and magnitude to those observed on proestrus in cycling female animals. In this model, the evening LH surge was associated with a twofold increase in α mRNA, but LH β mRNA was unchanged (42). Thus, in the presence of stable concentrations of E_2 and in the absence of P and ovarian peptides, a different pattern of α and LH β mRNA expression is found compared to proestrus. This suggests that the ovarian steroid/peptide milieu may effect differential modulation of subunit mRNA expression, but it remains unclear whether this reflects direct actions on the gonadotrope or effects exerted via changes in the pattern of GnRH secretion.

B.   GnRH PULSES AND THE REGULATION OF GONADOTROPIN SUBUNIT GENE EXPRESSION

The critical role of a pulsatile GnRH stimulus in maintaining gonadotropin secretion has been firmly established. In GnRH-deficient humans and in animal models, administration of GnRH pulses maintains LH and FSH secretion, whereas continuous GnRH infusions desensitize the gonadotrope, and LH and FSH secretion falls (43). In desensitized gonadotropes, α mRNA concentrations are increased, but LH β mRNA is unchanged or is decreased (44). The importance of GnRH in maintaining subunit gene expression is seen in studies utilizing GnRH antagonists or GnRH-deficient animal models. Blockade of GnRH action prevents the increase in mRNAs in castrate rats, and steady state mRNA concentrations fall to levels below those seen in intact animals (45,46). In GnRH-deficient sheep with surgical disconnection of the hypothalamus and pituitary, administration of GnRH pulses increased concentrations of α and LH β mRNA to values present in ovariectomized controls (47).

The role of the pattern of GnRH stimulation in the differential regulation of subunit gene expression was studied by administering GnRH pulses, at different amplitudes and frequencies, to GnRH-deficient rats. The model used for these studies was the castrate male rat replaced with physiologic concentrations of testosterone via subcutaneous silastic implants. When testosterone is replaced in this manner, the postcastration rise in serum gonadotropins and GnRH receptors is abolished, similar to the removal of endogenous GnRH secretion by hypo-thalamic lesions or administration of anti-GnRH serum (48). These data suggest that a constant concentration of plasma testosterone reduces GnRH secretion. Other studies have shown that LH pulses (by inference GnRH pulses) are either absent (75%) or occur infrequently (0–4 pulses/24 hours) in the remaining 25% of castrate rats with testosterone implants (49). Thus, the castrate testosterone-replaced rat is relatively GnRH deficient and provides a convenient model in which to study the effects of exogenous GnRH administration in the presence of a stable testosterone milieu.

1. Role of GnRH Pulse Amplitude

We have examined the effects of GnRH pulse amplitude on LH secretion, GnRH receptor number, and gonadotropin subunit mRNA concentrations (50,51). GnRH pulses were given to castrate T-replaced rats at intervals of every 30 minutes to mimic the frequency of endogenous GnRH secretion observed in castrate rats (49). The effects of different GnRH pulse amplitudes on subunit mRNA concentrations are shown in Fig. 5. After 24 or 48 hours of pulsa-tile GnRH administration, serum LH release to GnRH, the number of pituitary GnRH receptors, and LH β mRNA concentrations were all maximum after 25-ng GnRH pulses. This GnRH pulse dose produces plasma GnRH concentrations in the upper physiologic range [~200 pg/ml (48)] in pituitary stalk blood. Lower or higher amplitude pulses were less effective in increasing LH β mRNA concentra-tions. In contrast, all GnRH doses used (10–250 ng/pulse) increased α mRNA to a similar degree. Parallel to α responses, FSH β mRNA expression was not dependent on the amplitude of the GnRH pulse and was increased by all doses of GnRH used. These results suggest that the amplitude of the pulsatile GnRH stimulus can exert differential effects on subunit mRNA expression. α and FSH β expression appear to be relatively independent of dose, but LH β expression appears highly sensitive, and maximum mRNA concentrations are seen only in response to a narrow range of GnRH pulse amplitudes.

α subunit mRNA is expressed in both thyrotrope and gonadotrope cells, and the actions of GnRH in specifically regulating gonadotrope α mRNA were exam-ined in castrate T-replaced rats given triiodothyronine (T_3) to suppress thyrotrope function (51). T_3 treatment suppressed α mRNA by 50%, suggesting that half of pituitary α mRNA is present in thyrotrope cells. GnRH pulses increased α

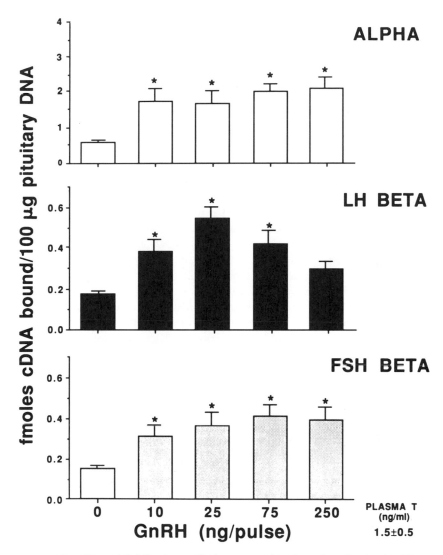

FIG. 5. The effects of GnRH pulse amplitude on expression of gonadotropin subunit mRNAs in male rats. Castrate testosterone-replaced rats were given GnRH pulses every 30 minutes for 24 hours. Pulse amplitude (ng/pulse) is shown. *$p < 0.05$ vs 0 ng GnRH (saline controls).

mRNA within 12 hours in T_3-replaced rats. The magnitude of the α response was similar in control and in T_3-treated animals, indicating that the increased α mRNA was of gonadotrope origin. These studies also revealed that the same pulsatile GnRH stimulus did not stimulate expression of gonadotropin subunit mRNAs on the same time course. Both FSH β and α mRNAs increased more rapidly (after 8–12 hours), whereas significant increases in LH β mRNA were not seen until after 24 hours of pulsatile GnRH stimulation (51).

Thus, in response to a physiologic GnRH pulse signal in male rats, FSH β and α mRNA expression occurs more rapidly than that of LH β, and both are relatively independent of the amplitude of the GnRH signal. In contrast, LH β mRNA expression is delayed until 24 hours, and is markedly dependent on GnRH amplitude.

Steroid hormones appear to act directly on the gonadotrope cell to modify subunit mRNA expression in response to GnRH pulse amplitude. The effects of GnRH pulse amplitude were examined in the presence of plasma testosterone concentrations of 1.5 ng/ml (low physiologic), 3.5 ng/ml, and 6 ng/ml (high physiologic). α mRNA expression tended to be lower in the presence of the highest testosterone concentration. FSH β mRNA responses were similar after all GnRH doses, at all three testosterone concentrations. In the presence of the high physiologic plasma testosterone, LH β mRNA expression was enhanced after the lower dose GnRH pulses (10 ng), whereas responses to the higher GnRH doses (75–250 ng) were suppressed (52). Similar response patterns of the three gonadotropin subunit mRNAs were observed in the presence of di-hydrotestosterone or estradiol replacement in male rats, although α mRNA concentrations were higher in the presence of estradiol. These data suggest that GnRH pulse amplitude may be a more important determinant of subunit mRNA responses than the gonadal steroid milieu, although physiologic changes in plasma T concentration can directly modulate LH β responses to GnRH pulse amplitude.

2. Role of GnRH Pulse Frequency

The effects of changing the frequency of the pulsatile GnRH secretion was also examined in the castrate T-replaced male rat model. GnRH pulses were administered at optimal amplitude (25 ng/pulse) at intervals between 8 and 480 min for a total duration of 24–48 hours. The number of pituitary GnRH receptors were increased by GnRH given at all pulse intervals less than 120 min, and acute LH release was maximal after pulses given every 15, 30, and 60 min. When GnRH pulses were administered at the frequency found in castrate rats (every 30 min), GnRH receptors, α, and LH β subunit mRNAs were increased to values similar to those present in castrate males (53,54). Thus, administration of exogenous GnRH at a frequency similar to that present after castration suggests that a 30-min GnRH frequency is optimal for expression of GnRH receptors, α, and LH β mRNAs.

We have also examined the effect of GnRH pulse frequency on FSH β mRNA expression. As seen in Fig. 6, fast-frequency GnRH pulses (every 8 min) maximally increased α and also increased LH β mRNA. GnRH pulses given every 30 min increased all three subunit mRNAs. In contrast, slower frequency pulses

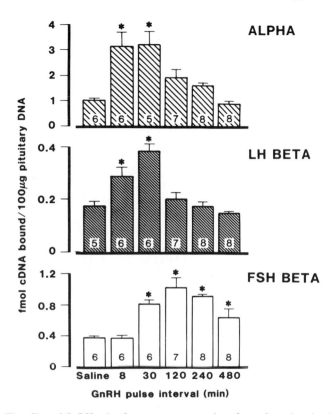

FIG. 6.   The effect of GnRH pulse frequency on expression of gonadotropin subunit mRNAs in castrate testosterone-replaced male rats. The dose/pulse of GnRH was constant (25 ng) and the pulse intervals are shown. *$p < 0.05$ vs saline. (Reproduced from Ref. 55 with permission.)

(every 120 min or longer) did not increase α or LH β mRNAs, but maintained the elevated level of FSH β mRNA expression (55). In the same study, the frequency dependence of subunit mRNA expression was examined by equalizing the total dose of GnRH given over 24 hours. GnRH pulses given every 8 min increased α mRNA, but the same total GnRH dose given every 480 min was ineffective. In contrast, pulses given every 480 min increased only FSH β mRNA, and faster frequency pulses were ineffective.

The effects of GnRH pulse frequency appear to be exerted at the level of subunit gene transcription (56). Using the same experimental paradigm, fast-frequency (8 min) GnRH pulses increased α gene transcription, while only 30-min pulses were effective in elevating LH β transcription rates (fourfold). Of interest, FSH β transcription was increased only after 120-min pulses, which is in accord with the above results where steady state mRNA concentrations were

GnRH PULSES

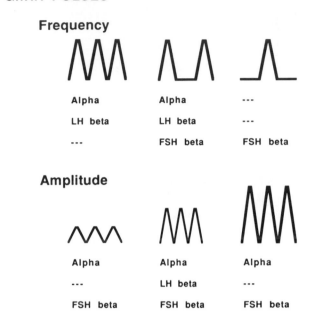

Frequency

Alpha	Alpha	---
LH beta	LH beta	---
---	FSH beta	FSH beta

Amplitude

Alpha	Alpha	Alpha
---	LH beta	---
FSH beta	FSH beta	FSH beta

TESTOSTERONE

slows GnRH frequency   ⎤ LH beta mRNA ↓
stabilizes FSH beta mRNA ⎦ FSH beta mRNA ↑

-favors FSH beta synthesis

FIG. 7.   Summary of the effects of the pattern of GnRH stimulation and of testosterone in regulating gonadotropin subunit mRNA expression in male rats. The subunit mRNA shown is expressed in response to the GnRH pulse stimulus shown.

measured. In contrast, the continuous administration of GnRH by infusion did not increase transcription rates of any of the gonadotropin subunit genes.

Similar results of GnRH pulse frequency have been observed in other animal models. In sheep (57), GnRH pulses every 30 min increased α mRNA, and pulses given every 60 min increased all three subunit mRNAs. Slower frequency GnRH pulses did not increase FSH β mRNA in the sheep model, however, and this difference may relate to the steroid milieu in the animal model used. Progesterone was used to inhibit endogenous GnRH release in sheep, and *in vitro* studies have suggested that progesterone can selectively inhibit FSH β gene transcription in ovine pituitary cells (34).

Overall, these results suggest that changes in the frequency of GnRH pulse stimulation of the gonadotrope may be physiologically important in differentially regulating gonadotropin subunit gene expression and gonadotropin secretion. In general, these results in the castrate T-replaced rat model are in accord with earlier studies in humans and primates, where slow-frequency GnRH stimuli favored FSH secretion and faster pulse stimuli were required to maintain LH release (58,59).

The effects of GnRH pulse frequency and amplitude on expression of subunit mRNAs are summarized in Fig. 7. The interactions of GnRH pulse frequency, amplitude, and plasma testosterone may be important in maintaining expression of FSH β mRNA, FSH secretion, and spermatogenesis in adult males. During pubertal maturation, the increased amplitude and frequency of GnRH stimulation would favor expression of the α and LH β genes. LH secretion would stimulate testosterone production by the testis, and the rising plasma testosterone would slow the frequency of endogenous GnRH secretion. This would reduce expression of the LH β gene, but maintain FSH β expression. The action of testosterone in stabilizing FSH β mRNA would act synergistically to maintain steady state FSH β mRNA concentrations, available for translation of FSH β subunit and subsequent FSH secretion. Thus, in the adult animal, these mechanisms would tend to maintain FSH β synthesis and secretion which may be important for maintenance of spermatogenesis. A decline in plasma testosterone would result in increased GnRH frequency, which would favor α and LH β mRNA expression, LH synthesis and secretion, with consequent restoration of the circulating adult testosterone milieu. Such feedback mechanisms remain speculative as physiologic mediators, but would allow maintained FSH synthesis and secretion, which might otherwise be compromised by the enhanced GnRH pulse frequency and amplitude present after pubertal maturation.

III. GnRH Pulses and the Regulation of Human Reproduction

A. PUBERTAL MATURATION

The reproductive system in humans and primates appears to be active during infancy and plasma gonadotropins are elevated during the first few months of life. Few data are available in humans (60), but in infant rhesus monkeys, pulsatile gonadotropin secretion occurs at this stage and the LH pulse patterns suggest that GnRH is being secreted at approximately 1 pulse/hour in males (61) and at a slower frequency (1 pulse/3–4 hours) in females (62). Subsequently, plasma gonadotropin levels decline (particularly that of LH, producing a high FSH-to-LH ratio), LH responses to exogenous GnRH fall, and gonadotropins remain low during the first decade of life. During this inhibitory phase, gonadotropins, presumably stimulated by GnRH, are secreted in a pulsatile manner in prepubertal children (63–65). Studies have suggested that GnRH pulse frequency is slow in prepubertal children. Gonadotropin pulses occur approximately every 3–4 hours during the day (66,67), with a small augmentation of both frequency and amplitude during sleep (68).

The resurgence of the reproductive system at puberty is initiated by a marked amplification of sleep-entrained secretion of gonadotropins. Work (66–69) has shown that this nocturnal increase in gonadotropin secretion is consequent on an increased frequency and amplitude of GnRH release. As shown in Fig. 8, in early adolescent boys LH pulse frequency and amplitude are highest with the onset of sleep and tend to decline during the night, perhaps consequent upon the increase in plasma testosterone (66).

The mechanisms underlying these changes in hypothalamic GnRH secretion remain uncertain. In some species, particularly sheep, hypothalamic opioid mechanisms inhibit pulsatile GnRH secretion during the prepubertal period as well as after puberty (70). In prepubertal children, however, administration of the opiate antagonist naloxone during the day does not increase LH secretion (71,72). This suggests that increased hypothalamic opioid tone is not the major limiting factor in reducing GnRH secretion before puberty in humans. GnRH is present in hypothalamic neurons in prepubertal animals, and administration of N-methyl-DL-aspartate (NMA, an analog of aspartate, a putative excitatory neurotransmitter) results in release of LH (73).

Thus, while the sleep-related onset of GnRH secretion in early puberty is well documented, the mechanisms involved in disinhibiting GnRH are unclear, but do not appear to involve opioids in humans. Similarly, little is known about the timing and effects of sex steroids and establishment of the inhibitory feedback mechanisms present in adults. Administration of testosterone during the evening does not inhibit the initial sleep-associated increase in GnRH pulses (66). However, when testosterone is administered by infusion earlier in the day, the frequency of nocturnal LH pulses is reduced, and nocturnal secretion resembles that seen in adult men (74). This suggests that the slowing of GnRH pulse frequency in the later hours of sleep during puberty may reflect the development of testosterone feedback to inhibit the rapid nocturnal GnRH secretion.

These data suggest that GnRH pulse patterns may regulate gonadotropin synthesis and secretion in humans. Pubertal maturation is associated with a change in gonadotropin responsiveness to exogenous GnRH, with the initial prepubertal predominance of FSH release diminishing, the LH responsiveness increasing, as puberty advances. By analogy to our data in rats, slow-frequency, low-amplitude GnRH stimuli would favor FSH β expression, and the increased amplitude and frequency occurring with the onset of puberty would favor expression of the α and LH β genes.

Once the rapid-frequency (~1 pulse/hour) high-amplitude GnRH secretion

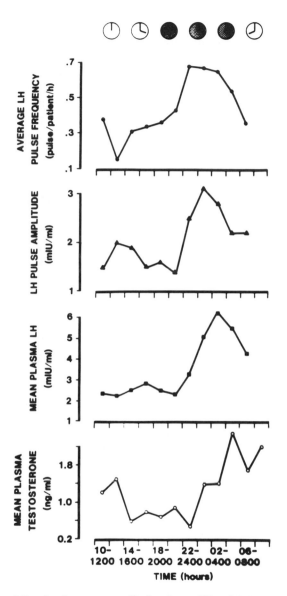

FIG. 8. Mean LH pulse frequency, amplitude, plasma LH and testosterone, in 2-hour time blocks, from data in 21 early pubertal boys. (Reproduced from Ref. 66 with permission.)

has been established at puberty, the subsequent transition to the adult reproductive state involves the establishment of adult patterns of steroid hormone feedback. We propose that in boys, the nocturnal secretion of testosterone dampens GnRH pulse frequency and amplitude during the later hours of sleep. As puberty progresses, testosterone inhibits the GnRH pulse discharge to every 90–120 min, a pattern which is maintained throughout the day, with little if any sleep-entrained differences. Castration at this time would remove testosterone feedback and restore the rapid (~1 pulse/hour) LH, and by inference GnRH frequency, which first appeared during the first few hours of sleep. In girls, similar mechanisms may be involved although fewer data are presently available. The initial amplification of sleep-entrained GnRH secretion increases FSH, which in turn is suppressed as ovarian estrogen secretion matures. Further increases in GnRH frequency and amplitude enhance LH secretion, leading to waves of incomplete follicular development. However, estradiol fails to slow GnRH pulse frequency, as has been described in the late follicular phase of the adult ovulatory cycle (75). As estradiol also augments LH responsiveness to GnRH, LH secretion increases. Anovulatory menstrual bleeding may occur at this juncture as ovarian steroid secretion fluctuates. However, full expression of the positive feedback of estradiol and ovulation is not possible until the ability to maintain a continuous hourly GnRH pulse frequency has been firmly established at midpuberty. Subsequent ovulation, and the actions of estradiol and progesterone, lead to the adult pattern of luteal slowing of GnRH secretion (see the next section).

B. OVULATORY CYCLES IN WOMEN

The availability of sensitive radioimmunoassays led to detailed studies of pulsatile gonadotropin secretion during the menstrual cycle. A series of studies using blood-sampling regimens of every 5–20 min have revealed a consensus as to the changes in GnRH pulsatile secretion during ovulatory menstrual cycles (75–79).

During the early follicular phase, LH (GnRH) pulses occur approximately every 90–100 min and as plasma estradiol and inhibin levels are low, the GnRH stimulus releases both FSH and LH. The half-life of FSH (~180 min) exceeds the interval between GnRH pulses and plasma FSH concentrations consequently rise and exceed those of LH. By the midfollicular phase, the frequency of GnRH pulses has increased to between every 60 and 90 min. The prior gonadotropin stimulus induces follicular maturation and elevation of serum estradiol, which selectively inhibits FSH release (possibly assisted by inhibin from the maturing follicle) and plasma FSH falls. The rise in estradiol does not appear to inhibit the increase in GnRH pulse frequency and, by the late follicular phase, GnRH frequency is approximately 1 pulse/hour. With continuing follicular maturation, estradiol rises further and initiates the positive feedback action to enhance LH responsiveness to GnRH. Plasma progesterone also increases just before the LH surge, and together with estradiol augments gonadotropin release, completing the preovulatory surges of LH and FSH (80,81). Thus, the LH surge results from the positive effects of ovarian steroids in augmenting gonadotrope responsiveness in the presence of a rapid frequency of GnRH stimulation. GnRH pulse amplitude may also be increased as has been shown in rats (82), but it is unclear whether this occurs in primates. LH surges can be produced following administration of a consistent dose of GnRH per pulse (83,84), but the surge is often of lower magnitude than that which occurs spontaneously. The LH surge is of limited duration, but the mechanisms involved in reducing the pulsatile release of LH in response to GnRH are unclear. In many studies, pulsatile LH release continues on the downslope of the LH surge, suggesting continuing pulsatile GnRH secretion with diminishing pituitary responsiveness. This may reflect the loss of estradiol positive feedback or, alternatively, may represent a degree of desensitization after a prolonged interval of rapid-frequency GnRH stimulation.

After ovulation, LH pulse frequency falls during the next 3–4 days and continues to fall, so that by the midluteal phase, LH pulse frequency is 1 pulse every 3–5 hours, less than half the frequency present during the follicular phase. The amplitude of LH pulses also varies, which probably reflects altered GnRH pulse amplitude, since LH responses to exogenous GnRH were consistent in the presence of luteal concentrations of ovarian steroids (85). The reduced GnRH pulse

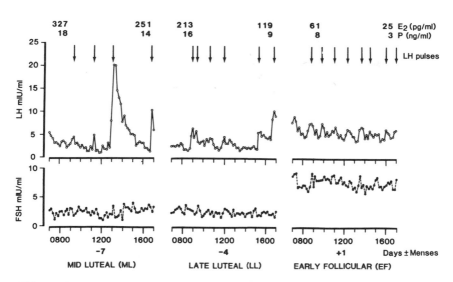

FIG. 9.   Gonadotropin secretion and gonadal steroids during the intercycle period of an ovulatory cycle. The transition of hormone secretion from the luteal to the follicular phase is shown. The arrows indicate LH (GnRH) pulses.

frequency reflects the actions of progesterone and estradiol acting at the hypothalamus (86,87). Estradiol and progesterone appear to slow GnRH pulse frequency by enhancing hypothalamic opioid activity and administration of naloxone increases GnRH pulse frequency during the luteal phase (88–90).

With the demise of the corpus luteum, estradiol, progesterone, and inhibin
concentrations fall and the frequency of GnRH secretion increases, returning to a
more rapid, consistent amplitude stimulus (see Fig. 9). The slow, irregular GnRH
stimulus during the luteal phase may not be optimal for stimulation of LH β
mRNA expression and, hence, LH synthesis may be impaired. This, together
with the intermittent release of LH, would deplete pituitary LH stores. FSH is not
released in response to the GnRH stimuli, due to the selective inhibitory effects
of estradiol and inhibin. As the slow irregular stimulus would be expected to
favor FSH mRNA expression, pituitary FSH stores would be maintained. Thus,
with the decline of estradiol and progesterone and the increase in GnRH pulse

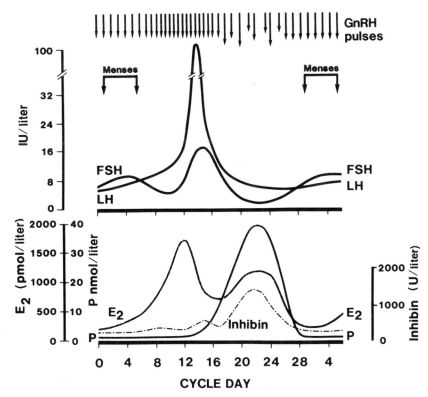

FIG. 10.   Schematic diagram of GnRH pulsatile stimuli, plasma gonadotropins, ovarian steroids,
and peptides during an ovulatory menstrual cycle.

frequency, the predominant gonadotropin secreted in the late luteal phase is FSH. This, together with a longer FSH half-life, produces the selective increase in FSH which appears essential for initiating the development of ovarian follicles destined to ovulate during the next cycle (91,92). A schematic diagram indicating the proposed changes in GnRH amplitude and frequency, plasma gonadotropins, ovarian steroids, and peptides is shown in Fig. 10.

The changes in pulsatile GnRH stimulation described above have been well documented, and would suggest an important role for the ability to regulate GnRH pulse frequency in the control of normal cyclical ovulation. As inferred from studies in rats, the increased GnRH frequency during the follicular phase would be expected to increase α and LH β mRNAs, leading to enhanced ability to synthesize LH prior to the LH surge. The irregular, slow-frequency midluteal GnRH stimulus would be anticipated to favor FSH β mRNA expression and FSH synthesis. As FSH is not released, FSH stores would be replete and, with the removal of selective inhibition, FSH would be released in response to the return of more rapid GnRH stimuli. The above suggestions are in accord with observations in women and primates, but it should be noted that the ability to change GnRH pulse frequency may not be a prerequisite for ovulatory cycles. Exogenous GnRH, given at fixed frequencies, can induce ovulation in GnRH-deficient women and primates, although in many cases the doses used were supraphysiologic. Slowing of GnRH stimulation in the luteal phase does appear to be important, however. Administration of GnRH at rapid frequencies in the luteal phase has led to deficient follicular development and corpus luteum function in subsequent cycles in both women and primates (93,94). Thus, the altered pattern of GnRH stimulation may be most important in the luteal phase; to allow FSH β synthesis and FSH secretion, which may have stimulatory actions on waves of developing ovarian follicles destined to ovulate several cycles in the future.

C.   GnRH PULSES IN ANOVULATORY CONDITIONS

Anovulation has been shown to be associated with abnormal patterns of LH (GnRH) pulsatile secretion in several clinical conditions.

1.   Hypothalamic Amenorrhea

Hypothalamic amenorrhea (HA) is a term used to describe a common disorder in which anovulation occurs in the absence of any demonstrable abnormalities of pituitary or ovarian function. The disorder is often preceded by conditions such as self-imposed weight loss, strenuous exercise such as gymnastics or competitive running, psychological stress, and on occasions the prior use of combined oral contraceptives (95,96). In approximately 70% of the patients, removal of the antecedent problem results in a return of ovulatory menses within a year, but in the remaining women, anovulation and amenorrhea persist. Plasma LH, FSH, and estradiol are low or low normal, prolactin is normal, and responsiveness to exogenous GnRH is preserved. Studies from several groups have revealed that the frequency of GnRH pulsatile secretion is markedly reduced in a majority (16 of 19 patients in our study, Ref. 95) of women with HA. In general, the GnRH frequency (1 pulse every 3–4 hours) resembles that seen in the luteal phase of an ovulatory cycle, and the irregular amplitude of LH pulsatile release also resembles that seen in the luteal phase. These similarities, together with the fact that normal suppression of GnRH frequency in the luteal phase is effected via increased hypothalamic opioid tone, led to the suggestion that the disorder may reflect abnormal suppression of GnRH frequency by hypothalamic opioids. Administration of naloxone to these women results in rapid (within 1–2 hours) restoration of normal frequency GnRH secretion in some 60–70% of women (97,98). This suggests that in a majority of women with this disorder, the anovulation reflects a slow frequency of GnRH pulsatile secretion which is inadequate to maintain the level of LH synthesis and secretion which is required for the ovulatory LH surge. This is in accord with data showing that administra-

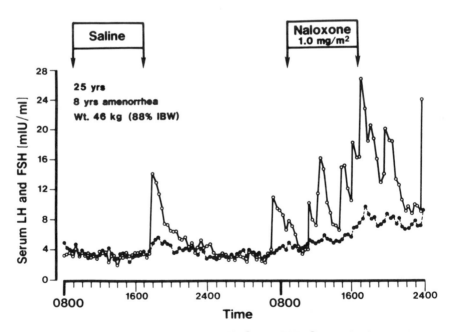

FIG. 11. Effects of naloxone on pulsatile LH (○) and FSH (●) secretion in a woman with hypothalamic amenorrhea. The patient was 25 years old with a history of weight loss and amenorrhea for 8 years. She had regained weight to 90% of ideal, 1 year prior to the study, but had remained amenorrheic. (Reproduced from Ref. 8 with permission.)

tion of slow-frequency (every 3 hours) GnRH pulses to GnRH-deficient primates did not maintain plasma LH concentrations (99). Figure 11 shows LH pulse patterns in a patient with hypothalamic amenorrhea before and during the administration of the opiate antagonist naloxone.

The mechanisms of amenorrhea in women who do not respond to opiate blockade remain uncertain. However, data have suggested that abnormalities of the hypothalamic–pituitary–adrenal axis may be involved. Stress is thought to elevate corticotropin-releasing hormone (CRH) and, in animal studies, CRH can directly inhibit GnRH secretion and reproductive function. Some women with HA have been shown to have elevated plasma cortisol levels and blunted cortisol responses to CRH (100), suggesting that stress-induced abnormalities of CRH secretion may be involved in the anovulation in women with AH. Thus, current evidence suggests that increased hypothalamic opioid activity is the underlying cause of abnormal GnRH secretion in most women with HA, but other abnormalities, perhaps involving CRH secretion, may be the predominant mechanism in other patients.

2. Hyperprolactinemia

An elevated serum prolactin, due to medications which reduce hypothalamic dopamine secretion or action, or to a prolactinoma of the pituitary gland, is commonly associated with amenorrhea. Studies of pulsatile release revealed slow irregular patterns of GnRH secretion, which were restored to normal follicular phase patterns after prolactin had been suppressed by the dopamine agonist bromocriptine (101–103). The mechanisms of reduced GnRH frequency and irregular amplitude in hyperprolactinemia also appear to involve excess opioid activity. In patients in whom prolactin remained elevated, administration of naloxone resulted in a rapid return of pulsatile GnRH secretion in a manner analogous to that seen in hypothalamic amenorrhea (104).

3. Polycystic Ovarian Disease

Polycystic ovarian disease (PCO) is a disorder of unknown etiology associated with anovulation, hirsutism, obesity, and multiple cysts in the ovaries. The excess androgen secretion in PCO has been shown to be predominantly of ovarian origin, but the exact mechanisms leading to excess androgen production remain unclear (105). While ovarian abnormalities, including abnormal ovarian steroidogenesis or follicular maturation, may be the primary cause of excess androgen secretion in some patients, increased stimulation of the ovary by LH appears to be the predominant mechanism in a majority of patients. Approximately 75% of patients with PCO have elevated mean serum LH values on repeated measurement. The increased LH, perhaps augmented by the hyperinsulinemia commonly associated with obesity in this disorder, would be expected to increase androgen synthesis by the ovary (106,107). The use of GnRH agonists to desensitize LH release results in reduced LH and ovarian androgen secretion (108). This strongly suggests that increased LH stimulation of the ovaries is causally involved in the syndrome.

More recently, studies have demonstrated that the frequency and amplitude of LH pulses are commonly increased in patients with PCO (109,110). Thus, the possibility exists that a persistent rapid frequency of GnRH secretion results in excess LH synthesis and secretion, which in turn is primarily involved in enhanced androgen production by the ovary. The mechanisms involved in producing abnormal GnRH secretion are unknown, and some authors view the GnRH abnormality as secondary to excess estrogen/androgen secretion from the ovaries (see Ref. 105 for review). An alternative view is that the underlying abnormality lies in the absence of normal regulation of pulsatile GnRH secretion. Specifically, a relative inability of the ovarian steroids, estradiol and progesterone, to suppress the frequency of pulsatile GnRH release could explain the observed changes in LH secretion. In many patients with PCO, a careful history reveals that the disorder began soon after pubertal maturation and, in some instances, regular menstrual cycles were never established. As discussed earlier, the attainment of increased frequency and amplitude of GnRH secretion at puberty is followed by ovulation and leads to the relationships whereby estradiol and progesterone regulate GnRH pulse frequency. If women destined to develop PCO were relatively resistant to the effects of estradiol and progesterone in slowing GnRH secretion, the normal slowing in the luteal phase may not occur. Thus, the events consequent on GnRH slowing in the luteal phase would not take place, and the relative increase in FSH to induce follicular maturation in the next cycle may be deficient—leading to irregular, infrequent ovulation. Over time, the continuing rapid GnRH pulse frequency and increased LH secretion would increase ovarian androgen production, and lead to hirsutism and cyst formation in the ovaries. Convincing data to support this proposed sequence of events are not available, but some evidence suggests that these events may occur. In 5 of 12 adolescent girls with anovulation, LH pulse amplitude and frequency were increased and this was associated with elevated plasma estradiol and testosterone (111). The subsequent clinical course in these patients remains to be established, but the hormonal changes are consistent with the later development of PCO. Additional support for the view that abnormalities of GnRH secretion may be a primary abnormality in this disorder is found in the results of administration of oral contraceptives to reduce hirsutism and regulate menses in PCO. Combined oral contraceptive preparations reduce LH and testosterone production and, not uncommonly, withdrawal of contraceptives is followed by the temporary return of ovulatory cycles for a few months. The estrogen/progesterone in the combined oral contraceptives may act to restore the normal luteal mechanisms and reduce the frequency of GnRH secretion.

We have begun to examine these possibilities in women with PCO and have administered physiologic concentrations of luteal steroids to assess regulation of GnRH pulse frequency (112). The data from one patient are shown in Fig. 12 and mean plasma gonadotropins from six patients in Fig. 13. As seen in Fig. 12, luteal phase concentrations of estradiol and progesterone initially reduced LH (GnRH) pulse frequency (by day 10), and subsequently pulse amplitude was lowered (day 20). Withdrawal of ovarian steroids was associated with increased GnRH pulse frequency and a selective increase in plasma FSH, which rose more

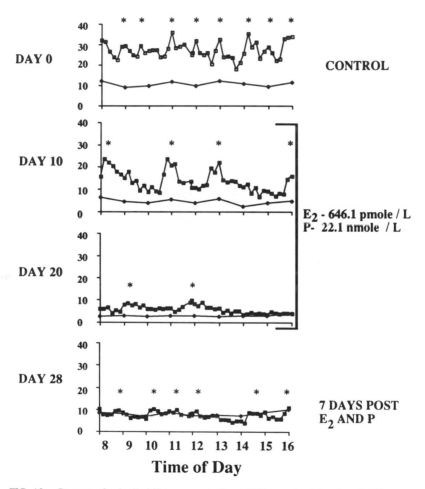

FIG. 12.   Patterns of pulsatile LH (open symbol) and FSH secretion (closed symbol) in a woman with polycystic ovarian disease, before, during, and 1 week after administration of estradiol (E$_2$) and progesterone (P). Estradiol was given by skin patch, and progesterone by vaginal suppository (three times daily), to reproduce midluteal plasma concentrations of these steroids for 3 weeks.

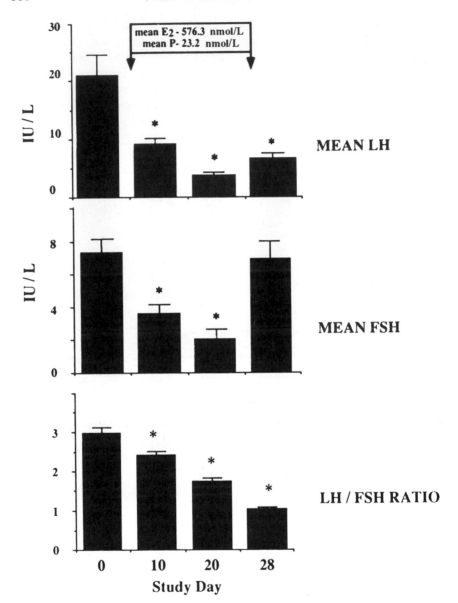

FIG. 13. Mean concentrations of plasma gonadotropins in women with polycystic ovarian disease (PCO) following administration of estradiol and progesterone at midluteal concentrations. Note the marked increase in FSH 1 week after withdrawal of ovarian steroids. *, $p < 0.05$ vs. Day.

rapidly than LH. Follicular maturation, as judged by ovarian ultrasound and plasma estradiol, occurred in all patients. These data measuring plasma hormones are in accord with our hypothesis that slowing of GnRH pulse frequency is not optimal for LH β mRNA expression and LH synthesis, but FSH synthesis is maintained. After estradiol and progesterone are withdrawn, the increase in GnRH frequency results in FSH release in a manner similar to that postulated to occur in the luteal-follicular transition in ovulatory cycles.

Thus, the etiology of PCO remains to be established, but the data reviewed above are consistent with an abnormality of GnRH regulation being an etiologic factor in many women with PCO. Further studies are required to determine the exact nature of the proposed abnormality, and to assess if this could reflect abnormalities of hypothalamic opioid regulation of GnRH pulse frequency.

IV. Summary

The data reviewed present evidence that the pattern of GnRH secretion is an important factor in the regulation of gonadotropin subunit gene expression, gonadotropin synthesis, and secretion. The information on regulation of mRNA expression by GnRH pulses should be considered with some caution, as the experiments were performed in male rats and may not accurately reflect events in female primates or humans. However, an overall pattern emerges which suggests that common factors may be involved in all mammalian species. If current evidence is correct, and only a single gonadotropin-releasing hormone exists, then mechanisms to differentially regulate the three gonadotropin genes may involve changes in GnRH secretion. Alterations in GnRH pulse frequency and amplitude are recognized by the pituitary gonadotrope cell and could be the mechanism used to effect differential expression of the gonadotropin subunit genes. Differential regulation of subunit gene expression would be expected to be critically important in the establishment of pubertal maturation, and subsequently in the maintenance of ovulatory cycles in women. Our hypotheses, proposing a major role of pulsatile GnRH secretion in the regulation of human reproduction, are summarized in schematic form in Fig. 14 for men and Fig. 15 for women.

In utero and during the first few months of life, GnRH is secreted at a relatively fast frequency (~ 1 pulse/hour). During the first year, GnRH secretion is inhibited and both the amplitude and apparent frequency of pulsatile release is markedly reduced. The mechanisms involved in inhibiting GnRH release remain unclear in humans. Similarly, the mechanisms involved in the disinhibition of GnRH secretion, which first occurs during sleep at the initiation of puberty, are unclear, but in humans do not appear to involve opiates. In males, the increased frequency and amplitude of GnRH secretion favor LH synthesis and release, which in turn stimulates testosterone secretion (Fig. 14). Testosterone acts at the hypothalamus, perhaps through opioid mechanisms, to inhibit GnRH pulse fre-

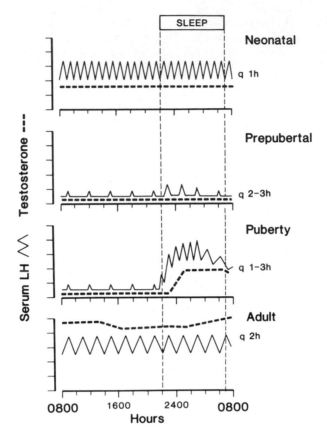

FIG. 14.   Schematic representation of the patterns of GnRH pulse secretion (inferred from LH pulses) during infancy, pubertal maturation, and adulthood in males. Dashed line indicates plasma testosterone.

quency and to maintain a regular pattern of pulses occurring approximately every 90–110 min in adult males.

In females, the mechanisms involving alterations in the patterns of GnRH secretion to regulate reproduction appear more complex. This may reflect the need to differentially synthesize and secrete FSH and LH at different times during reproductive cycles to allow orderly follicular maturation and ovulation. As shown in Fig. 15, we hypothesize that the events during the first decade of life and through the initiation of nocturnal GnRH secretion at puberty are similar in both sexes. With pubertal maturation, however, the increased LH synthesis and stimulation of estradiol secretion do not slow GnRH pulses and may augment pulsatile GnRH secretion, leading to intermittent partial follicular maturation and

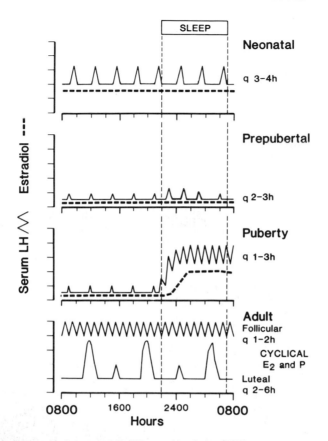

FIG. 15. Schematic diagram of GnRH secretion during childhood, pubertal maturation, and adulthood in females. The dashed line indicates plasma estradiol. The lower panel indicates the rapid GnRH pulses of the follicular phase and the slow, irregular pulses of the luteal phase.

partial luteinization. Over a period of several months, estradiol and progesterone effect a slowing of GnRH pulse secretion, which favors FSH synthesis and is a factor in allowing preferential FSH secretion in the luteal-follicular transition. This in turn stimulates several cohorts of ovarian follicles destined to mature and ovulate in subsequent cycles. We propose that at least two critical aspects of GnRH regulation are required for cyclic ovulation: (1) the individual has attained the ability to secrete GnRH at a rapid frequency ($\sim$1 pulse/hr), as occurs after puberty; (2) the mechanisms to suppress GnRH frequency in the luteal phase are effective. Abnormalities of the mechanisms involved in changing GnRH frequency would be expected to cause anovulation. Persistent slow GnRH stimuli would not effect LH secretion at midcycle and persistent rapid GnRH pulses would reduce luteal-follicular FSH secretion required for follicular maturation. Future studies will determine if these hypotheses prove to be correct. However, current data provide strong evidence that the pattern of GnRH secretion is a critical factor in the regulation of gonadotropin synthesis and secretion.

ACKNOWLEDGMENTS

We greatly appreciate the facilities of the Clinical Research Center at the University of Michigan and the skilled assistance of the nursing staff. The superb secretarial support of Linda McCrate is also recognized with sincere thanks. Supported by USPHS Grants HD11489 to J.C.M., HD23736 to J.C.M., HD16000 to R.P.K., and General Clinical Research Grant 5M01RR00042.

REFERENCES

1. Pierce, J. G., and Parsons, T. F. (1981). *Annu. Rev. Biochem.* **50,** 465.
2. Chin, W. W. (1987). *In* "Genes Encoding Hormones and Regulatory Peptides" (J. F. Habener, ed.), pp. 137–172. Humana Press, Clifton, New Jersey.
3. Santen, R. J., and Bardin, C. W. (1973). *J. Clin. Invest.* **52,** 2617.
4. Yen, S. S. C., Tsai, C. C., Naftolin, F., Vanderberg, G., and Ajabor, L. (1972). *J. Clin. Endocrinol. Metab.* **34,** 671.
5. Clarke, I. J., and Cummins, J. T. (1982). *Endocrinology (Baltimore)* **111,** 1737.
6. Belchetz, P. E., Plant, T. M., Nakai, Y., Keogh, E. G., and Knobil, E. (1978). *Science* **202,** 631.
7. Bergquist, C., Nillius, S. J., and Wide, L. (1979). *Contraception* **19,** 497.
8. Marshall, J. C., and Kelch, R. P. (1986). *N. Engl. J. Med.* **315,** 1459.
9. Godine, J. E., Chin, W. W., and Habener, J. F. (1982). *J. Biol. Chem.* **257,** 8368.
10. Jameson, J. L., Chin, W. W., Hollenberg, A. M., Chang, A. S., and Habener, J. F. (1983). *J. Biol. Chem.* **259,** 15474.
11. Maurer, R. A. (1987). *Mol. Endocrinol.* **1,** 717.
12. Papavasiliou, S. S., Zmeili, S., Herbon, L., Duncan-Weldon, J., Marshall, J. C., and Landefeld, T. D. (1986). *Endocrinology (Baltimore)* **119,** 691.
13. Shupnik, M. A., Gharib, S. D., and Chin, W. W. (1989). *Mol. Endocrinol.* **3,** 474.
14. Gharib, S. D., Bower, S. M., Need, L. R., and Chin, W. W. (1986). *J. Clin. Invest.* **77,** 582.
15. Gharib, S. D., Wierman, M. E., Badger, T. M., and Chin, W. W. (1987). *J. Clin. Invest.* **80,** 249.
16. Abbot, S. D., Docherty, K., and Clayton, R. N. (1988). *J. Mol. Endocrinol.* **1,** 49.
17. Dalkin, A. C., Haisenleder, D. J., Ortolano, G. A., Suhr, A., and Marshall, J. C. (1990). *Endocrinology (Baltimore)* **127,** 798.
18. Attardi, B., Keeping, H. S., Winters, S. J., Kotsuji, F., Maurer, R. A., and Troen, P. (1989). *Mol. Endocrinol.* **3,** 280.
19. Carroll, R. S., Corrigan, A. Z., Gharib, S. D., Vale, W., and Chin, W. W. (1989). *Mol. Endocrinol.* **3,** 1969.
20. Mercer, J. E., Clements, J. A., Funder, J. W., and Clarke, I. J. (1987). *Mol. Cell. Endocrinol.* **53,** 251.
21. Papavasiliou, S. S., Zmeili, S. M., Khoury, S., Landefeld, T. D., Chin, W. W., and Marshall, J. C. (1986). *Proc. Natl. Acad. Sci. U.S.A.* **83,** 4026.
22. Wierman, M. E., Gharib, S. D., LaRovere, J. M., Badger, T. M., and Chin, W. W. (1988). *Mol. Endocrinol.* **2,** 492.

23. Gharib, S. D., Wierman, M. E., Shupnik, M. A., and Chin, W. W. (1990). *Endocr. Rev.* **11**, 177.
24. Paul, S. J., Ellis, T. R., and Marshall, J. C. (1989). *Proc. 71st Annu. Meet. Endocr. Soc.* Abstr. No. 311, p. 270.
25. Paul, S. J., Ortolano, G. A., and Marshall, J. C. (1990). *Clin. Res.* **38**, 297A.
26. Perheentupa, A., and Huhtaniemi, I. (1990). *Endocrinology (Baltimore)* **126**, 3204.
27. Carroll, R. S., Corrigan, A. Z., and Chin, W. W. (1990). *Proc. 72nd Annu. Meet. Endocr. Soc.* Abstr. No. 777, p. 219.
28. Shupnik, M. A., Gharib, S. D., and Chin, W. W. (1988). *Endocrinology (Baltimore)* **122**, 1842.
29. Shupnik, M. A., Gharib, S. D., and Chin, W. W. (1989). *Mol. Endocrinol.* **3**, 474.
30. Mercer, J. E., Clements, J. E., Funder, J. W., and Clarke, I. J. (1988). *Neuroendocrinology* **47**, 563.
31. Shupnik, M. A., Wierman, C. M., Notides, A. C., and Chin, W. W. (1989). *J. Biol. Chem.* **264**, 80.
32. Leipheimer, R. E., Bona-Gallo, A., and Gallo, R. V. (1986). *Endocrinology (Baltimore)* **118**, 2083.
33. Soules, M. R., Steiner, R. A., Clifton, D. K., Cohen, N. L., Aksel, S., and Bremner, W. J. (1984). *Clin. Endocrinol. Metab.* **58**, 378.
34. Phillips, C. L., Lin, L. N., Wu, J. C., Guzman, K., Milsted, A., and Miller, W. L. (1988). *Mol. Endocrinol.* **2**, 641.
35. Simard, J., Labrie, C., Hubert, J. F., and Labrie, F. (1988). *Mol. Endocrinol.* **2**, 775.
36. Savoy-Moore, R. T., and Schwartz, M. B. (1980). *Int. Rev. Physiol.* **22**, 203.
37. Zmeili, S. M., Papavasiliou, S. S., Thorner, M. O., Evans, W. S., Marshall, J. C., and Landefeld, T. D. (1986). *Endocrinology (Baltimore)* **119**, 1867.
38. Ortolano, G. A., Haisenleder, D. J., Dalkin, A. C., Iliff-Sizemore, S. A., Landefeld, T. D., Maurer, R. A., and Marshall, J. C. (1988). *Endocrinology (Baltimore)* **123**, 2149.
39. Fox, S. E., and Smith, M. S. (1985). *Endocrinology (Baltimore)* **116**, 1485.
40. Levine, J. E., and Ramirez, V. D. (1982). *Endocrinology (Baltimore)* **111**, 1439.
41. Haisenleder, D. J., Ortolano, G. A., Jolly, D., Dalkin, A. C., Landefeld, T. D., Vale, W. W., and Marshall, J. C. (1990). *Life Sci.* **47**, 1769.
42. Haisenleder, D. J., Barkan, A. L., Papavasiliou, S., Zmeili, S. M., Dee, C., Jameel, M. L., Ortolano, G. A., El-Gewely, M. R., and Marshall, J. C. (1988). *Am. J. Physiol.* **254**, E99.
43. Belchetz, P. E., Plant, T. M., Nakai, Y., Keogh, E. G., and Knobil, E. (1978). *Science* **202**, 631.
44. Lalloz, M. R. A., Detta, A., and Clayton, R. N. (1988). *Endocrinology (Baltimore)* **122**, 1689.
45. Wierman, M. E., Rivier, J. E., and Wang, C. (1989). *Endocrinology (Baltimore)* **124**, 272.
46. Lalloz, M. R. A., Detta, A., and Clayton, R. N. (1988). *Endocrinology (Baltimore)* **122**, 1681.
47. Hammernik, D. L., and Nett, T. M. (1988). *Endocrinology (Baltimore)* **122**, 959.
48. Garcia, A., Schiff, M., and Marshall, J. C. (1984). *J. Clin. Invest.* **74**, 920.
49. Steiner, R. A., Bremner, W. J., and Clifton, D. K. (1982). *Endocrinology (Baltimore)* **111**, 2055.
50. Haisenleder, D. J., Katt, J. A., Ortolano, G. A., El-Gewely, M. R., Duncan, J. A., Dee, C. and Marshall, J. C. (1988). *Mol. Endocrinol.* **2**, 338.
51. Haisenleder, D. J., Ortolano, G. A., Dalkin, A. C., Paul, S. J., Chin, W. W., and Marshall, J. C. (1989). *J. Endocrinol.* **122**, 117.
52. Iliff-Sizemore, S. A., Ortolano, G. A., Haisenleder, D. J., Dalkin, A. C., Krueger, K. A., and Marshall, J. C. (1990). *Endocrinology (Baltimore)* **127**, 2876.
53. Haisenleder, D. J., Khoury, S., Zmeili, S. M., Papavasiliou, S., Ortolano, G A., Dee, C., Duncan, J. A., and Marshall, J. C. (1987). *Mol. Endocrinol.* **1**, 834.
54. Katt, J. A., Duncan, J. A. Herbon, L., Barkan, A., and Marshall, J. C. (1985). *Endocrinology (Baltimore)* **116**, 2113.

55. Dalkin, A. C., Haisenleder, D. J., Ortolano, G. A., Ellis, T. R., and Marshall, J. C. (1989). *Endocrinology (Baltimore)* **125**, 917.
56. Haisenleder, D. J., Dalkin, A. C., Ortolano, G. A., Marshall, J. C., and Shupnik, M. A. (1990). *Endocrinology (Baltimore)* **128**, 509.
57. Leung, K., Kaynard, A. H., Negrini, B. P., Kim, K. E., Maurer, R. A., and Landefeld, T. D. (1987). *Mol. Endocrinol.* **2**, 724.
58. Pohl, C. R., Richardson, D. W., Hutchison, J. S., Germak, J. A., and Knobil, E. (1983). *Endocrinology (Baltimore)* **112**, 2076.
59. Clarke, I. J., Cummins, J. T., Findlay, J. K., Burman, K. J., and Daughton, D. W. (1984). *Neuroendocrinology* **39**, 214.
60. Waldhauser, F., Weissenbacher, G., Frisch, H., and Pollack, A. (1981). *Eur. J. Pediatr.* **137**, 71.
61. Plant, T. M. (1982). *J. Endocrinol.* **93**, 71.
62. Plant, T. M. (1986). *Endocrinology (Baltimore)* **119**, 539.
63. Jakacki, R. I., Kelch, R. P., Sauder, S. E., Lloyd, J. S., Hopwood, N. J., and Marshall, J. C. (1982). *J. Clin. Endocrinol. Metab.* **55**, 453.
64. Penny, R., Olambiwonnu, N. O., and Frasier, S. D. (1977). *J. Clin. Endocrinol. Metab.* **45**, 307.
65. Wennink, J. M. B., Delemarre-Van deWaal, H. A., Van Kessel, H., Mulder, G. H., Foster, J. P., and Schoemaker, J. (1988). *J. Clin. Endocrinol. Metab.* **67**, 924.
66. Hale, P. M., Khoury, S., Foster, C. M., Beitins, I. Z., Hopwood, N. J., Marshall, J. C., and Kelch, R. P. (1988). *J. Clin. Endocrinol. Metab.* **66**, 785.
67. Wu, F. C. W., Borrow, S. M., Nicol, K., Elton, R., and Hunter, W. M. (1989). *J. Endocrinol.* **123**, 347.
68. Kelch, R. P., Khoury, S. A., Hale, P. M., Hopwood, N. J., and Marshall, J. C. (1987). *In* "The Episodic Secretion of Hormones" (W. F. Crowley and J. G. Hoffer, eds.), pp. 187–196. Churchill-Livingstone, New York.
69. Wu, F. C. W., Butler, G. E., Kelnar, C. J. H., and Sellar, R. E. (1990). *J. Clin. Endocrinol. Metab.* **70**, 629.
70. Ebling, F. J. P., Schwartz, M. L., and Foster, D. L. (1989). *Endocrinology (Baltimore)* **125**, 369.
71. Sauder, S. E., Case, G. D., Hopwood, N. J., Kelch, R. P., and Marshall, J. C. (1984). *Pediatr. Res.* **18**, 322.
72. Mauras, N., Veldhuis, J. D., and Rogol, L. (1986). *J. Clin. Endocrinol. Metab.* **62**, 1256.
73. Gay, V. L., and Plant, T. M. (1987). *Endocrinology (Baltimore)* **120**, 228.
74. Foster, C. M., Hassing, J. M., Mendes, T. M., Hale, P. M., Padmanabhan, V., Hopwood, N. J., Beitins, I. Z., Marshall, J. C., and Kelch, R. P. (1989). *J. Clin. Endocrinol. Metab.* **69**, 1213.
75. Backstrom, C. T., McNeilly, A. S., Leask, R. M., Baird, D. T. (1982). *Clin. Endocrinol. (Oxford)* **17**, 29.
76. Santen, R. J., and Bardin, C. W. (1973). *J. Clin. Invest.* **52**, 2617.
77. Reame, N., Sauder, S. E., Kelch, R. P., and Marshall, J. C. (1984). *J. Clin. Endocrinol. Metab.* **59**, 328.
78. Crowley, W. F., Filicori, M., Spratt, D. I., and Santoro, N. F. (1985). *Recent Prog. Horm. Res.* **41**, 473.
79. Yen, S. S. C., Tsai, C. C., Naftolin, S., Vandenberg, G., and Ajabor, L. (1972). *J. Clin. Endocrinol. Metab.* **34**, 671.
80. Liu, J. H., and Yen, S. S. C. (1983). *J. Clin. Endocrinol. Metab.* **57**, 797.
81. Hoff, J. D., Quigley, M. E., and Yen, S. S. C. (1983). *J. Clin. Endocrinol. Metab.* **57**, 792.
82. Sarkar, D. K., Chiappa, S. A., Fink, G., and Sherwood, N. M. (1975). *Nature (London)* **264**, 461.

83. Valk, T. W., Marshall, J. C., and Kelch, R. P. (1981). *Am. J. Obstet. Gynecol.* **141,** 842.

84. Leyendecker, G., Wildt, L., and Hansmen, M. (1980). *J. Clin. Endocrinol. Metab.* **51,** 1214.

85. Nippoldt, T. B., Khoury, S., Barkan, A., Kelch, R. P., and Marshall, J. C. (1987). *Clin. Endocrinol. (Oxford)* **26,** 293.

86. Soules, M. R., Steiner, R. A., Clifton, D. K., Cohen, N. L., Aksel, S., and Bremner, W. J. (1984). *J. Clin. Endocrinol. Metab.* **58,** 378.

87. Nippoldt, T. B., Reame, N. E., Kelch, R. P., and Marshall, J. C. (1989). *J. Clin. Endocrinol. Metab.* **69,** 67.

88. Wardlaw, S. L., Wehrenberg, W. B., Ferin, M., Antunes, J. L., and Frantz, A. G. (1982). *J. Clin. Endocrinol. Metab.* **55,** 877.

89. Van Vugt, D. A., Lam, N. Y., and Ferin, M. (1984). *Endocrinology (Baltimore)* **115,** 1095.

90. Quigley, M. E., and Yen, S. S. C. (1980). *J. Clin. Endocrinol. Metab.* **51,** 179.

91. Mais, V., Cetel, N. S., Muse, K. N., Quigley, M. E., Reid, R. L., and Yen, S. S. C. (1987). *J. Clin. Endocrinol. Metab.* **64,** 1109.

92. Filicori, M., Santoro, N., Merriam, G., and Crowley, W. F. (1986). *J. Clin. Endocrinol. Metab.* **62,** 1136.

93. Lam, N. Y., and Ferin, M. (1987). *Endocrinology (Baltimore)* **120,** 2044.

94. Soules, M. R., Clifton, D. K., Bremner, W. J., and Steiner, R. A. (1987). *J. Clin. Endocrinol. Metab.* **65,** 475.

95. Reame, N. E., Sauder, S. E., Case, G. D., Kelch, R. P., and Marshall, J. C. (1985). *J. Clin. Endocrinol. Metab.* **61,** 851.

96. Schwartz, B., Cumming, D. C., Riordan, E., Selye, M., Yen, S. S. C., and Rebar, R. W. (1981). *Am. J. Obstet. Gynecol.* **141,** 662.

97. Khoury, S. A., Reame, N. E., Kelch, R. P., and Marshall, J. C. (1987). *J. Clin. Endocrinol. Metab.* **64,** 755.

98. Wildt, L., and Leyendecker, G. (1987). *J. Clin. Endocrinol. Metab.* **64,** 1334.

99. Pohl, C. R., Richardson, D. W., Hutchison, J. S., Germak, J. A., and Knobil, E. (1983). *Endocrinology (Baltimore)* **112,** 2076.

100. Biller, B. M. K., Federoff, H. J., Koenig, J. I., and Klibanski, A. (1990). *J. Clin. Endocrinol. Metab.* **70,** 311.

101. Klibanski, A., Beitins, I. Z., Merriam, G. R., McArthur, J. W., Zervas, M. T., and Ridgway, E. C. (1984). *J. Clin. Endocrinol. Metab.* **58,** 1141.

102. Sauder, S. E., Frager, M., Case, G. D., Kelch, R. P., and Marshall, J. C. (1984). *J. Clin. Endocrinol. Metab.* **59,** 941.

103. Moult, P. J. A., Rees, L. H., and Besser, G. M. (1982). *Clin. Endocrinol. (Oxford)* **16,** 153.

104. Nippoldt, T. B., Cook, C. B., Kelch, R. P., and Marshall, J. C. (1988). *Proc. 70th Annu. Meet. Endocr. Soc.* Abstr. No. 880, p. 240.

105. Barnes, R., and Rosenfield, R. L. (1989). *Ann. Intern. Med.* **110,** 386.

106. Dunaif, A., and Graf, M. (1989). *J. Clin. Invest.* **83,** 23.

107. Erikson, G. F., Magoffin, D. A., Cragun, J. R., and Chang, R. J. (1990). *J. Clin. Endocrinol. Metab.* **70,** 894.

108. Chang, R. J., Laufer, L. R., and Meldrum, D. R. (1983). *J. Clin. Endocrinol. Metab.* **56,** 897.

109. Kazer, R. R., Kessel, B., and Yen, S. S. C. (1987). *J. Clin. Endocrinol. Metab.* **65,** 223.

110. Waldstreicher, J., Santoro, N. S., Hall, J. E., Filicori, M., and Crowley, W. F. (1988). *J. Clin. Endocrinol. Metab.* **66,** 165.

111. Porcu, E., Venturoli, S., Magrini, O., Bolzani, R., Gabbi, D., Paradisi, R., Fabbri, R., and Flagmigni, C. (1987). *J. Clin. Endocrinol. Metab.* **65,** 488.

112. Christman, G. M., Randolph, J. F., Kelch, R. P., and Marshall, J. C. (1989). *Proc. 71st Annu. Meet. Endocr. Soc.* Abstr. No. 990, p. 270.

DISCUSSION

A. Dunaif.   In the studies in which you showed estradiol and progesterone feedback, did you just study normal women? You speculate that such feedback might be abnormal in PCO. Have you studied any PCO women?

J. Marshall.   The data we presented were obtained in women with PCO. We have done similar studies in normal women, but with a different protocol. It is difficult to know what an appropriate "normal" control would be for studies in women with PCO. Some studies have used individuals with ovarian failure, but I feel that someone who has not been exposed to estrogen and/or progesterone for months or years is not necessarily a good control for a disorder such as PCO. What we did in normal women was to examine pulsatile LH secretion in individuals given estradiol or progesterone, or both hormones together to prolong the luteal phase. In this manner we selectively prolonged exposure to progesterone or estradiol. It is interesting that progesterone alone does not maintain the luteal slowing of LH pulses. Progesterone with estradiol is effective, and interestingly, estradiol alone maintains the slow LH pulse frequency. We do not know whether that reflects a direct effect of estradiol, or an action mediated by estradiol enhancing the number of progesterone receptors. If the latter is the case, the hypothalamus may recognize low concentrations (< 0.5 ng) of progesterone and thus the maintained LH slowing may still reflect an action of progesterone (*J. Clin. Endocrinol. Metab.* **69**, 67–76, 1989). These data showing that the luteal slowing of LH pulses can be prolonged by estrogen alone or by estrogen plus progesterone led us to address the issue of the ability of estrogen and progesterone to slow LH pulses in women with PCO. LH pulses slowed markedly in the PCO group. At present we are trying to assess the relative sensitivity of slowing LH pulses by estradiol and progesterone in both PCO and normal women. If our hypothesis is to be valid, the women with PCO must somehow be resistant to the effects of concentrations of estradiol and progesterone which are effective in slowing LH pulses in normal women. We hypothesized that in normal adolescents, during those early cycles after menarche, low levels of estradiol and progesterone from luteinized follicles will slow down LH pulsatile secretion and establish the normal patterns seen in the luteal phase. In women destined to develop PCO, LH pulses may not be slowed, leading to LH and ovarian androgen excess and anovulation. These studies comparing sensitivity to estradiol and progesterone in PCO and normal women are ongoing. I cannot present any data which will show whether we will find differences.

A. Dunaif.   Have you studied the effects of estradiol alone in PCO? We did some experiments and showed an enhancement of amplitude without changes in frequency.

J. Marshall.   No, we have not.

J. Funder.   Recently there were some data from Paul Stewart and colleagues who examined a subgroup of PCO and suggested that the problem was overactivity of 5α-reductase. They concentrated on liver and skin, but I think for the system to hold up it would clearly also need to be operant in the brain. If that were the case, then progesterone might really be much less effective if it were rapidly 5α reduced.

J. Marshall.   I have not seen those data, but that is an interesting thought.

N. Schwartz.   Regarding your data on the single dose of the GnRH antagonist, I was quite surprised at the rapidity with which the FSH message turned over. It seems particularly anomalous since we and others have shown that FSH in the blood, presumably representing secretion, is so slow in disappearing after antagonists. Do you have any explanation of that apparent paradox?

J. Marshall.   The half-time for the disappearance of FSH β mRNA was approximately 20 hours in the absence of testosterone and approximately 50 hours in the presence of testosterone, but even in the latter experiment, FSH β mRNA disappeared more rapidly than either α or LH β mRNAs. I think that regulation of FSH β mRNA is quite interesting in view of some data which I did not discuss. For example, if we use physiologic GnRH pulses and examine transcription after 24 hours, there is no sustained increase in β mRNA transcription even though we continued to give a physiologic stimulus that was initially effective. Interestingly, FSH β mRNA was increased after 24 hours of GnRH pulses, but was also increased to the same degree if we only gave GnRH pulses for the first 4 hours followed by 20 hours of saline pulses. This short burst of slow-frequency GnRH pulses increased FSH β mRNA for the subsequent 18–24 hours [*Endocrinology* **128**, 509 (1991)]. We have only recently obtained this data and at present do not know the mechanisms involved, but this was surprising in view of the apparent shorter half-life of FSH β mRNA.

W. F. Crowley, Jr. We have now reexamined our patients with polycystic ovarian disease with a second marker of gonadotrope function. Using the pulsatile secretion of the free α subunit of glycoproteins as an alternative marker of GnRH secretion, we find that, if anything, the frequencies that we have seen with LH are a little faster with the free α subunit. We also repeated these studies in the same patients with PCOD over several months and found that these abnormalities are sustained over time as you have stated. We also believe that this is the key to it (the etioloyg of PCOD). This is why I think your progesterone administration is such an important observation with therapeutic implications. How long can you sustain the progestin slowing? Have you studied this beyond the 20 days you mentioned? Will people eventually escape or maintain suppression?

J. Marshall. I do not know the answer to the duration for which LH pulse slowing can be maintained. In the present studies each of the women had evidence of follicular maturation, presumably in response to the "selective" FSH increase, or to the more rapid increase in plasma FSH. Only one of the six women actually ovulated. We felt this may have been due to the prolonged LH drive, high androgen levels, and disordered intraovarian biochemistry, which may take longer than 3 weeks to reverse after LH pulses are slowed. We have on-going studies which are aimed to cover a 6-week duration, which should answer the question of how long LH pulses will remain slow. The idea is that if the LH drive is shut down for that much longer, the androgen effects on the ovaries may no longer be a factor and follicular maturation may be more rapid, or more marked. Perhaps one would expect to see ovulation occurring in all patients after a longer duration of estradiol and progesterone suppression of LH pulse frequency.

W. F. Crowley, Jr. I would be surprised if PCOD patients had high 5α-reductase levels because they should have high levels of dihydrotestosterone, which should slow their LH pulse frequency. What would be interesting would be if they had an increase in their 3α metabolism in the CNS such that they were metabolizing androgens and thus lowering dihydrotestosterone levels. Such an effect might impair androgen slowing of GnRH secretion. This is just speculation.

J. Marshall. Yes I agree, though I think the other issue is whether testosterone acts via hypothalamic opioids to slow LH pulses. I was interested in some of Jon Levine's data in rats. Certainly in primates and humans it has been felt that both testosterone and estradiol plus progesterone produce their effects via increased hypothalamic opioids. This is supported by a lot of data. We have actually got some data from puberty studies done with Bob Kelch. While testosterone slows LH pulses in early pubertal boys, we cannot prevent this effect using opioid blockers. Thus it may be that testosterone can exert negative feedback actions on GnRH pulsatility by other mechanisms. The only area in which we have any data to support this would be from those studies with early pubertal boys. The question remains open at this time.

Steroids, Receptors, and Response Elements: The Limits of Signal Specificity

JOHN W. FUNDER

Baker Medical Research Institute, Prahran 3181, Victoria, Australia

This is an account of 25 years of scientific wandering and wondering. The wondering is necessarily personal; nevertheless, I hope to convey some of the sense of power and urgency that it can bring, even over 25 years. In terms of wandering, what I will recount is an Aeneid rather than an Odyssey; unashamedly *arma virosque cano*.

In the late 1960s, when I was a graduate student at the Howard Florey Institute in Melbourne, the institute produced a series of elegant papers on the control of aldosterone biosynthesis. Whereas most similar studies used rat adrenal cells *in vitro*, our studies used the autotransplanted adrenal gland in the conscious, undisturbed sheep, a piece of inspired improvization perfected by the late Professor R. D. "Panzee" Wright. This was a multistage preparation, involving repeat surgery and even in the best hands an ultimate success rate of only 50%: in those sheep in which it all came together, however, there was a possibility of an experimental animal able to be repeatedly used for a decade.

The model exploited several characteristics of the merino sheep. First, they have very large dewlaps, folds of redundant skin hanging down from the neck. Second, they are phlegmatic animals, tolerating surgery and experimental use with unusual sang-froid. Third, they eat dry food, do not bite, and do not pick out their sutures. In an adrenal autotransplant, the initial step is to prepare carotid and jugular loops, connected in the middle by a skin bridge; when the immediate postoperative edema and subsequent granulation have resolved, the overall effect is not unlike a H-shaped suitcase handle. Often, under the same anesthetic, a unilateral adrenalectomy was performed, with the kidney on that side untouched and intact.

The second stage in the procedure is contralateral nephrectomy, and transplantation of renal artery/adrenal artery/adrenal gland/adrenal vein to the skin pouch prepared in the neck. The renal artery, with one end tied off, is of a caliber sufficient to allow anastomosis to the carotid; the adrenal vein was harvested with a generous rim of inferior vena cava, again facilitating anastomosis to the jugular. When the local inflammation had settled postoperatively, the preparation offered unrivaled access to the adrenal, by the ability to cannulate either vessel, and to obstruct flow above and/or below the transplanted adrenal by inflating an external cuff to the appropriate pressure.

Using the preparation, for example, we showed that close intraarterial infusion of potassium chloride would selectively raise the aldosterone secretion rate, even when the amount infused was such that no difference in potassium concentration between adrenal venous effluent and peripheral blood could be determined by flame photometry (1). Routinely, adrenal venous samples were analyzed for aldosterone, corticosterone, and cortisol by double-isotope derivative dilution assays, and an estimate of secretion rate derived from concentrations and measured flow rate (by occluding the carotid beyond the transplant for the duration of the infusion, and the jugular on both sides of the gland). Laborious and time-consuming experiments they were; but in John Coghlan's words they were *in vivo* studies, which reflected what the gland did, not just what it could do.

Other experimental surgical techniques were similarly employed to address other questions. One such question was that of the relative roles of potassium and angiotensin as stimuli to aldosterone secretion; the ferocity with which one was held to be proximate, the other permissive, in retrospect was more suited to the Thirty Years' War than twentieth century science. Such partisanship notwithstanding, the last set of animals I prepared before thesis writing-up went as follows: parotid fistula, unilateral nephrectomy, contralateral indwelling renal arterial line connected externally to a syringe and pump. The rationale devised by John Coghlan for these studies was as follows: angiotensin is cleared by the kidney (*inter alia*), and angiotensin has a local feedback effect on renin release and thus endogenous angiotensin formation from renin substrate. Let us then infuse angiotensin to the kidney and block renin release while the sheep goes into sodium deficiency through uncompensated loss of salivary sodium bicarbonate. Peripheral angiotensin will not rise. Will aldosterone secretion increase in response to sodium deficiency under such circumstances? The answer is that it does; for 24 hours, such an infusion keeps peripheral angiotensin at basal (sodium replete) levels, despite which the aldosterone responds to the progressive sodium depletion altogether normally (2). Ergo, angiotensin is permissive rather than proximate; *cuius regio, eius religio*.

I have very deliberately introduced this article on receptors and signal specificity by a discussion of the control of signal generation, for a variety of reasons. The first is to pay personal tribute to Panzee Wright, Derek Denton, and John Coghlan; the second is to serve notice in terms of heuristic approach. Over the past decade, the contributions of molecular biology to endocrinology have been extraordinary, including providing answers to questions not yet framed; this said, a key component of endocrinology is communication between organs and tissues and systems, so that *in vivo* studies—experimental animal and clinical—are irreplaceable. They are irreplaceable not merely in testing new recombinant products in the crucible of the whole cell or living organism, but for setting the agenda for molecular biological studies, for ordering the context, for providing hypotheses to test. In brief, I believe that it is more likely that we will ask leading questions and do key experiments in the field of steroid action if we know something about the control of steroid hormone secretion—rather than merely regarding aldosterone as coming out of a bottle, and dexamethasone as out of a catalog.

The early studies in mineralocorticoid receptors provide a case in point. Gordon Tomkins' laboratory at the University of California, San Francisco was engaged in the late 1960s and early 1970s in pioneering studies on glucocorticoid action, and Isidore Edelman's laboratory on the next floor was conducting similar studies on mineralocorticoid action. Among the studies on HTC cell lines and toad bladders were the preliminary experiments on the binding of tritiated dexamethasone and tritiated aldosterone. John Baxter and Guy Rousseau had shown that aldosterone could optimally induce tyrosine aminotransferase in HTC cell lines, though high concentrations were required; presumably, therefore, it had some ability to activate glucocorticoid receptors. We also knew, from physiological studies and observations, that there are glucocorticoid effects on the kidney—gluconeogenesis, changes in free water clearance—quite distinct from mineralocorticoid effects on sodium balance. If aldosterone was glucocorticoid only at very high doses, it seemed likely that its mineralocorticoid effect might be mediated via a distinct class of receptors for which it has much higher affinity. And so the study was done, over a few weeks, between the two laboratories—and aldosterone indeed was shown to bind displaceably to two distinct classes of sites in adrenalectomized rat kidney cytosols (3), later shown to represent mineralocorticoid and glucocorticoid receptors (4–6). At the same time, both classes of aldosterone binding site were documented in slices of rat parotid gland (7), echoing the use of the changes in parotid salivary $Na^+:K^+$ ratio as an index of aldosterone action in the sheep (2).

Though the first studies distinguishing two classes of aldosterone binding sites were on renal cytosols, such preparations commonly yielded very inconsistent results until the introduction of sodium molybdate as a constituent of homogenization buffers almost a decade later. This inconsistency reflected the lability of the high-affinity aldosterone binding sites in broken cell preparations, particularly in the absence of steroid, i.e., when the receptors are unoccupied. To overcome this problem, most studies on renal mineralocorticoid receptors used conditions in which binding took place in whole cells (either *in vivo*, or in kidney slices *in vitro*), following which the cells were fractionated, and cytoplasmic and nuclear uptake and retention of steroid determined.

The first of these kidney slice studies (4) focused squarely on what was to emerge as one of the two dominant questions in the area over the next 15 years, that of the specificity of these high-affinity (type 1) aldosterone binding sites. In kidney slice preparations, corticosterone (the physiological glucocorticoid in rats) was approximately one-tenth as potent as unlabeled aldosterone in displacing tritiated aldosterone from the totality of its displaceable binding sites, type I (mineralocorticoid) and type II (glucocorticoid) receptors together. This overall level of competition, however, reflected very unequal contributions in terms of the relative ability of corticosterone to compete for aldosterone binding to the two types of receptors. From a variety of rather indirect studies, we estimated that whereas corticosterone had a substantially higher affinity than aldosterone for glucocorticoid receptors, its affinity for mineralocorticoid receptors was only 2–4% that of aldosterone.

Even despite this 25- to 50-fold advantage in affinity, aldosterone occupancy of mineralocorticoid receptors *in vivo* remained problematical, inasmuch as circulating aldosterone concentrations are two to three orders of magnitude lower than those of the physiological glucocorticoids (corticosterone in rat and mouse; cortisol in man, dog, sheep, and most other experimental species). Clearly, as mineralocorticoid and glucocorticoid secretion respond to different stimuli, the exact ratio will vary with sodium balance, extent of stress, and to a lesser extent time of day; on the other hand, even at two orders of magnitude difference in concentration, a receptor with a 25:1 or 50:1 ratio of preferring signal to noise is a remarkably imprecise detection device.

What appears to be a crucial specificity-conferring mechanism in such circumstances was also described in this initial paper (4), that of the preferential plasma binding of the physiological glucocorticoids compared with aldosterone. Most steroids circulate largely bound to plasma protein; aldosterone is unusual, in that although it is modestly bound by serum albumin, more than half of the circulating levels are free. In contrast, the physiological glucocorticoids are bound not only to albumin but to a specific globulin (transcortin, or corticosteroid-binding globulin, CBG), so that in most species only approximately 5% circulates as free steroid.

To examine this possible specificity-conferring mechanism, David Feldman and I compared the ability of corticosterone to compete for tritiated aldosterone binding sites *in vivo* and *in vitro*. *In vivo*, we gingerly anesthetized adrenalectomized rats, and after a loading dose gave them a constant infusion for 20 minutes of tritiated aldosterone, alone and with a range of doses of nonradioactive aldosterone or corticosterone. *In vitro*, we incubated kidney slices from adrenalectomized rats with tritiated aldosterone, alone and with a range of concentrations of nonradioactive aldosterone or corticosterone, for 20 minutes at 37°C. *In vitro*, corticosterone was one-tenth as potent as aldosterone at displacing tritiated aldosterone, and *in vivo* only one-hundredth, consistent with a 10-fold increase in aldosterone potency *in vivo* as a consequence of the preferential plasma binding of corticosterone, denying it access to the intracellular compartment and thus the ability to bid for receptor occupancy.

In 1973, then, the specificy of mineralocorticoid action was assumed to reflect an apparent inherent (25- to 50-fold) receptor selectivity for aldosterone over the physiological glucocorticoids, coupled with a 10-fold advantage conferred by the preferential plasma binding of glucocorticoids. A decade later, in 1983, it finally became clear (8–10) that the first of these premises was untrue, and that mineralocorticoid receptors had identical intrinsic affinity for aldosterone and corticosterone [and later, in man (11,12) for cortisol]. Fifteen years later, in 1988, the observations in 1973–*in vivo*, and in kidney slices—and those of 1983—*in vitro*, on molybdate-stabilized cytosols—were reconciled, by the demonstration of the crucial, prereceptor specificity-conferring role of the enzyme 11β-hydroxysteroid dehydrogenase, *in vivo* (13,14) and in kidney slices (15). These developments will be described, in detail, later in this article: for the moment, however, I want to focus on the second of the major areas of uncertainty in the decade post-1973, that of the anatomical distribution of mineralocorticoid or type I receptors.

The initial studies, already described, were on two classical mineralocorticoid target tissues, the kidney (3,4) and the parotid (7). Subsequently, high-affinity aldosterone binding sites were shown to be present in whole-cell preparations from small and large intestine (16), but absent in liver cytosols and heart slices (17,18); in retrospect, the failure to demonstrate the low levels more recently reported for both these latter tissues (19,20) almost certainly reflects the methodological inadequacies of our earlier studies. At the time, however, there was a certain internal consistency in the findings: mineralocorticoid receptors appeared confined to tissues which were classical aldosterone targets, as defined by the criterion of steroid-induced unidirectional transepithelial sodium transport.

From other laboratories, however, there came a series of studies describing high-affinity aldosterone binding in tissues that were clearly not sites of mineralocorticoid action in any classic sense. From Pat Mulrow's laboratory, for example, the binding of tritiated deoxycorticosterone in rat brain cytosol was reported, better displaced by aldosterone than by glucocorticoids, and differing between normotensive and hypertensive rats (21). In Darrell Fanestil's laboratory, at the same time, studies on the ability of various steroids, including the mineralocorticoid antagonist spirolactone, to compete for tritiated aldosterone binding in rat brain preparations were similarly interpreted as evidence for the existence of mineralocorticoid receptors in this tissue (22).

Given the physiological plausibility on aldosterone-specific effects on salt appetite, later demonstrated by intracerebroventricular infusion studies (23), the existence of aldosterone-specific binding sites in the brain was therefore not surprising, even if it did necessitate widening of the definition of mineralocorticoid action to allow the possibility of secondary effects on the machinery of sodium homeostasis. On the other hand, various groups were showing indisputable high-affinity binding of aldosterone in a variety of tissues which were only remotely if at all implicated in any sodium homeostatic feedback mechanism. In addition, binding in such tissues appeared to differ from that in classical mineralocorticoid target tissues, in that deoxycorticosterone, corticosterone, and aldosterone had equivalent affinity for the high-affinity aldosterone binding sites; for renal mineralocorticoid receptors, the hierarchy was clearly aldosterone > deoxycorticosterone ≫ corticosterone, happily consistent with their established *in vivo* mineralocorticoid potency.

In cultured rat aortic vascular smooth muscle cells, Walter Meyer and colleagues pioneered the description of such corticosteroid receptors (24); subsequently, this group has shown aldosterone-induced changes in β-adrenoceptor levels via such receptors in fibroblast cell lines (25). Nancy Lan, in John Baxter's laboratory, defined these "mineralocorticoid-like receptors" in GH_3 cells, a cultured rat somatotrope/lactotrope cell line (26); again, as for the vascular smooth muscle and fibroblast cells in culture, there was essentially no difference in affinity of these sites for aldosterone or corticosterone, in contrast with renal mineralocorticoid receptors. In addition, these authors characterized the receptors' affinity for a range of natural and synthetic steroids, which clearly distinguished them from classical glucocorticoid (type II) receptors.

At the same time, Bruce McEwen and colleagues embarked on a series of ground-breaking studies, in which they described "corticosterone-preferring" glucocorticoid binding sites in the hippocampus of the adrenalectomized rat (27). As for the initial studies on renal mineralocorticoid receptors, the crucial observations came from *in vivo* experiments, in this case comparing the brain localization of radioactivity after injection of various steroid tracers. Tritiated dexamethasone was found at highest concentrations in the pituitary and hypothalamus, not surprisingly in view of their known roles as glucocorticoid target tissues. In contrast, however, tritiated corticosterone was far better taken up and retained than dexamethasone in the hippocampus and septum.

Though with characteristic prudence Bruce McEwen termed these hippocampal binders "corticosterone-preferring" sites rather than receptors, their receptor status has since been established unequivocally, initially by collaborative studies between McEwen and Paul Greengard (28), and subsequently in a series of studies from Ron de Kloet's laboratory (29,30). In the initial study, levels of synapsin, an 80K phosphoprotein, were shown to fall in the hippocampus following adrenalectomy, and to be restored by administration of a low dose of corticosterone, but not by the same dose of dexamethasone (28). De Kloet and colleagues extended this preliminary observation, by showing that corticosterone, but not dexamethasone, produced both acute and chronic changes in levels of hippocampal serotonin receptors, when given to adrenalectomized rats. Aldosterone was without agonist effect, but when given with corticosterone substantially antagonized the effect of the latter. The logical interpretation of these data is that such effects are mediated by corticosterone-preferring sites acting as receptors, for which corticosterone has much higher affinity than dexamethasone, and for which aldosterone also has substantial affinity but acts as an antagonist.

That renal mineralocorticoid receptors and hippocampal corticosterone-preferring glucocorticoid receptors are similar if not identical was very strongly suggested by a trio of studies in 1983 (8–10), and supported by the subsequent cloning of the human renal mineralocorticoid receptor (12) and the rat hippocampal species (31). In the first of the 1983 studies, Wrange and Yu (8) showed identical isoelectric focusing patterns for tryptic fragments of the aldosterone-binding species from rat kidney and hippocampus, in contrast with rat renal dexamethasone binding (classic glucocorticoid) receptors. Beaumont and Fanestil subsequently reported surprisingly high affinity of aldosterone for hippocampal tritiated corticosterone binding sites (9), presaging de Kloet's demonstration of the antagonist action of aldosterone on serotonin receptor response to corticosterone (30). Finally, in a comprehensive study of the binding of various mineralocorticoids and glucocorticoids in rat thymus, hippocampus, and kidney, Zygmunt Krozowski in this laboratory (10) made two crucial observations. The first was that when steroid binding to classical glucocorticoid receptors was blocked by excess RU26988 (a highly specific synthetic glucorticoid from Roussel-Uclaf, Romainville, France), the hippocampal binding species has identical, very high (K_d 0.3 nM) affinity for corticosterone, deoxycorticosterone, aldosterone, and 9α-fluorocortisol. Second, these studies showed that hemoglobin-free cytosols, prepared from exhaustively perfused adrenalectomized rat kidneys, bound tritiated aldosterone with "classical" mineralocorticoid receptor specificity (aldosterone > corticosterone). When, however, the receptors were adsorbed onto hydroxylapatite and washed free of transcortin, and then eluted for binding studies, the affinity of binding was indistinguishable from that seen in hippocampal cytosols. The "classical" mineralocorticoid receptor specificity thus presumably reflects contamination of renal cytosols by extravascular transcortin, a presumption confirmed by the demonstration of "mineralocorticoid" specificity in hippocampal cytosols made 1% with adrenalectomized rat plasma.

A decade after the initial description of mineralocorticoid receptors (3,4), then, the intrinsic affinity of such receptors for a range of natural and synthetic steroids was finally established. Given that such receptors include physiological mineralocorticoid receptors in kidney and high-affinity glucocorticoid receptors in the hippocampus, the 1973 terminology (type I receptors) appears amply justified. To refer to type I sites in the hippocampus as "mineralocorticoid receptors" when corticosterone restores the deficit postadrenalectomy, and aldosterone is without agonist action but antagonizes that of corticosterone, is to beggar adrenal physiology.

The demonstration that renal type I receptors had equivalent intrinsic affinity for aldosterone and corticosterone—and that the previous apparent aldosterone selectivity of the receptors reflected cytosol contamination by transcortin—thus placed into even more exquisite relief the conundrum of *in vivo* specificity in mineralocorticoid target tissues. Gabriela Stephenson in this laboratory showed that a gradient existed for extravascular transcortin between outer cortex and inner medulla–papilla (32), allowing the possibility of a countercurrent mechanism of reversible corticosterone sequestration in the inner zones of the kidney, where transcortin concentrations were 8–10 times higher than in plasma. Such a system, if operant, might confer up to an order of magnitude advantage to aldosterone over corticosterone in terms of binding to mineralocorticoid receptors in the inner zones of the kidney: it would not, however, be able to contribute to aldosterone selectivity in the cortical collecting tubule and connecting segment, prime sites of aldosterone action to modulate sodium retention. Though such a countercurrent mechanism is conceptually attractive, and may modulate aldosterone effects on proton secretion further down the nephron, it clearly cannot explain the observed aldosterone selectivity of cortical type I sites.

That this was experimentally the case was demonstrated by Karen Sheppard, in studies on 1-day adrenalectomized, 10-day-old rats (33,34). From around 4 to 14 days of age, rats have vanishingly low levels of both plasma and extravascular transcortin. Renal cytosols from such rats bind aldosterone and corticosterone equivalently, as would be predicted. When, however, such rats are injected with tritiated aldosterone or tritiated corticosterone, clear differences in the pattern of uptake and retention are seen. In the hippocampus, *in vivo*, there is equivalent binding of tritiated aldosterone and tritiated corticosterone (in the presence of excess RU28362, the second-generation type II receptor-specific glucocorticoid ligand); in the physiological mineralocorticoid target tissues, however,—kidney, parotid, and colon—the uptake and retention of tritiated aldosterone was over an order of magnitude higher than that of corticosterone.

There are two possible, nonexclusive mechanisms for such selectivity. First, it is possible that the intrinsic specificity of the receptors is identical for aldosterone and corticosterone, but that a postreceptor tertiary interaction with the chromatin somehow "locks" aldosterone but not corticosterone into the receptor, thus increasing its apparent affinity *in vivo* (35). When Jeff Arriza and Ron Evans cotransfected human renal mineralocorticoid receptors and mouse mammary tumor virus long-terminal-repeat luciferase reported sequences, they found that aldosterone was an order of magnitude more potent than cortisol (36), in spite of the fact that the expressed receptor has identical affinity for the two steroids (12); one interpretation of this finding is that a chromatin-induced increase in receptor affinity for aldosterone is in fact the case in whole cells.

Whether such a mechanism is operant, there clearly are prereceptor mechanisms serving to exclude corticosterone and cortisol from mineralocorticoid receptors in classical aldosterone target tissues. As previously detailed, transcortin is an insufficient explanation, inasmuch as kidneys from 10-day-old rats are still highly aldosterone specific *in vivo*. The other prereceptor specificity-conferring mechanism, responsible at least in large part for the aldosterone selectivity of type I receptors in mineralocorticoid target tissues, is the enzyme 11β-hydroxysteroid dehydrogenase, which converts cortisol and corticosterone (but not aldosterone) to receptor-inactive 11-dehydro metabolites.

The stimulus to the studies demonstrating the crucial physiological role of 11β-hydroxysteroid dehydrogenase came from a series of clinical observations in patients with the rare syndrome of apparent mineralocorticoid excess. Maria New and Stanley Ulick were the first to document such a patient, a young native American (37). They showed that the syndrome of marked salt retention and hypertension was accompanied by suppressed renin and aldosterone, normal plasma cortisol levels, and very unusual patterns of urinary cortisol metabolites—in particular, ratios of cortisol to cortisone metabolites ranging from 10:1 to 50:1, rather than the normal value of unity, and clear evidence for 5α rather than 5β reduction of cortisol.

Initially, the implications of these findings for the pathogenesis of the salt retention and blood pressure elevation were not clear; subsequently, however, both New and Ulick (38) and Chris Edwards and colleagues (39) focused on the possibility that such patients lack renal 11β-hydroxysteroid dehydrogenase, and that their syndrome thus reflects local, renal hypercortisolism. Edwards developed the notion of the physiological importance of the cortisol–cortisone shuttle, reflecting the different activity in different tissues of 11β-hydroxysteroid dehydrogenase on the one hand, and of 11-ketosteroid reductase on the other. In patients with apparent mineralocorticoid excess, he suggested (40) the abnormally high intrarenal cortisol levels overcome the normal aldosterone selectivity-conferring mechanisms, and thus caused sodium retention and blood pressure elevation by inappropriately occupying renal type I receptors.

From the vantage point of normal physiology, it appeared to us that 11β-hydroxysteroid dehydrogenase might itself be the physiological aldosterone selectivity-conferring mechanism, given that the hydroxyl in the 11-position in aldosterone is cyclized with the very reactive aldehyde group at the 18-position, making it an unlikely substrate for enzymatic attack, and the knowledge that cortisone is itself inactive as a glucocorticoid, needing to be reduced to cortisol for receptor binding.

Our ability to test this hypothesis stemmed from an elegant study by Paul Stewart in Chris Edwards' laboratory, which illuminated the pathogenesis of licorice-induced hypertension and sodium retention (40). Though glycyrrhetinic acid, the active principle in licorice, is a very weak mineralocorticoid agonist (41), it is now clear that its principal mechanism of action is via blockade of renal 11β-hydroxysteroid dehydrogenase. Stewart *et al.* fed seven normal volunteers 250 g of licorice per day, and noted a change in the ratio of urinary cortisol to cortisone metabolites from unity to 2:1; in three subjects, infusion studies with radioactive cortisol showed a more than doubling in half-time, again consistent with enzyme blockade.

In our *in vivo* studies to test the hypothesis—that 11β-hydroxysteroid dehydrogenase was the mechanism allowing aldosterone occupancy of type I receptors in physiological mineralocorticoid target tissues—we used carbenoxolone sodium, a drug widely used in various countries in the 1970s for the treatment of peptic ulcers. Carbenoxolone sodium is the hemisuccinate of glycyrrhetinic acid; though a very effective antiulcer drug (by mechanisms yet to be established), it notoriously produces side effects of marked salt and water retention, and blood pressure elevation. Given this clinical observation, it seemed not inappropriate to pretreat adrenalectomized rats with carbenoxolone sodium, as a candidate blocker of 11β-hydroxysteroid dehydrogenase, before injecting them with tritiated aldosterone or tritiated corticosterone. As in the previous *in vivo* binding studies, excess RU28362 was simultaneously injected to exclude injected tracer from type II glucocorticoid receptors; similarly, 1-day adrenalectomized, 10-day-old rats were used, to avoid the problem of transcortin binding of corticosterone.

Under such conditions, we showed (13) that carbenoxolone sodium pretreatment had no significant effect on the binding of either steroid in the hippocampus (or in the heart, included in case the blood–brain barrier excluded the drug). Whereas in these two tissues the extent of binding of tritiated aldosterone and tritiated corticosterone was identical, in the kidney, parotid, and colon in the absence of carbenoxolone sodium, the same high degree of aldosterone selectivity as previously noted was found. When, however, the rats were pretreated with carbenoxolone sodium to block 11β-hydroxysteroid dehydrogenase, the extent of corticosterone binding rose dramatically in all three mineralocorticoid target tissues—in kidney and parotid to levels equal to those of aldosterone, and in the colon approaching those of aldosterone. What this study shows, therefore, is that the *in vivo* aldosterone selectivity of type I receptors is clearly mediated at the prereceptor level, in that blockade of 11β-hydroxysteroid dehydrogenase reduced or abolished the preferential uptake and retention of aldosterone over corticosterone.

If indeed the enzyme 11β-hydroxysteroid dehydrogenase is the crucial specificity-conferring mechanism in mineralocorticoid target tissues, a number of things are entailed. First, the dose of carbenoxolone sodium used in the *in vivo* blockade of binding studies should be sufficient to block conversion of corticosterone to 11-dehydrocorticosterone; in fact, kidneys taken from carbenoxolone sodium-loaded rats show a $> 90\%$ reduction in ability to metabolize corticosterone to 11-dehydrocorticosterone *in vitro*, at 37°C over 2 hours, suggesting that the drug is essentially a suicide substrate for the enzyme (13).

Second, whereas kidney minces from 1-day adrenalectomized, 10-day-old rats metabolize tritiated cortisol to tritiated cortisone with high efficiency, we found essentially no conversion by hippocampal minces from the same animals. The heart similarly showed absent or low activity; activity in the parotid was high, but in the colon much lower than predicted, although still present (13). These studies were done with tritiated cortisol, given the unavailability of authentic tritiated 11-dehydrocorticosterone to serve as a marker for corticosterone metabolism. In subsequent unpublished studies we and others (C. Monder, personal communication) have shown modest but clear 11β-hydroxysteroid dehydrogenase activity in rat hippocampus when corticosterone rather than cortisol is used as substrate, a finding at variance with published data from Edwards and colleagues (14).

Third, we have formally demonstrated (13) that 11β-hydroxysteroid dehydrogenase activity, to convert corticosterone to 11-dehydrocorticosterone, lowers the affinity of type I receptor binding of the product to approximately 0.3% that of its parent steroid. Though the parent corticosterone may circulate at 100–1000 times higher concentration than aldosterone, the combination of preferential plasma binding and higher levels of 11β-hydroxysteroid dehydrogenase activity together are sufficient to allow aldosterone rather than corticosterone to occupy type I receptors.

The relative affinities of corticosterone and 11-dehydrocorticosterone for type I receptors were established by classical displacement studies, determining the ability of progressively increasing concentrations of both steroids to compete with tritiated aldosterone binding in kidney cytosol from adrenalectomized rats, in the presence of excess RU28362. Almost as an afterthought, Paul Pearce also determined the relative affinities of corticosterone and 11-dehydrocorticosterone for classical type II glucocorticoid receptors by comparing their ability to compete for tritiated dexamethasone binding in thymic cytosol preparations from adrenalectomized rats (13). Not surprisingly, perhaps—given the known glucocorticoid receptor inactivity of cortisone—11-dehydrocorticosterone was also essentially receptor inactive, with an affinity clearly < 1% that of corticosterone for classical glucocorticoid receptors.

What this finding underlines is the possibility that 11β-hydroxysteroid dehydrogenase activity may not merely limit glucocorticoid access to type I receptors in mineralocorticoid target tissues, but may also locally modulate levels of signal available to occupy glucocorticoid receptors. In this context, the clinical studies (42) by Maria New and colleagues on their patient with apparent mineralocorticoid excess come into particular focus. Such patients are more sensitive to administered cortisol than aldosterone, in terms of sodium retention and hypertension, an observation difficult to reconcile with the proposed etiology of the syndrome, that of inappropriate occupancy of renal type I receptors by the high intrarenal levels of cortisol. Second, when such patients are treated with dexamethasone, to suppress adrenocorticotropin (ACTH) and thus cortisol secretion, their blood pressure does not fall, again an unexpected finding if the syndrome simply reflects inappropriate occupancy of mineralocorticoid receptors by cortisol. If, on the other hand, the role of the enzyme in the kidney is to protect both type I and type II receptors from inappropriate occupancy by cortisol, then both of these unexpected clinical observations might be accommodated in terms of pathogenesis.

To test the null hypothesis—that the clinical syndrome of apparent mineralocorticoid excess reflects absent or low 11β-hydroxysteroid dehydrogenase activity, allowing cortisol inappropriate access to mineralocorticoid (type I) receptors—Paul Roy and I studied four patients with pseudohypoaldosteronism. Pseudohypoaldosteronism—hyponatremia, hyperkalemia, elevated renin and aldosterone, and unresponsiveness to administered mineralocorticoids—was first described as a syndrome over 30 years ago by Don Cheek and Don Perry from the Royal Children's Hospital in Melbourne (43). Much more recently, Decio Armanini made the seminal observation that normal peripheral monocytes contain low but clearly detectable levels of type I receptors (11), a finding subsequently exploited in a joint study (44) to show that patients with the syndrome have absent receptors, on the criterion of lack of aldosterone binding by isolated monocytes *in vitro*.

If the pathogenesis of apparent mineralocorticoid excess solely reflects inappropriate type I receptor occupancy by cortisol, then the patients with pseudohypoaldosteronism (i.e., no type I receptors) should be unaffected by licorice or carbenoxolone. To test this hypothesis, we studied four patients with pseudohypoaldosteronism—a mother and her three surviving children—in the Children's Hospital in Sydney, for 3 weeks (45). After 4 days for equilibration, the subjects began carbenoxolone sodium, 50–150 mg/day, for a total of 15 days. Over this period, a marked fall in urinary sodium to creatinine, and sodium to potassium, ratio was seen; in addition, a prompt elevation in plasma bicarbonate was noted. In the presence of carbenoxolone sodium, then, these subjects manifested a prototypic mineralocorticoid response—despite the absence of mineralocorticoid receptors, and presumably reflecting cortisol occupancy of normally forbidden glucocorticoid receptors.

The immediate implications of this study are profound. First, newborns with pseudohypoaldosteronism should be given carbenoxolone sodium, as well as the often poorly tolerated 2-hourly 2 N sodium chloride solution. Second, it would appear clear that apparent mineralocorticoid excess involves inappropriate cortisol occupancy of receptors other than and/or in addition to type I sites. Finally, and in a sense most importantly, these studies suggest that just as there is no mineralocorticoid/glucocorticoid discrimination at the level of the type I receptor, there may in addition be no discrimination at the level of the nuclear response element. Otherwise stated, an activated glucocorticoid receptor in the cortical collecting tubules may be able to trigger the identical genomic response as an activated mineralocorticoid receptor, further underlying the crucial importance of prereceptor specificity-conferring mechanisms in determining not merely which steroid will occupy a type I receptor, but whether glucocorticoids will occupy any receptors in a mineralocorticoid target tissue.

To extend these clinical observations, we used a modification of the classic Kagawa bioassay, designed to explore the effect of administered steroids on urinary sodium to potassium ratios in adrenalectomized rats, as an index of their mineralocorticoid activity (45). Four groups of rats were given vehicle, carbenoxolone sodium, RU28362, or carbenoxolone sodium plus RU28362, and the effects on urinary electrolyte excreting determined. The dose of carbenoxolone sodium used (1 mg) had no mineralocorticoid activity in the adrenalectomized rat; in contrast, however, a modest dose (10 μg) of the pure glucocorticoid RU28362 was followed by an electrolyte effect half that of a maximal dose of aldosterone, and a maximal effect in the presence of 1 mg carbenoxolone sodium. It should be emphasized that RU28362 has vanishingly low affinity for type I receptors, such that it is used at 100- to 1000-fold excess in binding studies to exclude tracer from type II sites, with minimal effects on type I receptor occupancy. Similarly, in the cotransfection studies of Arizza and Evans previously cited (36), RU28362 over a range of doses was ineffective in human renal mineralocorticoid receptor-expressing transfects. Our interpretation of these studies is that RU28362 can occupy type II glucocorticoid receptors in the cortical collecting tubule—better in the presence of carbenoxolone, as in the clinical studies—and that agonist occupancy of type II receptors is equally effective in evoking the benchmark mineralocorticoid effect on urinary electrolytes as is agonist occupancy of type I receptors. In futher support of this interpretation, the type I selective agonist RU28318 substantially reversed the effect of aldosterone on urinary electrolytes, but did not alter the response to concurrently administered RU28362.

Interpretation of *in vivo* studies, clinical or in experimental animals, must necessarily be careful, inasmuch as the interventions in most cases affect a series of interactive organs and systems, raising the possibility of indirect and on occasion reverberating effects. Carbenoxolone sodium is clearly a "dirty" drug, with effects on receptors and a variety of enzymes at moderate to high doses (46); similarly, administration of RU28362 to adrenalectomized animals is followed by a variety of metabolic, circulatory, and renal effects. In such circumstances, though effects of cortisol and RU28362 via cortical collecting tubule type II receptors may be an attractive hypothesis to explain the findings, it is clearly not an exclusive one: such is the nature of many *in vivo* studies.

Confirmation of the hypothesis, however, comes from a complementary series of *in vitro* studies by Aniko Naray-Fejes-Toth and Geza Fejes-Toth on cortical collecting tubule cells in culture (47). These workers immunodissect kidneys, after raising panels of monoclonal antibodies to a range of proteins in renal extracts. By exposing kidney sections to each antibody, they are then able to select those antibodies recognizing epitopes expressed solely by cortical collecting tubules. When collagenase-perfused kidneys are exposed to such antibodies linked to a solid phase, only cortical collecting tubules are retained; plated out on grids in a double chamber, the cells form a monolayer, seal around the rim, and begin pumping sodium from one chamber to the other, and potassium in the other direction. This is reflected as a potential difference between chambers, which can be measured as a short circuit current, by techniques similar to those classically used to study ion transport by toad bladder or frog skin.

When the Fejes-Toths add aldosterone at 5 nM to their system, they find an almost threefold increase in sodium and potassium flux, and a commensurate increase in potential difference and short circuit current. When they add either dexamethasone or RU28362 at 50 nM—a concentration at which RU28362 would occupy few if any type I receptors—they obtain identical effects on ion flux and electrical parameters as seen with aldosterone. As in the *in vivo* studies on adrenalectomized rats given RU28362, the electrolyte effects of the glucocorticoids are not blocked by type I-specific antagonists, further evidence that the response is mediated by type II receptors in the cultured cortical collecting tubule cell.

When the clinical studies on patients with pseudohypoaldosteronism, the Kagawa bioassay studies on RU28362, and the *in vitro* cortical collecting tubule cell culture studies are considered together, they provide very strong evidence for the physiological promiscuity of the putative "mineralocorticoid response element(s)" responsible for the genomic effects of mineralocorticoids. Such promiscuity has previously been noted in *in vitro* transfection studies, in which mineralocorticoid, glucocorticoid, androgen, and progesterone receptors are all capable, for example, of activating a protypic "glucocorticoid" response element. The extent to which such *in vitro* studies, commonly with considerable receptor excess and very possibly in the absence of proteins modifying receptor function, can be extrapolated to the *in vivo* situation remains moot; currently, therefore, the triad of studies—clinical, *in vivo*, and *in vitro*—described above provides an unusually integrated insight into endocrine physiology.

What they also do is to underline the crucial specificity-conferring role of two enzymes in terms of specifying mineralocorticoid action—18 methyloxidase in the adrenal glomerulosa (48) and 11β-hydroxysteroid dehydrogenase (49) in physiological aldosterone target tissues. As such, they leave us with an intriguing physiological question: given the scope and versatility of the steroid/thyroid/retinoic acid/orphan receptor family of transcriptional regulators, where is the gain in not only having a mineralocorticoid receptor with equal affinity for mineralocorticoids and glucocorticoids, but also mineralocorticoid-regulated genes which appear equivalently responsive to mineralocorticoid or glucocorticoid receptors? Put another way, why have we not evolved a specific mineralocorticoid receptor? It is possible that the answer to this rather impertinent question may be found to lie within the conceptual framework of classical endocrine control systems; it is also possible that there are other determinants which have currently not even been considered for their possible relevance. We may never find the final answer to these sorts of questions; but if we are to have even partial success, in the long haul, it will surely come from observations and experiments made on patients, on animals *in vivo*, and on cells and subcellular systems *in vitro*.

REFERENCES

1. Funder, J. W., Blair-West, J. R., Coghlan, J. P., Denton, D. A., Scoggins, B. A., and Wright, R. D. (1969). Effect of plasma (K⁺) on the secretion of aldosterone. *Endocrinology* **85**, 381–384.
2. Blair-West, J. R., Coghlan, J. P., Cran, E., Denton, D. A., Funder, J. W. and Scoggins, B. A. (1973). Increased aldosterone secretion during sodium depletion with inhibition of renin release. *Am. J. Physiol.* **224**, 1409–1414.
3. Rousseau, G., Baxter, J. D., Funder, J. W., Edelman, I. S., and Tomkins, G. M. (1972). Glucocorticoid and mineralocorticoid receptors for aldosterone. *J. Steroid Biochem.* **3**, 219–227.
4. Funder, J. W., Feldman, D., and Edelman, I. S. (1973). The roles of plasma binding and receptor specificity in the mineralocorticoid action of aldosterone. *Endocrinology* **92**, 994–1004.
5. Funder, J. W., Feldman, D., and Edelman, I. S. (1973). Glucocorticoid receptors in the rat kidney: The binding of tritiated dexamethasone. *Endocrinology* **92**, 1005–1013.
6. Marver, D., Stewart, J., Funder, J. W., Feldman, D., and Edelman, I. S. (1974). Renal aldosterone receptors: Studies with (³H) aldosterone and the anti-mineralocorticoid (³H) spirolactone (SC-26304). *Proc. Natl. Acad. Sci. U.S.A.* **71**, 1431–1435.
7. Funder, J. W., Feldman, D., and Edelman, I. S. (1972). Specific aldosterone binding in rat kidney and parotid. *J. Steroid Biochem.* **3**, 209–218.
8. Wrange, O., and Yu, Z.-Y. (1983). Mineralocorticoid receptor in rat kidney and hippocampus: Characterization and quantitation by isoelectric focusing. *Endocrinology (Baltimore)* **113**, 243–250.
9. Beaumont, K., and Fanestil, D. D. (1983). Characterization of rat brain aldosterone receptors reveals high affinity for corticosterone. *Endocrinology (Baltimore)* **113**, 2043–2051.
10. Krozowski, Z. S., and Funder, J. W. (1983). Renal mineralocorticoid receptors and hippocampal corticosterone binding species have identical intrinsic steroid specificity. *Proc. Natl. Acad. Sci. U.S.A.* **80**, 6056–6060.
11. Armanini, D., Strasser, T., and Weber, P. C. (1985). Characterization of aldosterone binding sites in cirulatory human mononuclear leukocytes. *Am. J. Physiol.* **248**, E388–E390.
12. Arriza, J. L., Weinberger, C., Cerelli, G., Glaser, T. M., Handelin, B. L., Housman, D. E., and Evans, R. M. (1987). Cloning of human mineralocorticoid receptor complementary DNA: Structural and functional kinship with the glucocorticoid receptor. *Science* **237**, 268–275.
13. Funder, J. W., Pearce, P. T., Smith, R., and Smith, A. I. (1988). Mineralocorticoid action: Target-tissue specificity is enzyme, not receptor-mediated. *Science* **242**, 583–585.
14. Edwards, C. R. W., Stewart, P. M., Burt, D., McIntyre, M. A., de Kloet, E. R., Brett, L., Sutanto, W. S., and Monder, C. (1988). Localization of 11β-hydroxysteroid dehydrogenase— tissue specific protector of the mineralocorticoid receptor. *Lancet* **2**, 986.
15. Doyle, D., Smith, R., Krozowski, Z. S., and Funder, J. W. (1989). Mineralocorticoid specificity of renal Type I receptors: Binding and metabolism of corticosterone. *J. Steroid Biochem.* **32**(2), 165–170.

16. Pressley, L. A., and Funder, J. W. (1975). Glucocorticoid and mineralocorticoid receptors in gut mucosa. *Endocrinology (Baltimore)* **97,** 588–596.
17. Duval, D., and Funder, J. W. (1974). The binding of tritiated aldosterone in the rat liver cytosol. *Endocrinology* **94,** 575–579.
18. Funder, J. W., Duval, D., and Meyer, P. (1973). Cardiac glucocorticoid receptors: The binding of tritiated dexamethasone in rat and dog heart. *Endocrinology* **93,** 1300–1308.
19. Zaini, A., Pearce, P., and Funder, J. W. (1987). High-affinity aldosterone binding in rat liver—a re-evaluation. *Clin. Exp. Pharmacol. Physiol.* **14,** 39–45.
20. Pearce, P., and Funder, J. W. (1987). High affinity aldosterone binding sites (type I receptors) in rat heart. *Clin. Exp. Pharmacol. Physiol.* **14,** 859–866.
21. Lassman, M. N., and Mulrow, P. J. (1974). Deficiency of deoxycorticosterone binding protein in the hypothalamus of rats resistant to deoxycorticosterone induced hypertension. *Endocrinology (Baltimore)* **94,** 1541–1549.
22. Anderson, N. S., III, and Fanestil, D. D. (1976). Corticoid receptors in rat brain: Evidence for an aldosterone receptor. *Endocrinology (Baltimore)* **98,** 676–684.
23. McEwen, B. S., Lambdin, L. T., Rainbow, T. C., and Denicola, A. F. (1986). Aldosterone effects on salt appetite in adrenalectomized rats. *Neuroendocrinology* **43,** 38–43.
24. Meyer, W. J., and Nichols, N. R. (1981). Mineralocorticoid binding in cultured smooth muscle cells and fibroblasts from rat aorta. *J. Steroid Biochem.* **14,** 1157–1168.
25. Jazayeri, A., and Meyer, W. J. (1989). Mineralocorticoid-induced increase in β-adrenergic receptors of cultured rat arterial smooth muscle cells. *J. Steroid Biochem.* **33,** 987–991.
26. Lan, N. C., Matulich, D. T., Morris, J. A., and Baxter, J. D. (1981). Mineralocorticoid receptor-like aldosterone-binding protein in cell culture. *Endocrinology (Baltimore)* **109,** 1963–1970.
27. De Kloet, E. R., Wallach, G., and McEwen, B. S. (1975). Differences in binding of corticosterone and dexamethasone to rat brain and pituitary. *Endocrinology (Baltimore)* **96,** 598–611.
28. Nestler, E. J., Rainbow, T. C., McEwen, B. S., and Greengard, P. (1981). Corticosterone increases the level of protein 1, a neuron-specific protein in rat hippocampus. *Science* **212,** 1162–1164.
29. De Kloet, E. R., Kovacs, G. L., Szabo, G., Telegdy, G. Bohus, B., and Versteeg, D. H. G. (1982). Decreased serotonin turnover in the dorsal hippocampus of rat brain shortly after adrenalectomy: Selective normalization after corticosterone substitution. *Brain Res.* **239,** 659–663.
30. de Kloet, E. R., Versteeg, D. H. G., and Kovacs, G. L. (1983). Aldosterone blocks the response to corticosterone in the raphe-hippocampal serotonin system. *Brain Res.* **264,** 323–327.
31. Patel, P. D., Sherman, T. G., Goldman, D. J., and Watson, S. J. (1989). Molecular closing of a mineralocorticoid (type I) receptor complementary DNA for rat hippocampus. *Mol. Endocrinol.* **3, 11,** 1877–1885.
32. Stephenson, G., Krozowski, Z. S., and Funder, J. W. (1984). Extravascular CBG-like sites in rat kidney and mineralocorticoid receptor specificity. *Am. J. Physiol.* **246,** F227–F233.
33. Sheppard, K., and Funder, J. W. (1987). Mineralocorticoid specificity of renal type I receptors: In vivo binding studies. *Am. J. Physiol.* **252,** E224–E229.
34. Sheppard, K., and Funder, J. W. (1987). Type I receptors in parotid, colon and pituitary are aldosterone-selective *in vivo. Am. J. Physiol.* **253**(3), E467–E471.
35. Funder, J. W., and Sheppard, K. (1987). Adrenocortical steroids and the brain. *Annu. Rev. Physiol.* **49,** 397–412.
36. Arriza, J., Simerly, R. B., Swanson, L. W., and Evans, R. M. (1988). Neuronal mineralocorticoid receptor as a mediator of glucocorticoid response. *Neuron* **1,** 887–900.
37. Ulick, S., Levine, L. S., Gunczler, P., Zanconato, G., Ramirez, L. C., Rauh, W., Rösler, A., Bradlow, H. L., and New, M. I. (1979). A syndrome of apparent mineralocorticoid excess

associated with defects in the peripheral metabolism of cortisol. *J. Clin. Endocrinol. Metab.* **49,** 757–764.

38. New, M. I., Oberfield, S. E., Carey, R., Greig, F., Ulick, S., and Levine, L. S. (1983). A genetic defect in cortisol metabolism as the basis for the syndrome of apparent mineralocorticoid excess. *In* "Endocrinology of Hypertension" (F. Mantero, E. G. Biglieri, and C. R. W. Edwards, eds.) Serono Symp. 50, pp. 85–101. Academic Press, New York.

39. Stewart, P. M., Shackleton, C. H. L., and Edwards, C. R. W. (1987). The cortisol–cortisone shuttle and the genesis of hypertension. *In* "Corticosteroids and Peptide Homones in Hypertension" (F. Mantero and P. Vecsei, eds.), Serono Symp., pp. 163–177. Raven Press, New York.

40. Stewart, P. M., Valentino, R., Wallace, A. M., Burt, D., Shackleton, C. H. L., and Edwards, C. R. W. (1987). Mineralocorticoid activity of licorice: 11β-hydroxysteroid dehydrogenase activity comes of age. *Lancet* **2,** 821–824.

41. Armanini, D., Karbowiak, I., and Funder, J. W. (1983). Affinity of licorice derivatives for mineralocorticoid and glucocorticoid receptors. *Clin. Endocrinol. (Oxford)* **19,** 609–612.

42. Oberfield, S. E., Levine, L. S., Carey, R. M., Greig, F., Ulick, S., and New, M. I. (1983). Metabolic and blood pressure responses to hydrocortisone in the syndrome of apparent mineralocorticoid excess. *J. Clin. Endocrinol. Metab.* **56,** 332–339.

43. Cheek, D. B., and Perry, J. W. (1958). A salt-washing syndrome in infancy. *Arch. Dis. Child.* **33,** 252–256.

44. Armanini, D., Kuhnle, U., Strasser, T., Dorr, H., Weber, P. C., Stockigt, J. R., Pearce, P., and Funder, J. W. (1985). Aldosterone-receptor deficiency in pseudohypoaldosteronism. *N. Engl. J. Med.* **313,** 1178–1181.

45. Funder, J. W., Pearce, P., Myles, K., and Roy, L. P. (1990). Apparent mineralocorticoid excess, pseudohypoaldosteronism and urinary electrolyte excretion: Towards a redefinition of "mineralocorticoid" action. *FASEB J.* **4,** 3234–3238.

46. Latif, S. A., Conca, J. J., and Morris, D. J. (1990). The effect of the licorice derivative, glycyrrhetinic acid, on hepatic 3α- and 3β-hydroxysteroid dehydrogenase and 5α- and 5β-reductase pathways of metabolism of aldosterone in male rats. *Steroids* **55,** 52–58.

47. Naray-Fejes-Toth, A., and Fejes-Toth, G. (1990). Glucocorticoid receptors mediate mineralocorticoid-like effects in cultured collecting duct cells. *Am. J. Physiol.* **259,** F672–F678.

48. Imai, M., Shimada, H., Okada, Y., Matsushima-Hibiya, Y., Ogishima, T., and Ishimura, Y. (1990). Molecular cloning of a cDNA encoding aldosterone synthase cytochrome P_{450} in rat adrenal cortex. *FEBS Lett.* **263,** 299–302.

49. Agarwal, A. K., Monder, C., Eckstein, B., and White, P. C. (1989). Cloning and expression of rat cDNA encoding corticosteroid 11β-dehydrogenase. *J. Biol. Chem.* **264,** 18939–18943.

DISCUSSION

B. O'Malley. I can accept everything you said so far as salt retention and the processes that you are measuring. However, there are two other factors that could play a role and supply a need for two separate receptors. First, a careful developmental study of these receptors may reveal differential expression. There might be some combination of receptor and hormone, plus or minus contributions from steroid-metabolizing enzymes, that provides certain selective advantages to the embryo. The second possibility to consider is that either receptor may activate gene sets differentially depending on what ligand occupies it. You may not see this if you choose one or a limited number of model target genes to test in cell transfection studies. This concept has been validated in two instances to date and should be considered in your system. It is quite possible that if there were different ligands on the molecules, the C-terminal domains would be induced into different allosteric configurations, would expose different activation domains, and would induce differential gene activation.

J. Funder.   There have been a number of studies on the ontogeny of 11β-hydroxysteroid dehydrogenase, for example, by Beverley Murphy. The balance between dehydrogenase and reductase activity varies among fetus, neonate, and adult, and these developmental changes can be seen in a number of tissues. Much less well studied is the ontogeny of mineralocorticoid receptors. I cannot think of any studies specifically addressing that question. Perhaps a parallel study would be to do an evolutionary scan. Aldosterone seems to have emerged as the mineralocorticoid with lung fish, and before that 1α-hydroxycorticosterone, for example, appeared to fill the same role. 11β-Hydroxysteroid dehydrogenase has been here for millions of years; Carl Monder has fascinating data on its activity in a variety of species. In terms of your second possibility all I can do is agree that maybe the types of things you mentioned underlie the differences alluded to between the commonality of aldosterone and corticosterone actions in the kidney and their having agonist and antagonist roles in the hippocampus.

M. New.   It is clear that this mechanism must be vital to life because the patients who lack 11β-hydroxysteroid dehydrogenase are affected by a fatal disease. I only know of Chris Edwards' patient who has reached adulthood. I am reminded of something that Gordon Tomkins once told me, which was that too much cortisol is bad for you. Apparently the conversion of cortisol to cortisone is a very important shuttle to keep people alive. You have to buffer yourself against the production of cortisol, and this is perhaps the most important function 11β-dehydrogenase serves, that plus the fact that aldosterone is in the lactone.

I am troubled by two things. If apparent mineralocorticoid excess were an enzyme defect you would predict that the genetic defects of enzyme deficiency would be inherited as a recessive disorder. I have tested family members of patients with apparent mineralocorticoid excess (AME) and have never found any evidence by biochemical testing of the defect or of heterozygosity. Also, there is only one sib pair reported in the world's literature. Most have been sporadic cases. Thus I am not convinced that AME is a genetic defect, yet most enzyme defects are. This is worrisome. I think once Perry White has the probe for the human enzyme we will be able to test for genetic defects. Finally, the administration of aldosterone to AME patients produces no response. My interpretation of that is that the patients are in aldosterone escape, i.e., the mineralocorticoid receptors are filled. We gave patients as much as 2 mg of aldosterone and there was no effect whatsoever. On the other hand, 10 mg of cortisol raised the blood pressure and aggravated hypokalemia.

J. Funder.   In terms of your last point, I think we are in fact in agreement. If these patients are in mineralocorticoid escape because their type I receptors are filled, then added cortisol at low doses is raising blood pressure through receptors other than mineralocorticoid receptors, so the whole pathology would appear to be a combination of inappropriate mineralocorticoid receptor and glucocorticoid receptor occupancy. I do not think we differ on that point. In terms of an organism protecting itself from cortisol, cortisol production is not the problem. The design problem, if you wanted to play God or Charles Darwin, is that essentially every tissue in the body has glucocorticoid receptors. Therefore, it seems not unreasonable that we have evolved mechanisms to fractionate the signal. One of them clearly is transcortin in that the binding of cortisol to transcortin is markedly temperature dependent, so at skin temperature of 28°C, two to three times more cortisol is bound to transcortin than at the core temperature of 37°C. This is a very simple and an almost mechanical way of fractionating glucocorticoid signal, but the variable expression of 11β-hydroxysteroid dehydrogenase and reductase in a whole range of tissues makes for a system with the possibility of much finer tuning. The 11β-hydroxysteroid dehydrogenase that has been cloned, incidentally, is not the one that protects mineralocorticoid receptors. It is in high abundance in testis, lung, liver, and the renal proximal tubule, but not in the cortical collecting tubule, salivary glands, or colon. There clearly will be another member of the enzyme family in these tissues. The role of the one that has been cloned and characterized appears to be to fractionate the glucocorticoid signal in the tissues I have mentioned; to provide the possibility of cell- and tissue-specific differential responses to the same level of steroid. Can you repeat your second point?

M. New.   My second point is just a classical genetic one: an enzyme deficiency should be genetic.

J. Funder.   I think that even though enzyme activity is down in these patients it may not be a primary enzyme defect. Even though the problem I have highlighted is 11-oxidoreduction of steroids, these patients also have changes in their ratios of 5α to 5β reduced urinary steroids, and in 3α- and 3β-hydroxysteroid dehydrogenase activity. Second, there are two patients who are a sib pair. . .

M. New.   The sib pair that I know of is mine. I do not know whether there are others; I do not think so.

J. Funder.   . . . of Franco Mantero's whose steroid profiles (worked up by Stan Ulick) have an apparent deficiency of both dehydrogenase and reductase, normal urinary E to F metabolite ratios, but the clinical stigmata of the apparent mineralocorticoid excess syndrome. The evidence for reduced activity of both dehydrogenase and reductase is that they have very prolonged half-lives of infused radioactive E and F. These patients are said to have type II apparent mineralocorticoid excess to distinguish them from the classical patients (type I) in whom the lowered activity appears confined to the dehydrogenase. The very interesting point is that David Morris and colleagues have reproduced both the type I and type II pattern of enzyme deficiency in rats by giving them either glycyrrhetinic acid, in one instance, or carbenoxolone, in the other. Now 5β-reductase and 3β-dehydrogenase are very different; the reductase is microsomal and the dehydrogenase cytosolic. The chances that the two enzymes have a genetically linked deficiency appears to be vanishingly low. The fact that the pattern of apparent enzyme defect of type I apparent mineralocorticoid excess can be reproduced with one drug and the type II defect, where conversion is impaired in both directions, can be blocked with another makes me think that one group is making an "endogenous glycyrrhetinic acid" and the other an "endogenous carbenoxolone" to produce the complex pattern of apparent enzyme defects. Thus, I think, we should not necessarily think of these syndromes as a genetic defect of 11β-hydroxysteroid dehydrogenase. For this reason I prefer to say that these patients have reduced enzyme activity, rather than that the enzyme is congenitally impaired or absent.

W. Bardin.   Dr. Funder has the ability to synthesize data that in isolation do not seem to make sense. We are still left with an important conundrum that is not only important for these observations but also for almost all steroid-responsive systems. What we refer to is the apparent tissue difference in response between various tissues. What are your thoughts on brain versus the kidney?

J. Funder.   There is one small piece of evidence that may be potentially relevant from a study by Damian Doyle in my laboratory, who was trying to purify type I receptors from hippocampus and kidney. Clearly there are no differences between tissues at the level of message. In 1983, Wrange and Yu, in an article published in *Endocrinology*, showed that on electrofocusing of tryptic digests the receptor in the two tissues was the same; the binding specificity, as you saw in the steroid hierarchy shown for the two tissues, is the same. When Damian subjected hippocampal and renal receptors to FPLC, under conditions in which they still bound ligand, he consistently found a difference of about 15K in the apparent molecular weight. Whether this reflects a posttranslational modification, or perhaps, more likely, an associated relatively low-molecular-weight protein present in hippocampus but not present in kidney, I do not know. I suspect, however, that it will be by following up this type of study that we may be able to find the subtle difference between tissues.

W. Bardin.   Are there studies on tissue differences in receptors? For example, are there studies evaluating on rates and off rates? One such example for this appears to be the tissue-dependent differences in the androgen receptor where the off rate for testosterone in the prostate receptor is very fast, whereas in most other tissues, including the pituitary, the brain, and the kidney, the off rate for testosterone is very slow.

J. Funder.   To my knowledge these studies have never been done for mineralocorticoid receptors. It was not until 1980 and the use of molybdate in buffers that broken-cell studies on mineralocorticoid receptors were really possible. Until then, studies had to be in whole cells or in tissue slices, because the receptors were really very labile. By the 1980s, people were no longer interested in doing those types of studies.

V. Lakshmi.   Regarding the experiments you showed on the inhibition of 11β-hydroxysteroid dehydrogenase, is it possible that carbenoxolone could not reach the brain because of the blood–brain barrier?

J. Funder.   Yes, that is possible, and that is the reason I included the heart as an additional control since I do not know whether carbenoxolone crosses the blood–brain barrier.

J. Baxter.   Why do you think dexamethasone does not elicit mineralocorticoid-like actions when given to humans? Might it be a partial mineralocorticoid agonist?

J. Funder.   Certainly in cotransfection studies it acts as a full mineralocorticoid agonist. I suspect that *in vivo* it does not act as a mineralocorticoid because its 11β-hydroxyl group is labile in the kidney.

J. Baxter.   This has never been shown to be the major metabolic fate of dexamethasone.

J. Funder.   In this regard an interesting question is why adding a 16α-methyl group to a 9α-fluoroprednisolone, which is an extremely active mineralocorticoid, alters the susceptibility of the 11β-hydroxyl to enzymatic attack.

N. Schwartz.   Some of the earlier studies from McEwen's laboratory on pituitary receptors may have come to the wrong conclusions. Have there been new studies on anterior pituitaries with respect to suppression by glucocorticoids and to the receptors present?

J. Funder.   So far as we can tell corticotropes do not have type I (mineralocorticoid) receptors; the pituitary complement of these appears confined to cells of the somatomammotropic lineage. At the pituitary level you are therefore looking at a regular type II receptor glucocorticoid effect. At the hypothalamic level the effect is similarly via type II glucocorticoid receptors. There are, however, compelling data, predominantly from Mary Dallman's laboratory, showing that type I receptors in the hippocampus are intimately involved in the control of ACTH secretion, particularly in the evening. Mary has recently done some very elegant studies comparing the combination of very low doses of dexamethasone and corticosterone, and has shown that together they completely abolish ACTH secretion in adrenalectomized rats in the evening. So type I receptors in the hippocampus and type II receptors in the hypothalamus and pituitary together provide a complex pattern of network for steroid control of proopiomelanocortin (POMC) synthesis and ACTH secretion.

N. Schwartz.   Dallman has shown that there is some suppression at the level of the pituitary of ACTH secretion and POMC synthesis.

J. Funder.   POMC synthesis can certainly be turned off by a direct pituitary action. I think, however, that there are really serious doubts about rapid, direct effects of glucocorticoids on ACTH secretion and that there are differences between fragments, dispersed cells, and the whole pituitary. I am sure that the predominant effect of an intravenous bolus of cortisol in causing a fall in ACTH levels in plasma is at the level of the hypothalamus, and not a direct effect on the pituitary.

Molecular Biology of Human Renin and Its Gene

John D. Baxter,* Keith Duncan,* William Chu,*
Michael N. G. James,† Robert B. Russell,†
Mohammad A. Haidar,* Frances M. DeNoto,* Willa Hsueh,‡
and Timothy L. Reudelhuber§

*Metabolic Research Unit and the Department of Medicine, University of California at San Francisco, San Francisco, California, †Medical Research Council Group in Protein Structure and Function, Department of Biochemistry, University of Alberta, Edmonton, Alberta, Canada, ‡Department of Medicine, University of Southern California, Los Angeles, California, and § Clinical Research Institute of Montreal, Montreal, Quebec, Canada

I. Introduction

Human renin is the rate-limiting component of the renin–angiotensin system (RAS; reviewed in Baxter et al., 1987; Sealey and Laragh, 1990). Renin cleaves angiotensinogen to release the decapeptide angiotensin I (AI). The latter appears to be inactive, but is cleaved to angiotensin II (AII) by the actions of converting enzyme. Angiotensin II increases the blood pressure, stimulates the release of aldosterone that causes sodium retention and potassium loss, and regulates growth and other processes related to the control of fluid homeostasis and cardiovascular function. Abnormal function of the RAS participates in the pathogenesis of hypertension, heart failure, and renal failure (Hollenberg, 1985; Gavras, 1988; Lancet, 1987; Laragh, 1989; Williams, 1988). The importance of this system in these pathological processes is underscored by the widespread use of converting enzyme inhibitors in treating them (Gavras, 1988; Feig and Rutan, 1989; Lancet, 1987, 1990; Williams, 1988).

The expression of renin and other components of the RAS is a complex process involving multiple tissues. First, the renin gene is expressed in multiple tissues, including the kidneys, adrenals, gonads, placenta, pituitary, brain, and hypothalamus (Deschepper et al., 1986; Dzau et al., 1987a,b; Ekker et al., 1989; Miller et al., 1989; Mullins et al., 1989). Other components of the system, angiotensinogen and converting enzyme, are also widely expressed (Campbell, 1989; Dzau et al., 1987a,b; Ehlers and Riordan, 1989; Erdos, 1990; Jin et al., 1988; Lever, 1989; Unger and Gohlke, 1990). The levels of renin expression are regulated, and overall they tend to change more than those of angiotensinogen and converting enzyme. Renin mRNA levels and renin release appear to be controlled by multiple factors (Baxter et al., 1987; Barrett et al., 1989; Dzau et

211

al., 1987b; Everett *et al.*, 1990; Iwao *et al.*, 1988; Makrides *et al.*, 1988; Moffet *et al.*, 1986; Nakamura *et al.*, 1985; Sealey and Laragh, 1990; Toffelmire *et al.*, 1989). In essential hypertension, plasma renin levels can be low, normal, or high, implying possible differences in the control of this gene between people (Kotchen *et al.*, 1982; Laragh, 1989).

Prorenin is the precursor of renin and proteolytic cleavage by tissues such as the kidney results in removal of a 43-amino acid prosegment and in active renin (reviewed in Hsueh and Baxter, 1991; Lenz and Sealey, 1990; Sealey *et al.*, 1980; Shier *et al.*, 1989). Prorenin has only about 1–8% of the enzyme activity of renin at pH 7.4 and 37°C, but its activity can be increased nonproteolytically by a number of manipulations, including lowering the pH or temperature, and by lipids (Glorioso *et al.*, 1989; Higashimori *et al.*, 1989; Hseuh *et al.*, 1986; Hsueh and Baxter, 1991; Lenz and Sealey, 1990; Sealey *et al.*, 1980). Since many tissues produce prorenin but not renin, there is great interest in whether the intrinsic renin activity of prorenin, perhaps stimulated by cellular factors, may be significant in AI and ultimately AII production. By contrast, the kidney, adrenals, and perhaps a few other tissues cleave prorenin to renin (Hsueh and Baxter, 1991; Lenz and Sealey, 1990; Shier *et al.*, 1989). The factors that regulate renal renin release control the levels of renin mRNA, the conversion of prorenin to renin, and the release of renin. In diabetes mellitus there can be elevated prorenin levels with normal renin levels early in the course of the disease (Wilson and Luetscher, 1990) and in some cases there can be hyporeninemic hypoaldosteronism with normal to elevated prorenin levels in the circulation (Luetscher *et al.*, 1985). Renin and prorenin may also be taken up from the plasma into peripheral tissues where prorenin may be activated (Kim *et al.*, 1987; Lever, 1989; Skeggs and Dorer, 1989; Skinner *et al.*, 1986).

There is currently great debate as to the role of renal vs extrarenal RASs in the overall function of the system. The kidney appears to be responsible for essentially all of the renin activity in the circulation. Nephrectomy results in a complete disappearance of renin in the circulation with a consequent marked reduction in plasma AII levels (Campbell, 1989; Derkx *et al.*, 1978; Frohlich *et al.*, 1989; Jin *et al.*, 1988; Lever, 1989; Sealey and Rubattu, 1989). Renal defects in renin release lead to hyporeninemic hypoaldosteronism with low levels of AII and hyperkalemia (Luetscher *et al.*, 1985). These findings and others demonstrate the critical role of the kidney in the production of active renin.

In spite of this evidence for a central role of renal renin, other evidence suggests a potential function of extrarenal RASs. First, a significant proportion of the total AI in tissues and in the circulation may be derived outside the plasma (Admiraal *et al.*, 1990; Campbell, 1989; Lever, 1989; Swales and Samani, 1990). This AI may be generated by renal-derived renin that has been taken up by the peripheral tissues, since plasma AI generation drops to very low levels following nephrectomy (Campbell, 1989; Lever, 1989), and renin and prorenin can be taken up by peripheral tissues where, in some cases (e.g., liver), there can be substantial conversion of prorenin to renin (Kim *et al.*, 1987; Lever, 1989; Skeggs and Dorer, 1989). Nevertheless, AII has been found in plasma and tissues of nephrectomized rats and humans (Frohlich *et al.*, 1989; Itoh *et al.*, 1988; Jin *et al.*, 1988; Lever, 1989). In addition, as addressed later, prorenin and renin in extrarenal tissues appear to reside mostly in cells/tissues where there is renin mRNA and the potential to produce prorenin and renin. Further, whereas plasma prorenin levels decrease following nephrectomy, low levels of prorenin persist (Derkx *et al.*, 1978; Sealey and Rubattu, 1989; Hsueh and Baxter, 1991). It is conceivable that low RAS activity following nephrectomy or with renal disease involves not only decreased renal renin but a suppression of extrarenal RAS activity. For example, adrenal renin production responds to some of the same stimuli as renal renin and its release and levels are suppressed by salt retention that may occur following nephrectomy (Doi *et al.*, 1984; Inagami *et al.*, 1989; Shier *et al.*, 1989). Second, studies with renin (Boger *et al.*, 1990; Blaine *et al.*, 1985; Campbell, 1989; Corvol *et al.*, 1990; Kleinert *et al.*, 1987; Verburg *et al.*, 1989) and converting enzyme (Brunner *et al.*, 1989; Lancet, 1990; Lees *et al.*, 1990; Unger and Gohlke, 1990) inhibitors (or with infused renin; Swales and Samani, 1990) are best explained by proposing that important effects of those inhibitors are on RAS components outside the circulation; this is based on analyses of the dose response and the kinetics of both the onset of effects of treatment and the reversal of effects when the compounds are withdrawn. Third, there is evidence for a renal-independent RAS in the brain (Jin *et al.*, 1988), adrenals (Yamauchi *et al.*, 1990), vascular endothelial cells (Kifor and Dzau, 1987), and other tissues (Admiraal *et al.*, 1990). Fourth, there is some evidence that RAS activity outside the circulation controls blood pressure (Jin *et al.*, 1988; Lever, 1989; Oliver and Sciacca, 1984), and two recent studies using transgenic animals suggest that excessive extrarenal RAS activity can cause hypertension. Mullins *et al.* (1990) prepared transgenic rats that expressed the renin gene in several extrarenal tissues, especially the adrenal. Renal renin and plasma renin and AII levels were low in the face of severe hypertension that was correctable by treating the rats with converting enzyme inhibitors. Ohkubo *et al.* (1990) also produced transgenic mice that were hypertensive when they simultaneously expressed both the rat renin and angiotensinogen genes (predominantly in the liver).

AII and analogs can be generated by alternate means. Thus, other enzymes such as tonin and cathepsin D can produce angiotensin peptides from angiotensinogen (Campbell, 1989; Dzau, 1986; Hirata *et al.*, 1986; Rosenthal *et al.*, 1990). Other enzymes can generate angiotensin(1–7), containing the seven amino-terminal amino acids of AI that has AII-like activity in the brain but not on peripheral vasoconstrictor activities (Ferrario *et al.*, 1990). The qualitative and quantitative importance of these processes is currently not understood.

All of these considerations and others make it imperative to understand the mechanisms that control the expression of the renin gene in both renal and extrarenal tissues. This control is exerted not only at the level of expression of the gene, which differs markedly in various tissues, but also at other levels, including the processing of prorenin to renin, release of renin, and, possibly, renin activity of prorenin.

An additional reason to study renin and the control of the system is to develop alternative means to block it. The major success of the converting enzyme inhibitors in treating the conditions listed above demonstrates the importance of blocking the system. However, it is possible that blockade of the system at other levels could either complement or be superior to blockade of converting enzyme for all or selected aspects of the therapy (Mento *et al.*, 1989; Corvol *et al.*, 1990; also discussed below).

To understand the role of the RAS in health and disease and when and how to inhibit this system, we have been investigating a number of aspects of this system, including the mechanisms controlling tissue-specific and regulated activity of the renin promoter, the mechanisms of intracellular sorting of prorenin and renin to constitutive or regulated secretory pathways, the structures on prorenin and renin that both direct and affect the efficiency of these processes, the regulation of renin release, the mechanisms of prorenin prosegment inactivation of renin and of proteolytic and nonproteolytic activation of prorenin, and the structures of renin and prorenin. This article summarizes a number of aspects of our work in this area.

II. Cloning and Structure of the Human Renin Gene

The cDNA complementary to human renin mRNA was initially cloned by Imai *et al.*, (1983), and this followed the cloning of the cDNAs for mouse renin (Panthier *et al.*, 1984; Rougeon *et al.*, 1981). We used the mouse cDNA to clone the human renin chromosomal gene sequences (Hardman *et al.*, 1984) and later the human renin cDNA (Fritz *et al.*, 1986). The cDNA sequence agreed with that of Imai *et al.* (1983) except for one nucleotide difference. Analysis of the gene revealed it to span over 12 kilobases (kb) of DNA, and to contain 10 exons separated by 9 introns. Only one human renin gene was found (Hardman *et al.*, 1984), like the case with the rat and some but not all mouse strains (Fukamizu *et al.*, 1988; Holm *et al.*, 1984; Mullins *et al.*, 1982; Piccini *et al.*, 1982). The overall structure of the gene agreed with that reported by Hobart *et al.* (1984) and Miyazaki *et al.* (1984).

The renin-coding sequences showed homology with those of pepsinogen, one of the aspartyl proteases to which renin is related (Hardman *et al.*, 1984; Sogawa *et al.*, 1983). The gene also showed organizational similarity with the pepsinogen gene in terms of placement of introns within the exons. Based on their structures it can be speculated that the two exons containing each of the two Asp residues of the active site were derived from a duplication of an ancestral gene that led to the bilobal aspartyl proteases (Hardman *et al.*, 1984; Holm *et al.*,

```
             -100         -90         -80         -70         -60         -51
hRenin       GCCCTGCCATCTACCCCAGGGTAATAAATCAGGGGAGAGCAGAATTGCAA
rRenin       ..CCTGCCAT..ACCC..GGGTAATAAATCAG...AG.GC..
mRen 1 and 2      ..CC..GGGTAATAAATCA..GCAGA.C..


             -50         -40         -30         -20         -10         -1
hRenin       TCACCCCATGCATGGAGTGTATAAAGGGGAAGGGCTAAGGGAGCCACAG
rRenin       ..GTGTATAAA.G...
mRen 1 and 2      AT.CATGG.GTGTATAAA.G...
```

FIG. 1.   Homology in the 5′-flanking DNA between human (h) and rat (r) renin and mouse (m) renin 1 (*Ren 1*) and renin 2 (*Ren 2*) gene. Data from Hardman *et al.* (1984), Panthier *et al.* (1984), Soubrier *et al.* (1986), and Tronik *et al.* (1988).

1984; Sogawa *et al.*, 1983). Repetitive DNA was located flanking the gene. The gene structure provided an accurate prediction of the transcriptional start site; there was also a suggestion of a second TATA box upstream from the major promoter (Hardman *et al.*, 1984), although there is no evidence that this is functional in either the kidney (Soubrier *et al.*, 1986) or in an extrarenal tissue, the placenta (Duncan *et al.*, 1990). There were several regions in the 5′-flanking DNA of homology with consensus steroid hormone response elements (Hardman *et al.*, 1984), although as addressed below, we have been unable to obtain evidence that they are functional. Overall, there is little homology between the 5′-flanking DNA regions of the human renin gene with the mouse renin 1 and renin 2 gene (Soubrier *et al.*, 1986) and the rat renin gene (Tronik *et al.*, 1988), with the exception of four regions centered around the TATA box ($-45/-18$), the upstream TATA boxlike structure mentioned above ($-98/-62$) and a region from -365 to -333 in the human gene with homology to regions from -11 to $+9$, -450 to -433 and -671 to -654 in the mouse genes. The homologous sequences in the first 100 bp of 5′-flanking DNA of the human, rat, and mouse genes are shown in Fig. 1. Of note is that whereas there is homology between these genes in the -98 to -67 region, such homology does not extend to the human pepsinogen gene (Hardman *et al.*, 1984).

III.   Transcriptional Control of Renin Gene Expression

A.   TISSUE/CELL-TYPE EXPRESSION OF THE RENIN GENE

As mentioned in Section I, it is particularly important to know where the renin gene is expressed to understand the role of extrarenal RASs and the importance of local generation vs uptake of renin or prorenin in tissue renin activity. Studies using AI generation as an assay for renin do not distinguish between these possibilities. To obtain more information on this point, we analyzed rat tissues using mouse anti-renin antisera and mouse renin cDNA as probes (Deschepper *et al.*, 1986). We found renin mRNA in the renal juxtaglomerular cells, intermediate pituitary lobe, adrenal glomerulosa, scattered anterior pituitary cells, and testicular Leydig cells. The antisera detected renin protein in the same tissues; however, whereas all of our antisera recognized renal renin, and at least one batch of antiserum recognized renin in each of the tissues that contained renin mRNA, individual batches failed to recognize the renin in some of the extrarenal tissues. These results could imply that there are differences in posttranslation processing of rat renin, possibly in glycosylation in different tissues.

We also screened a number of cultured cell lines for renin, prorenin, and renin mRNA in an effort to find a cultured cell line to study the expression of the renin promoter. As shown in Table I, most cell lines produced no detectable renin or

TABLE I
Cell Lines Tested for Renin mRNA[a]

| | | Renin activity (ng AI/ml/hour) | | | |
Cell line	Origin	Total	Active	Prorenin	mRNA
SKUT-1	Uterine leiomyosarcoma	3.6	3.6	0	−
SKUT-1B	Uterine leiomyosarcoma	5.4	5.4	0	−
HeLa	Cervical carcinoma	3.6	3.6	0	−
C33A	Cervical carcinoma	3.2	3.2	0	−
BeWo	Choriocarcinoma	10.8	10.8	0	−
JAR	Choriocarcinoma	0	0	0	−
JEG-3	Choriocarcinoma	4.0	4.0	0	−
SKOV-3	Ovarian adenocarcinoma	0.2	0.2	0	−
CaOV-4	Ovarian adenocarcinoma	0	0	0	−
NIH OVCAR3	Ovarian adenocarcinoma	4.0	3.0	1.0	−
SK-LMS-1	Vulvar leiomyosarcoma	165.1	12.0	153.0	+
Choriodecidual	Primary choriodecidual cultures	1500	50	1450	+
SK-NEP-1	Wilm's tumor	0	0	0	−
G401	Wilm's tumor	10.8	7.2	3.6	−
G402	Renal leiomyoblastoma	9.0	5.4	3.6	−
Caki-2	Clear cell carcinoma (renal primary)	0.2	0.2	0	−
A-498	Kidney carcinoma	4.0	4.0	0	−
293	Transformed embryonal kidney	9.0	3.6	5.4	−
A-704	Renal adenocarcinoma	3.2	1.2	2.0	−

[a]The cell lines were obtained from the American Type Tissue Culture Association (Rockville, MD); mRNA analysis was by Northern blotting. Renin and prorenin activity were measured using angiotensin I (AI) generation with or without trypsin treatment (Duncan *et al.*, 1990) and are expressed as nanograms AI per hour per milliliter of supernatant from approximately 3×10^6 cells in 2-ml incubations. Total, Total active plus trypsin-activatable renin. + or − under mRNA indicate whether renin mRNA was detected.

prorenin and did not have detectable renin mRNA. However, we did find renin mRNA in SK-LMS-1 cells derived from a vulvar carcinoma and in human placental cells (discussed in the next section). In addition, we ultimately found renin mRNA under conditions of cAMP stimulation in a Simian virus 40 (SV40)-transformed human lung fibroblast cell line (discussed in Section III, E). These results provide additional evidence that the renin gene is not expressed at very high levels in most tissues, even though it is expressed in a number of tissues.

Overall these results have been confirmed by other studies (Dzau *et al.*, 1987a,b; Ekker *et al.*, 1989; Miller *et al.*, 1989). Since we and others have not found tissues with high levels of prorenin or renin that did not contain renin mRNA, it may be that in general most extrarenal prorenin and renin is produced in the tissues where it is found. However, a quantitative assessment of this issue in terms of AI production will require knowing in addition the activity of the "renin" proteins in these tissues, since much of the protein is likely to be prorenin rather than renin. The renin activity of this prorenin may be greater in these tissues than it is in the plasma due to specific cellular influences (discussed in Section VII).

More recently, Ekker *et al.* (1989) and Miller *et al.* (1989), using polymerase chain reaction (PCR) technology and RNase protection assays, respectively, have found lower levels of renin mRNA in several other tissues, including heart (only by Miller *et al.*, 1989), liver, spleen, lung, thymus, prostate, and brain; some of these tissues are implicated as being important in extrarenal/extraplasma RAS activity. The lower levels of expression raise the question of the quantitative importance of any renin produced, but overall the pattern suggests that the renin promoter is widely expressed with levels of activity that vary greatly from very low in many tissues to high levels in others.

B. EXPRESSION OF THE RENIN GENE IN HUMAN PLACENTAL CELLS

Several groups have identified human placenta as a source of prorenin in pregnancy (Acker *et al.*, 1982; Pinet *et al.*, 1988; Poisner *et al.*, 1981; Poisner and Poisner, 1987; Shaw *et al.*, 1989). This was not a surprising finding, since levels of prorenin in amniotic fluid exceed by up to 10-fold those in plasma; uterine–placental-derived prorenin probably accounts in part for the rise in maternal plasma prorenin levels that occurs in pregnancy (Hsueh and Baxter, 1991; Lumbers, 1971; Sealey *et al.*, 1980; Shaw *et al.*, 1989). Although the role of this prorenin is not known, theories include the possibilities that it may act through either AI and subsequent AII generation, or conceivably in some other manner as a growth factor or as a regulator of placental blood flow (Hsueh and Baxter, 1991). AII influences growth of vascular smooth muscle cells (Geistefer *et al.*, 1988), and there is evidence that converting enzyme inhibitors can block aspects of pathological processes such as atherosclerosis and hypertension-related vascular hypertrophy that involve proliferation of smooth muscle cells (Chobanian *et al.*, 1990). Abnormalities in the RAS appear not to be primary in the development of preeclampsia or eclampsia; however, this system may exacerbate or in other ways contribute to the pathology of these conditions (Hsueh and Baxter, 1991).

For all these reasons it is of interest to understand the mechanisms by which uterine–placental prorenin production is controlled. In addition, we are interested in uterine–placental cells as a suitable human cell type to study regulation of renin gene expression using the transfected renin promoter (discussed later). In our screening of a large number of cell lines in culture these cells produced more prorenin and contained more renin mRNA than other cell types examined (Table I). The cells we have used were prepared and propagated as described by Poisner *et al.* (1981). This group and others (Pinet *et al.*, 1988; Poisner *et al.*, 1981; Poisner and Poisner, 1987; Shaw *et al.*, 1989) have shown that these cells produce mostly prorenin and convert very little of this to renin.

There is current controversy as to which cells in this preparation are actually synthesizing the prorenin. The predominant cell type is chorion derived, and after some time most of the cells in the culture appear to be making prorenin (Pinet *et al.*, 1988). Thus, most groups have concluded that chorion and not decidual cells are the prorenin-producing cells (Pinet *et al.*, 1988; Poisner *et al.*, 1981). However, Shaw *et al.* (1989) concluded after examining decidual cells that were not contaminated with chorion (from ectopic pregnancies) and chorion without decidua (intertwin and basal plate chorion) that the decidual cells produce most of the prorenin. We have not addressed this controversy. For our purposes the cells in the culture produce prorenin and therefore are likely to contain factors that are important for renin promoter activity that can be studied. We now call these cultures "choriodecidual" cells.

We first confirmed previous results (Poisner and Poisner, 1987) that the cells do produce prorenin and that prorenin production can be increased by forskolin and by a calcium ionophore (Duncan *et al.*, 1990). For these studies renin levels are ordinarily measured by assaying for AI-generating activity using a radioimmunoassay for AI. Prorenin is measured by the amount of AI-generating activity that results after its activation with limited trypsin digestion (Duncan *et al.*, 1990).

We next examined the control of renin mRNA in these cells (Duncan *et al.*, 1990). We found that the renin mRNA start site was the same as that used by the kidney (Duncan *et al.*, 1990; Soubrier *et al.*, 1986). Figure 2 shows that renin mRNA levels were unaffected by $10^{-6}\,M$ forskolin (although $10^{-5}\,M$ forskolin does increase renin mRNA, not shown), and were increased by a calcium ionophore, and that the two stimulators together can act synergistically. These changes in renin mRNA paralleled changes in prorenin released into the medium (Duncan *et al.*, 1990), indicating that prorenin production can be regulated by

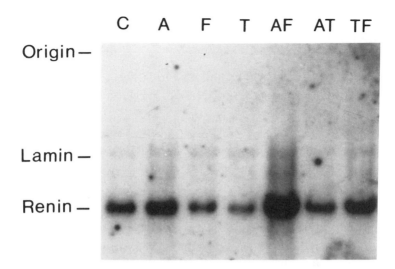

Rel. Amt. hR RNA: 1.0 2.1 0.8 0.5 6.3 1.3 3.1

FIG. 2.   Regulation of renin mRNA levels in primary cultures of human choriodecidual cells. Shown is the autoradiogram of a blot of renin mRNA and for comparison, lamin mRNA. Shown at the bottom are the relative renin mRNA levels assessed by densitometric scanning of the autoradiogram. C, Control; A, calcium ionophore A23186; F, 10^{-6} M forskolin; T, phorbol ester, PMA (phorbol 12-myristate 13-acetate). Reprinted from Duncan et al. (1990).

controlling the levels of renin mRNA. These cells also appear not to store prorenin, but release it continuously in response to a secretagogue, suggesting that they contain a constitutive but not a regulated release pathway. The stimulating effects of cAMP on renin mRNA are similar to those observed with rat kidney microvessels (Everett et al., 1990). The stimulating effects of calcium ion are also similar to those found with adrenal renin (Inagami et al., 1989), but they differ from those on the kidney, since increased calcium appears to block renal renin release (Churchill, 1985). The effects of calcium ion on renal renin mRNA have not, to our knowledge, been assessed; however, systems that respond to calcium ion and cAMP can respond in the same or in opposite directions to the two second messengers (Nishizuki, 1983). In some cases these differences can be due to variations in the way extracellular calcium affects intracellular calcium ion levels (Churchill, 1985; Hano et al., 1990; Kotchen and Guthrie, 1988; Park et al., 1986). However, there may be other differences. Calcium ion has been shown to activate transcription of the c-fos gene, the product of which acts through the AP-1 transcription factor complex through sequences that resemble a cAMP-responsive element (Fisch et al., 1987; Sheng et al., 1988). However, because jun and fos, which form AP-1, can have complex interrelationships with a variety of transcription factors (Yang-Yen *et al.*, 1990; Schule *et al.*, 1990), there is room for complex stimulatory and inhibitory actions of calcium ion.

C. CELL TYPE-SPECIFIC EXPRESSION OF THE TRANSFECTED HUMAN RENIN PROMOTER

To understand how the human prorenin and/or renin are produced in certain tissues and are regulated at the level of transcription it will be critical to define the DNA elements on the renin gene and the protein factors that interact with them to control renin promoter activity. To date, little work has been done to define such elements; the available data mostly come from experiments with transgenic rats and mice and the mouse renin promoter (Jones *et al.*, 1990; Miller *et al.*, 1989; Mullins *et al.*, 1989; Sigmund *et al.*, 1990; Tronik *et al.*, 1987). Overall, these studies show that the mouse renin gene 5'-flanking DNA can direct tissue-specific expression of this gene in tissues such as kidney and adrenal (Jones *et al.*, 1990; Sigmund *et al.*, 1990). Studies using the mouse renin promoter linked to the coding sequences for the simian virus 40 (SV40) T antigen, the product of which causes tumors, also reveal renin promoter activity in certain subcutaneous tissues and portions of the adrenal, where high levels of expression are not found in the adult (Jones *et al.*, 1990; Sigmund *et al.*, 1990). These results may imply that the renin promoter is more active in these tissues at some time during development than in the adult. However, in another study with such contructs, tumors developed in tissues reflecting the tumorigenic spectrum of the SV40 early region gene, implying a lack of reninlike tissue-specific expression (Sola *et al.*, 1989). When the transgenic mice containing the intact human renin gene were prepared the gene was expressed specifically in the kidney (Seo *et al.*, 1990).

Because our interest is mostly in how the human renin promoter is expressed in human tissues, the choriodecidual cells described above seem to be ideal candidates to study this promoter, and to characterize and ultimately isolate key DNA elements and transacting factors. They are human, nontransformed cells, and they express the endogenous renin gene. Whereas a first impression might be to focus on renal cells [cultured juxtaglomerular cells have not as yet been successfully used with gene transfer (Pinet *et al.*, 1985)], the considerations outlined earlier underscore the immense importance of understanding extrarenal as well as renal renin gene expression.

We therefore transfected renin gene DNA fragments linked to "reporter" gene-coding sequences, those encoding chloramphenicol acetyltransferase (CAT) into placental and other cell types. Initial efforts using CAT assays were frustrating, due to high background levels of nonspecific promoter activity in the plasmid. However, by using RNase protection assays and measuring start sites at the renin promoter and comparing the expression with that of a cotransfected

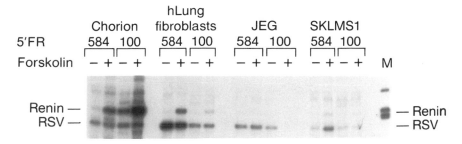

FIG. 3.   RNase protection assays of transcripts from transfected hR (−584/+11) CST and hR (−102/+11) CAT vectors cotransfected along with an RSV hGH gene into the indicated cell types by calcium phosphate precipitation and treated with (+) or without (−) forskolin. For methods, see Duncan *et al.* (1990).

Rous sarcoma virus (RSV) promoter (Fig. 3) or thymidine kinase promoter (not shown), or by using CAT assays and normalizing the expression to that from a cotransfected gene containing the human metallothionein 2a (hMT-2a) promoter and human growth hormone (hGH)-coding sequence (Table II), the renin promoter was found to be active in these cells and to use the correct start site.

Figure 3 shows results where RNase protection was used with constructs containing 584 or 102 base pairs (bp) of the human renin promoter. These renin

TABLE II

Relative Human Renin Pomoter Activity in Various Human Cell Lines in the Absence and Presence of Forskolin Stimulation[a]

Cell line			Activity	
Designation	Origin	Assay	Basal	Plus forskolin
Choriodecidual	Placenta	RNase/CAT	100	910
WI26VA4	Lung fibroblasts	RNase	0.2	31
JEG-3	Choriocarcinoma	RNase	1.7	6.5
SK-LMS-1	Vulvar leiomyosarcoma	RNase	4.6	6.3
SK-OV-3	Ovarian carcinoma	CAT	4.0	—
CaKi-2	Renal carcinoma	CAT	4.0	—
HeLa	Cervical carcinoma	CAT	1.1	—

[a]Cell lines from the American Type Tissue Culture Association were assayed for expression of a transfected gene containing 102 bp of the human renin gene 5′-flanking DNA linked to chloramphenicol acetyltransferase (CAT)-coding sequences assessed under basal conditions and following treatment of the cultures for 24 hours with 10^{-5} M forskolin. Expression was assayed by nuclease protection or CAT assays as described in Duncan *et al.* (1990, 1991), from which the data were taken. For the RNase protection, the expression is normalized to that of a cotransfected Rous sarcoma virus (RSV) promoter. For CAT assays the results were normalized to those from a cotransfected gene containing the hMT-2a promoter and hGH-coding sequences.

promoters show high activity relative to the cotransfected RSV promoter. Further, the renin promoter is more active in the choridecidual cells than in several other human cell types, including smooth muscle cells of a vulvar carcinoma (SK-LMS-1), choriocarcinoma cells (JEG-3), and cultured lung fibroblasts (WI26VA4). Table II shows these results and those using CAT assays, where several additional human cell types were examined. Again, the renin promoter was more active than the cotransfected hMT-2a promoter. These results indicate that the placental cells contain factors that act on the first 102 bp of the human renin promoter that direct its expression in a cell type-specific fashion. These factors have not yet been characterized, but the data indicate that this system should be useful for this purpose.

D.   cAMP CONTROL OF THE TRANSFECTED RENIN PROMOTER

Sine placental renin mRNA levels are regulated by cAMP (discussed above) and β-adrenergic stimuli that activate adenylate cyclase stimulate renal renin release (Baxter *et al.*, 1987; Churchill, 1985; Itoh *et al.*, 1985; Matsumura *et al.*, 1988; Park *et al.*, 1986; Pinet *et al.*, 1987), cAMP is a candidate to activate the renin promoter. As Fig. 3 and Table II show, 10^{-5} M forskolin increases the activity of the renin promoter in the choriodecidual cells by about 10-fold. This occurs with constructs containing either 584 or 102 bp of the promoter and the response-fold in both cases is the same (Fig. 3). These results indicate that DNA sequences within the first 102 bp of the promoter respond to cAMP. This response is specific for the renin promoter, since the cotransfected RSV promoter did not respond to forskolin. In other experiments (not shown), similar results were obtained using CAT assays. In these cases we cotransfected an hMT-2a promoter that did not respond to forskolin. These results are in general agreement with those using the RNase protection assays. Of note in this respect is that the hMT-2a promoter is known to respond to cAMP through the transcription factor AP-2 (Imagawa *et al.*, 1987). These results suggest that the placental cells do not have AP-2 activity under the conditions used. They also imply that AP-2 is not mediating the cAMP influence on the renin promoter. This point is worth emphasis, since there is a structure 5'-ACCCCAGGGT-3' in the renin gene 5'-flanking DNA (-88 to -79) that contains 8 of 10 bases with homology to the proposed AP-2 consensus structure 5'-TCCCCANGGG-3' (Imagawa *et al.*, 1987).

cAMP is also known to regulate transcription by modifying the transcription factor CREB (cyclic AMP response element-binding protein), which binds to a specific sequence (cyclic AMP response element or CRE) located in many cAMP-responsive genes (Hoeffler *et al.*, 1988; Yamamoto *et al.*, 1988) by induction (e.g., through AP-2 or CREB) of transcription factors that act on the target gene (Hjorth *et al.*, 1989), and possibly by effects on other proteins (Cherry *et al.*, 1989; Weih *et al.*, 1990). The first 102 bp of the renin promoter do not

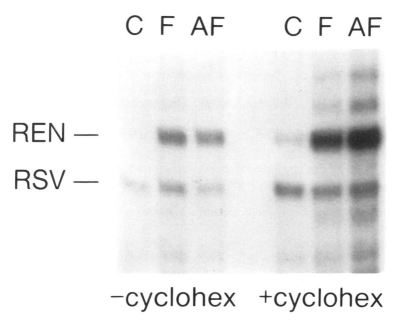

FIG. 4. Effect of cycloheximide on forskolin induction of renin promoter activity in choriodecidual cells. The cultures were transfected with the same gene described in Fig. 2 with 102 bp of human renin gene 5'-flanking DNA. Following the glycerol shock the cells were incubated without (− cyclohex) or with (+ cyclohex) $10^{-4} M$ cycloheximide and with or without $10^{-5} M$ forskolin (F) or $10^{-5} M$ forskolin plus 1 μM calcium ionophore A23187 (AF) for 8 hours and transcripts from the renin (REN) and a cotransfected RSV promoter (RSV) were measured. The conditions for transfection were as described in Duncan *et al.* (1990).

contain sequences that are homologous to the CRE, making it unlikely that this factor regulates the renin promoter. We examined the possibility that cAMP may regulate renin promoter activity by inducing the synthesis of a transcription factor that in turns acts on the renin promoter. As Fig. 4 shows, when we blocked protein synthesis with cycloheximide, the cAMP response could still be elicited. Thus, this possibility appears to be excluded. Experiments are in progress to determine how the cAMP responsiveness of this promoter is mediated, and it is possible that a novel mechanism is involved. Given that cAMP-activated kinases phosphorylate a large number of substrates and that cAMP regulates other cellular processes by acting on a number of different proteins it would be expected that it regulates transcription by affecting more than one transcription factor.

Burt *et al.* (1989) and Nakamura *et al.* (1989) have also reported cAMP responsiveness of the human renin 5'-flanking DNA in JEG-3 cells using the renin 5'-flanking DNA fragments linked to the herpes simplex virus thymidine kinase (HSVTK) promoter. The effects were modest, 1.1- to 3-fold, and the localization of the responsive sequences was not clear. These workers predicted that cAMP responsiveness resides between $+3$ and -149, but their best responsiveness was obtained with larger 5′-flanking DNA fragments. At present we cannot reconcile these differences. Although we were unable to observe cAMP responsiveness of similar constructs in JEG cells over that observed with the TK promoter alone, it is possible that subtle variations in the experimental conditions account for the differences. These results, if confirmed, could imply the existence of more than one cAMP-responsive element in the renin promoter.

E.   cAMP ACTIVATION OF RENIN PROMOTER ACTIVITY IN HUMAN LUNG FIBROBLASTS

As Fig. 3 and Table II show, the renin promoter shows negligible activity in human lung fibroblasts (WI26VA4 cells) in the absence of cAMP, but shows activity that is about one-third that of the RSV promoter in the presence of forskolin. Figure 3 suggests that the induction is greatest with the -584 human renin promoter, but an analysis of three experiments suggests that the induction is similar with both -584 and -100 constructs. The induction, up to 150-fold, is comparable to gene activation. An analogous situation exists in newborn rat kidney microvesicles, in which forskolin increases renin mRNA levels by increasing the number of renin-secreting cells without changing the amount of renin secreted by individual cells (Everett et al., 1990). This response may also be compared to the activation of renin gene expression in smooth muscle cells along the afferent renal arteriole with severe salt restriction and blockage of converting enzyme action (Gomez et al., 1988; Taugner et al., 1982). Ordinarily only the renal juxtaglomerular cells of the adult produce significant quantities of renin. There is no information about the mechanisms of the smooth muscle activation, but cAMP appears to be capable of eliciting a similar phenomenon.

These results also raise the issue of whether cAMP is involved in the high basal activity of the renin promoter in the choridecidual cells. To test this hypothesis, we cotransfected the renin promoter along with a vector that expresses a mutant protein kinase A regulatory subunit (Duncan et al., 1991) that ordinarily binds cAMP (Clegg et al., 1987). By forming heterodimers with native wild-type kinase regulatory subunits, this mutant subunit inactivates the native kinase. This transfection abolished the forskolin stimulation of renin promoter activity, but did not affect basal transcription (Fig. 5; Duncan et al., 1991). By contrast, transfection of a gene that expressed the wild-type regulatory subunit did not affect basal promoter activity or forskolin stimulation (Fig. 5; Duncan et al., 1991). Thus, elements other than cAMP-responsive ones probably direct basal renin promoter activity in placental cells. These results also imply that there may be differences in the way this promoter is regulated in placental cells vs human lung fibroblasts.

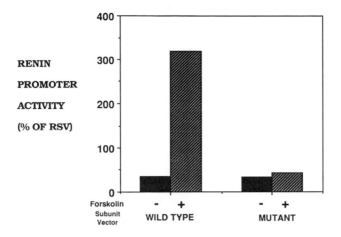

FIG. 5. Effect of cotransfection of a plasmid containing 102 bp of the renin promoter (Fig. 3) with a vector that expresses either a wild-type or mutant cAMP-responsive protein kinase regulatory subunit into choriodecidual cells on basal and forskolin (10^{-5} M, 24 hours)-stimulated transcription. Expression was assessed by RNase protection and activity expressed relative to RSV promoter expression, as described in Fig. 3.

F. A NEGATIVE CONTROL ELEMENT IN THE $-584/-102$ REGION

The data of Fig. 3 show that the -102 renin promoter is more active than the -584 promoter, implying the presence of a negative element(s) between -584 and -102. An examination of additional fragments reveals that a negative element resides in the $-420/-275$ region (Duncan et al., 1991) and that there may also be a negative element in the $-142/-102$ region (Duncan et al., 1991). This element(s) appears to function in both the choriodecidual cells and in human lung fibroblasts, but had less activity in several other cell types, implying that it has some tissue specificity in its action (Duncan et al., 1991).

G. DOES THE HUMAN RENIN PROMOTER RESPOND TO STEROID HORMONES?

Renal renin release is known to be negatively regulated by mineralocorticoid hormones (Baxter et al., 1987). This control is thought to be due dominantly to the mineralocorticoid-induced sodium chloride retention rather than to direct actions of the steroids (Baxter et al., 1987; Kotchen & Guthrie, 1988). However, there have been suggestions that mineralocorticoids, glucocorticoids, and sex steroids also regulate renin production through more direct mechanisms (Rubattu et al., 1989; Makrides et al., 1988; Welch et al., 1983). For example, plasma prorenin levels in pregnancy increase with the elevations of plasma estradiol, although changes in other hormones such as chorionic gonadotropin could also account for these influences (Itskovitz *et al.*, 1987). Dexamethasone blocks ketamine (a centrally acting anesthetic) induction of prorenin release (Rubattu *et al.*, 1989). Glucocorticoids enhance renin responsiveness to NaCl (Welch *et al.*, 1983). Additional reasons to think that the renin promoter may respond to steroids is that sequences with some weaker homology to the steroid response elements are present in the renin gene 5'-flanking DNA (Hardman *et al.*, 1984; Fukamizu *et al.*, 1986).

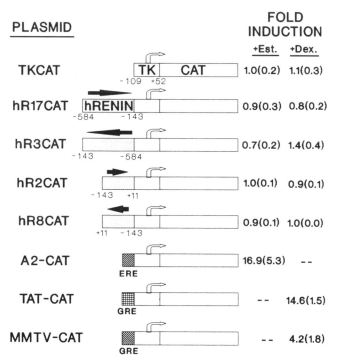

FIG. 6.   Lack of steroid regulatory elements in renin gene 5'-flanking DNA fragments. HeLa cells were transfected with vectors that contained the indicated human renin gene sequences, or sequences of an estrogen regulatory element from the chicken vitellogenin gene (A$_2$–CAT), a glucocorticoid regulatory element from the tyrosine aminotransferase (TAT–CAT), or a mouse mammary tumor virus (MMTV–CAT) gene. These sequences were placed upstream from a gene containing the indicated sequences of the herpes simplex virus thymidine kinase (TK) promoter and CAT gene-coding sequences. They were cotransfected with vectors that express the glucocorticoid or estrogen receptors and estradiol (Estr.) (10^{-7} M) or dexamethasone (Dex.) (10^{-7} M), respectively, and the cells assayed for CAT activity. Methods are given in Duncan *et al.* (1990), from which these results are taken.

Our data to date fail to provide evidence that the renin promoter is steroid responsive. First, several different classes of steroids failed to increase endogenous renin mRNA levels in the placental cells (Duncan *et al.*, 1990). Although these cells may not be steroid responsive, the placenta is known to be a rich source of steroid receptors (Ville, 1979). Second, when renin gene 5'-flanking DNA fragments were placed upstream of a HSVTK–CAT gene and cotransfected into HeLa cells along with vectors that expressed estrogen or glucocorticoid receptors, CAT activity was not increased by the steroids, whereas it was increased when known steroid response elements were placed upstream of this construct (Fig. 6). These results argue against strong estrogen or glucocorticoid response elements in the first 584 bp of the renin promoter. They also argue against the presence of mineralocorticoid, progesterone, or androgen response elements, as these are usually the same as glucocorticoid response elements (Beato, 1989). Of course, they do not exclude the existence of steroid response elements elsewhere on the renin gene. Nevertheless, it appears possible that direct actions of steroids on the promoter may not be significant for the renin gene.

IV. Posttranscriptional Control of Human Prorenin and Renin

Our studies of placental cells suggest that all of the changes in prorenin release are due to influences on renin mRNA, implying that control in this case is dominantly transcriptional in nature. As discussed above, these cells differ from renal cells, for example, in that they do not appear to store prorenin and do not process prorenin to renin. Renal renin release appears to be acutely regulatable by a variety of stimuli, including β-adrenergic activation (increased release), sodium chloride (suppression), and AII (suppression; Barrett *et al.*, 1989; Baxter *et al.*, 1987; Churchill, 1985; Dzau *et al.*, 1987b; Everett *et al.*, 1990; Iwao *et al.*, 1988; Makrides *et al.*, 1988; Nakamura *et al.*, 1985; Sealey and Laragh, 1990; Toffelmire *et al.*, 1989). The kidney also stores renin in dense secretory granules, and processes prorenin to renin (Taugner *et al.*, 1985a, 1986). The effects of sodium chloride, probably mediated by the chloride component, appear to be mediated through influences on the macula densa cells that release adenosine, which in turn regulates renin release from the juxtaglomerular cells (Baxter *et al.*, 1987). The ratio of prorenin to renin released by the kidney varies significantly (Toffelmire *et al.*, 1989) and, as mentioned earlier, in some circumstances (diabetes) the proportion of prorenin released is much greater (Wilson and Leutscher, 1990). The major difference between the kidney and the placenta in terms of prorenin and renin may be due to the fact that the kidney contains a regulated secretory pathway (Fritz *et al.*, 1987) that possesses the prorenin processing enzyme (PPE) and the capability to sequester renin in secretory granules and release it in response to stimuli.

To obtain a model system to better understand how prorenin is processed to renin and how renin release is regulated, we transfected a gene containing the hMT-2a promoter and preprorenin-coding sequences into cultured mouse pituitary AtT-20 cells that contain a regulated secretory pathway (Fritz et al., 1987). These cells express the endogenous proopiomelanocortin (POMC) gene, process POMC into corticotropin (ACTH) and other peptides, and release them in response to provocative stimuli (Moore et al., 1983). They also process proinsulin and other prohormones expressed from transfected genes (Moore et al., 1983).

We found that the transfected cells produce prorenin, process it to renin, and release renin in response to the secretogogue, 8-Br-cAMP (Fritz et al., 1987). The site of cleavage of prorenin to renin was the same as that used by the kidney, and the secretogogue preferentially stimulated renin vs prorenin release. Thus, enzymes that process prorenin to renin exist in extrarenal tissues, and these cells are suitable for examining certain aspects of renin release with respect to a regulated secretory pathway. Our findings with AtT-20 cells have since been confirmed by two other groups (Nakayama et al., 1989; Pratt et al., 1988).

We also transfected the preprorenin expression vector into Chinese hamster ovary (CHO) cells that have a constitutive but not a regulated secretory pathway (Fritz et al., 1987). These cells expressed prorenin and did not release detectable renin; prorenin release was not enhanced by stimulation with a secretogogue. Thus prorenin-to-renin conversion does not occur in either of the two constitutive pathway-only cell types, choriodecidual and CHO cells. This supports the notion that processing of prorenin to renin may occur only in cells that contain a regulated secretory pathway. The findings also raise the question of whether processing of prorenin to renin occurs in all tissues that contain a regulated secretory pathway. As noted earlier, after some period of time following nephrectomy in man, plasma prorenin levels are significant, whereas plasma renin levels remain undetectable (Derkx et al., 1978; Sealey and Rubattu, 1989; Hsueh and Baxter, 1991). Thus the extrarenal sources that release this prorenin into the circulation clearly do not process the prorenin to renin. Further, in most extrarenal tissues it appears that prorenin rather than renin is the major product, although in some cases (adrenal, pituitary) renin is present (Doi et al., 1984; Inagami et al., 1989; Shier et al., 1989). It appears that this prorenin could come from cells with either regulated or constitutive secretory pathways, since in two cases now prorenin has been expressed in cells that contain a regulated secretory pathway; the prorenin was found to be sorted to this pathway but not cleaved to renin. These results come from the published studies of Chidgey and Harrison (1990), using neuronal PC-12 cells, and from work with GH_4 cells (Hatsuzawa et al., 1990). In the latter case these workers found that prorenin was also not cleaved in GH_4 cells by furin expressed endogenously or from a transfected vector. This enzyme is homologous to the yeast Kex2 protein, which processes a number of proteins (Hatsuzawa et al., 1990).

V. Recombinant Human Prorenin and Renin

A. RECOMBINANT PRODUCTION OF THE PROTEINS

Due to its low abundance in plasma and even kidney it has been difficult to obtain large quantities of human prorenin and renin for structural and other analyses. We therefore used the human renin gene sequences for recombinant production of prorenin and renin. Our initial attempts to produce these proteins in bacteria and yeasts were unsuccessful, but we did program cultured CHO cells to overexpress prorenin (Fritz *et al.*, 1986). These cells produced 0.1–0.7 mg of prorenin/liter. Prorenin overexpression has also been achieved by other groups (Poorman *et al.*, 1986; Harrison *et al.*, 1989).

B. PROPERTIES OF RECOMBINANT PRORENIN AND RENIN

The prorenin produced by the CHO cells was either purified as such or converted to renin by limited trypsin treatment and then purified (Carilli *et al.*, 1988a; Shinagawa *et al.*, 1990). Recombinant prorenin appeared to be identical to purified renal prorenin in a number of properties, including the sequence of the amino terminus, activation by acid treatment or low temperature, binding to Cibacron Blue, and inhibition by pepstatin (Hsueh *et al.*, 1986; Carilli *et al.*, 1988a). These studies provided formal proof that the material previously called "inactive renal renin" is in fact prorenin (Hsueh *et al.*, 1986).

C. ACTIVITY OF INFUSED RECOMBINANT PRORENIN

It has been generally thought that the renin activity of prorenin in human plasma is negligible, since measurements of this activity at pH 7.4 and 37°C show it to be 1–8% that of renin (Hsueh *et al.*, 1986; Hsueh and Baxter, 1991; Glorioso *et al.*, 1989; Lenz and Sealey, 1990; Scalcy *et al.*, 1980). However, the fact that prorenin can be activated nonproteolytically leaves open the possibility that tissue or other factors could lead to renin activity of plasma prorenin. In addition, prorenin can be taken up by tissues and converted to renin (discussed earlier). To address this issue, we infused our recombinant-produced prorenin into rhesus monkeys (Table III; Lenz *et al.*, 1990) and measured renin activity, plasma aldosterone levels, and AII effects such as a rise in blood pressure. As shown, plasma prorenin levels were elevated by about threefold by the infusion. However, renin and aldosterone levels did not increase significantly and blood pressure did not increase. These results indicate that short-term modest elevations of plasma prorenin do not lead to physiologically significant consequences in terms of the parameters studied, assuming that the monkey can handle human prorenin in a manner similar to the human. These results are also consistent with

TABLE III

Response to Prorenin Infusion in Rhesus Monkeys[a]

	Response		
Parameter	Before	During[b]	After[c]
Plasma prorenin (ng/ml/hour)	70	220	180
Plasma renin (ng/ml/hour)	25	31	28
Plasma aldosterone (ng/ml)	22	24	21
Mean blood pressure (mmHg)	118	111	115
$\mu_{Na}V(\mu Eq/minute)$	3.1	2.9	4.8

[a]Data taken from Lenz *et al.* (1990); rhesus monkeys were infused for 40 minutes with 400 ng of prorenin/minute.

[b]Mean of 20- and 40-minute determinations.

[c]Mean of response 20 and 40 minutes following discontinuing the infusion.

the findings that diabetics with high levels of circulating prorenin but low levels of active renin have hyporeninemic hypoaldosteronism (Luetscher *et al.*, 1985) and that patients with tumors that release prorenin but not renin, leading to high circulating prorenin (but not renin) levels, do not develop hypertension (Hsueh and Baxter, 1991). These results, however, do not exclude an important role for plasma prorenin. Higher levels of prorenin or more time than allowed during the 40-minute infusion may be required for the prorenin to be taken up by the peripheral tissues to exert significant influences. As discussed earlier, clinically important actions of inhibitors of the RAS on blood pressure control occur outside the circulation, and these effects require a number of hours to days to be manifest and persist for similar periods of time following withdrawal of the agent.

D. PURIFICATION, IDENTIFICATION, AND PROPERTIES OF THE RENAL PRORENIN PROCESSING ENZYME

The availability of purified prorenin gave us a substrate for isolating a candidate renal prorenin processing enzyme (PPE). To date, this enzyme has not been unambiguously identified. This is important for an understanding of regulation of the RAS. In addition, blockage of the prorenin-to-renin conversion could be a conceivable target for pharmaceutical inhibition of the RAS. A number of enzymes have been reported to be capable of activating prorenin (Hsueh and Baxter, 1991; Inagami *et al.*, 1982; Imagami and Murakami, 1980; Lenz and Sealey, 1990; Sealey *et al.*, 1980); therefore, it is possible to isolate a nonspecific protease. Several criteria can be established as necessary properties of the authentic renal prorenin processing enzyme (Shinagawa *et al.*, 1990). The enzyme must (1) activate prorenin, (2) cleave the 43-amino acid prosegment of prorenin, (3) not degrade prorenin, (4) be present in the renal juxtaglomerular cells, and (5) have a pH optimum appropriate for activity in the secretory vesicles of the juxtaglomerular cells. Even these criteria are correlative; definitive proof will require experiments in which the putative enzyme is specifically inhibited or mutated with a resulting blockage of prorenin cleavage. Nevertheless, these criteria can be most helpful in excluding certain enzymes as renal PPEs and we would be encouraged to pursue characterization of candidate PPEs that fulfill these criteria.

We began by purifying the prevalent PPE activity from homogenates of human kidney using the purified recombinant prorenin as a substrate and assaying for generation of renin activity and changes in the migration of the prorenin on sodium dodecyl sulfate (SDS) gels (Shinagawa et al., 1990). Initially, the preparations both activated prorenin and degraded it, but as purification proceeded the latter activity decreased and nearly disappeared at pH 6.0. The steps used in purification of this enzyme are shown in Table IV. This resulted in a 550-fold purification (Wang et al., 1991).

The purified enzyme cleaves prorenin accurately as demonstrated by microsequencing of the amino terminus of the product (Shinagawa et al., 1990). It has a pH optimum in the 4–6 range; at pH 6.0 it docs not significantly degrade renin further, although some degradation did occur in the pH 4–5 range. Sensitivity to a panel of enzyme inhibitors showed it to be a thiol protease. Of the enzymes known to cleave prorenin to renin, this enzyme most closely resembles cathepsin B. The molecular weight of the purified enzyme is 30,000 (Wang et al., 1991), which is identical to that of cathepsin B. The purified enzyme was microsequenced (Wang et al., 1991) and the amino-terminal sequence is identical to that of cathepsin B, implying that it is cathepsin B.

This enzyme is relatively abundant in kidney and is likely to be present in renal cells outside the juxtaglomerular apparatus. However, previous studies of Taugner et al. (1985a), using immunohistochemistry, suggest that cathepsin B is

TABLE IV
Purification of Renal Prorenin Processing Enzyme[a]

Step	Specific activity (GU of renin/hour/mg)	Purification (−fold)
Kidney homogenate	37	1
$NH_4)_2SO_4$ precipitation	53	1.4
Affinity chromatography[b]	20,301	550

[a]Purification of the enzyme was performed as described by Wang et al. (1991), using purified prorenin as a substratate for assay.

[b]Affinity chromatcgraphy employed Sepharose Leu-Leu-arginyl.

present in renal juxtaglomerular cells. We also performed immunohistochemical analyses using sheep anti-human liver cathepsin B antibodies (Wang *et al.*, 1991). The predominant staining was in the proximal tubules of the kidney; however, the renal juxtaglomerular cells also showed staining (confirming previous results). Electron microscopy using the immunogold staining technique revealed that the cathepsin B was colocalized with renin in the dense secretory granules of the juxtaglomerular cells (Wang *et al.*, 1991). These results, taken together, suggest that cathepsin B is the renal PPE. Studies are in progress using GH_3 and PC-12 cells that do not process prorenin (discussed above) to determine whether expression of a transfected cathepsin B gene can confer processing in these cells.

VI.   Three-Dimensional Structures of Renin and Prorenin

A.   X-RAY CRYSTALLOGRAPHIC STRUCTURE OF HUMAN RENIN

The availability of milligram quantities of human renin provided the opportunity to crystallize it and determine its three-dimensional structure by X-ray diffraction. The major stimulus to do this was to use the structure to design inhibitors of human renin that have clinical utility (Corvol *et al.*, 1990; Wood *et al.*, 1987). As discussed in Section I, the enormous success of the converting enzyme inhibitors in treating hypertension, heart failure, and renal failure raises the question of whether blockade of the RAS at an alternative site might even be better than inhibiting converting enzyme. Such alternative blockade could be of renin activity, AII action, or prorenin-to-renin conversion. The currently available information does not allow us to know whether a renin inhibitor would be better than a blocker of other steps, although there are reasons to consider that this might be the case. For example, whereas renin has only one known substrate (Wood *et al.*, 1987), converting enzyme has many substrates, including bradykinin, for which it has a higher affinity than for AI (Ehlers and Riordan, 1989). Certain side effects of converting enzyme inhibitors, such as coughing, may be due to drug-induced elevations of bradykinin or substance P that occur as a consequence of blocking the breakdown of these peptides (Morice *et al.*, 1987; Ogihara *et al.*, 1991). Such effects might not be observed with a renin inhibitor. There are also studies that suggest that combined renin and converting enzyme inhibitor therapy might be more effective than the use of either of the inhibitors alone (Mento *et al.*, 1989) and there may be special situations where renin inhibition may be preferable (Corvol *et al.*, 1990). In addition, the use of structure for drug design is growing in popularity and renin appears to be an excellent model system for such design.

There were several other reasons to obtain a three-dimensional structure of human renin. Because of an interest in drug design, several groups had developed theoretical models of the renin three-dimensional structure based on the known structures of the renin-related enzymes, penicillopepsin and *Rhizopus* pepsin (Akahane *et al.*, 1985; Blundell *et al.*, 1983; Evin *et al.*, 1988; Sibanda *et al.*, 1984). The solution of the renin three-dimensional structure would help evaluate how well this could be done. This was particularly suited for our purposes, because James and Sielecki (1986) had solved the three-dimensional structure of porcine pepsinogen, which is more closely related to renin than these other aspartyl proteases. This structure could be used to prepare a model for human renin by substituting the amino acids of renin for those of pepsinogen and using molecular dynamics simulations. In addition, knowledge of the structure of renin would be helpful in understanding the role of surface structures in intracellular sorting of renin, and in developing a model for prorenin that would facilitate understanding the nonproteolytic and proteolytic activation of prorenin and the conformational changes that occur with the prorenin-to-renin transition.

The purified renin was prepared in milligram quantities and about 1500 crystallization trials were performed (Sielecki *et al.*, 1989). Microcrystals were first formed from which larger crystals suitable for X-ray diffraction were obtained. These crystals had tetragonal symmetry and space group I4.

Approximately 23,000 X-ray diffractions of the crystals were obtained and analyzed from which electron density maps were derived. Heavy metal-binding sites were not found that would have assisted in orienting the structure. To solve this problem, we developed a model for renin using the porcine pepsinogen structure as described above. When this model was compared with the electron density maps of renin, it was revealed that the model was accurate for portions of renin that were highly conserved between pepsinogen and renin, in the hydrophobic core of the protein that contains a β-pleated sheet. By contrast, there were marked discrepancies between the model and the electron density maps for surface loops and key structures near the active site (Fig. 7). These probably include the same structures that dictate the different substrate specificities of renin and pepsin. Clinically useful renin inhibitors should be specific for renin and not affect pepsin. For drug design, the most accurate structural information is needed for precisely those portions of the molecule where modeling gave the poorest results. Thus, theoretical molecular modeling techniques alone cannot be relied on for deciphering the detailed types of differences required in this case, although clearly these methods have utility in situations where high-resolution structural information is not required.

To refine the structure of human renin, and resolve the differences between the electron density maps and the model, we used molecular dynamics simulations and classical refinement techniques to fit the primary renin structure into the

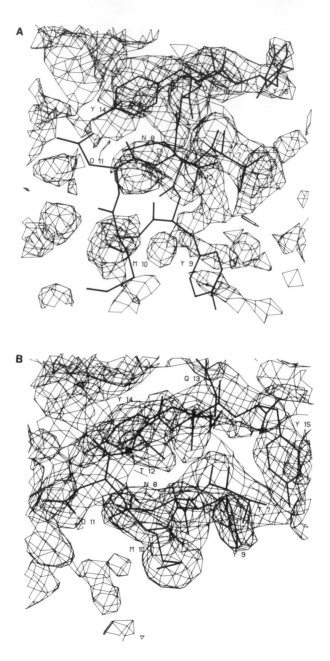

FIG. 7. Comparison of the X-ray diffraction data with the model for renin. (A) Electron density distribution and initial model (heavy lines) in the region of the NH$_2$-terminal strand. (B) Same region and orientation as in (A) but with the current electron density and the superimposed refined model. Comparison of (A) and (B) shows that model bias has little influence on these electron density maps. Residues are denoted by the one-letter code. Reprinted from Sielecki *et al.* (1989), by permission of the AAAS.

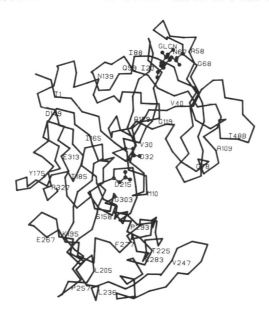

FIG. 8.   Three-dimensional structure of renin. Shown is the Cα-atom representation of the renin molecule (numbering derived from our alignment with pepsin). The side-chain atoms of Asp-32, Asp-215, and Asn-67 (Asn-75 in renin numbering) with an attached carbohydrate moiety (GLCN) are indicated with filled circles. There is no electron density associated with the first four of the six residues preceding Thr-1 (T1), the residue aligned to the NH₂ terminus of pepsin. The COOH terminus of the molecule is labeled R327. Reprinted from Sielecki *et al.* (1989), by permission of the AAAS.

electron density maps. The final structure was at a resolution of 2.5 Å for core structures, with lower resolution in the 8-Å range for certain loops. The electron density maps of some surface loops were too fuzzy to define the structure, suggesting that they display significant flexibility in the crystal lattice and perhaps also in solution.

The overall shape of the renin molecule (Fig. 8) is bilobar and can be likened to a croissant. The active site aspartate residues, positions 38 and 226, are located in the cleft of the croissant. The amino terminus is located on the "back" of the molecule and forms a β strand that is part of the six-stranded β-pleated sheet. This point is relevant for both the activation of renin and possibly the sorting of renin discussed in Section VII below. Renin and pepsin also differ in the charged residues; in renin the ratio of negative to positive charges is 1:1, whereas in pepsin it is 8.5:1. Whether this in part accounts for the increased pH optimum for renin relative to the other aspartyl proteases is not known.

B.   MODEL FOR HUMAN PRORENIN

Having the coordinates for human renin prompted us to develop a model for human prorenin (Baxter *et al.*, 1989; James *et al.*, 1991). This seemed plausible, since we needed only to add the 43 amino acids of the prorenin prosegment. We were also assisted by the structure of porcine pepsinogen (James and Sielecki, 1986) and an unpublished structure determined by Sielecki and James of porcine pepsin that provided detailed information about the placement of the prosegment of pepsinogen and the transition that occurs with prosegment cleavage and pepsin formation. Pepsinogen differs from prorenin in that it is autocatalytic. The prosegment of pepsinogen forms a β strand that is part of the β-pleated sheet mentioned above. This β strand fits into the same hydrophobic groove in the back of the molecule that is occupied by the amino terminus of pepsin in that molecule. Thus, when the pepsinogen prosegment is cleaved, the protein undergoes an enormous conformational change and the amino terminus of pepsin moves by about 40 Å to the location previously occupied by the prosegment of pepsinogen. By comparing the three-dimensional structures and the primary prorenin prosegment sequence, it appears that analogous changes occur when prorenin is cleaved to form renin.

Thus, with this information and the use of molecular replacement and molecular dynamics simulation methodologies, we prepared a model for human prorenin (Fig. 9). As said, the amino terminus extends out from the back of the

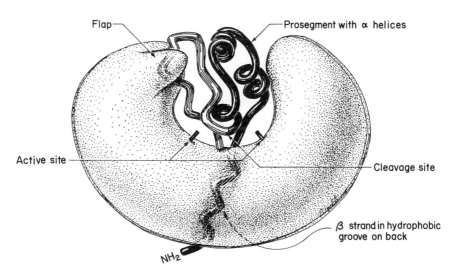

FIG. 9.   Model of prorenin. Reprinted from Hsueh and Baxter (1991), by permission of the American Heart Association, Inc.

molecule and is followed by the β strand that forms part of the β sheet. This is followed by a turn and then three α helices. The second amino acid (Lys-2) of the prosegment that forms part of the dibasic pair (Lys-2/Arg-1) at the prosegment cleavage site participates in ionic interactions with the aspartate groups of the active site; these interactions may be involved in the stabilization of the prosegment in inhibiting renin activity. This model makes several predictions that can be tested.

First, there must be conformational changes in the molecule in order for the prorenin processing enzyme to cleave prorenin at the diabasic residues, since the prosegment Lys-Arg $(-2/-1)$ at the cleavage site is buried in the active site. Second, several interactions may be involved in stabilizing the prosegment so it can block the catalytic activity of renin. These include the interactions of Lys-2 with the active site Asp residues of renin (at positions 38 and 226), hydrophobic effects due to fitting of the β strand into the hydrophobic groove, and several other ionic interactions, including bonds between the prosegment Arg-34, Arg-29, Arg-24, and Lys-20 residues with various amino acids in renin. Third, as mentioned above, the prosegment must undergo enormous conformational changes in order for prorenin to have renin activity. It is important to note that whereas this structure is useful for these predictions, more functional studies are required to demonstrate their importance. Some of these issues are addressed in the following section.

VII. Structures of Prorenin and Renin Involved in Activation, Processing, and Sorting

A. NONPROTEOLYTIC ACTIVATION OF PRORENIN

At 37°C and pH 7.4 prorenin has about 1–8% the renin activity of renin (Glorioso *et al.*, 1989; Higashimori *et al.*, 1989; Hsueh *et al.*, 1986; Hsueh and Baxter, 1991; Lenz and Sealey, 1990; Sealey *et al.*, 1980). This activity of prorenin can also be increased by several maneuvers, including lowering the pH, lowering the temperature, adding certain lipids, and probably others. These activations are reversible, implying that prorenin is in an equilibrium between active and inactive forms and that at pH 7.4 and 37°C the prosegment interactions with the body of renin favor the inactive form.

The model for prorenin described above depicts the molecule in the inactive configuration that is stabilized by the hydrophobic effects and electrostatic interactions already mentioned. We do not have a model for the structure of active prorenin, but in this state the prosegment amino acids in the vicinity of the active site must be moved away enough from the active site to allow the renin substrate, angiotensinogen, to fit into the active site. This substrate–enzyme interaction must be highly specific, given the unique substrate specificity of renin. In order for the prosegment amino acids to leave the active site, the β strand fitted into the groove on the back of the molecule (Fig. 9) would have to dissociate from the groove. We do not know whether the amino terminus of renin in active prorenin, with its attached prosegment amino acids, undergoes the same conformational change that occurs after the prosegment is cleaved (described above), but this seems likely given the critical interactions between the substrate and renin that are required for catalytic activity.

We have begun to ask what interactions favor the inactive form of prorenin as depicted by the model. To do this, we have mutated various amino acids of the prosegment (Chu *et al.*, 1991). These mutated prorenin genes, contained in a preprorenin expression vector described by Chu *et al.* (1990), were transfected into CHO cells and the renin activity of the secreted prorenin measured (Table V). As shown, mutation of any of the carboxyl-terminal amino acids of the prosegment, including Lys−2 (−2 to the amino terminus of renin) to Ala, did not increase the renin activity of prorenin. Thus, in spite of the possible electrostatic interactions between this residue and the Asp groups of the active site, these interactions are not essential for holding the prosegment in place. Mutation of +1 Leu (+1 to the amino terminus of renin) increased the prorenin activity to about 18% (Table V), suggesting potential importance of the hydrophobic side chain or some other feature of this amino acid. Mutation of amino acids +2 to +5 of the renin amino terminus individually did not have a significant effect. Mutation of −34 Arg, predicted to interact with an Asp residue in the body of renin, to Ala did result in an increase in the activity of the prorenin to about 14%, suggesting that this interaction helps to stabilize the prosegment. This Arg was also changed to Asp to create potential repulsive forces. This mutation resulted in 79% activation of the resulting prorenin, implying that the predicted interaction does occur, and that the interactions between the β strand and the body of renin are important for stabilizing the prosegment in the configuration that inactivates prorenin. This conclusion is also consistent with the data of Yamauchi *et al.* (1990), who found that mutation of the −34 Arg, −29 Arg, and −24 Arg residues to Glu partially activated prorenin. However, mutating the −12 Arg, −35 Lys, or −30 Lys resulted in no to minimal activation (Yamauchi *et al.*, 1990). Studies of Heinrickson *et al.* (1989) can also be interpreted in terms of the model. These workers found that deletions of as few as 10–14 amino acids of the amino terminus of prorenin resulted in active prorenin. These deletions extend just into the β strand and would disrupt the interactions with the hydrophobic groove. We also found that some preparations of CHO cell and placental-derived prorenin prepared by ammonium sulfate precipitation contained irreversibly activated prorenin (Shinagawa *et al.*, 1991); amino-terminal sequencing of the CHO

TABLE V
Renin and Prorenin Activity of Native and Mutated Prorenins
Expressed in CHO Cells Transfected with Various
Preprorenin Expression Vectors[a]

Preprorenin	Activity (ng AI/nl/hour)		Active (total)
	Active	Prorenin	
Native	75	1755	0.04
−5 Gln → Ala	73	1434	0.04
−4 Pro → Ala	162	3767	0.04
−3 Met → Ala	101	1679	0.05
−2 Lys → Ala	15	968	0.01
−1 Arg → Ala	79	2139	0.03
+1 Leu → Ala	346	1520	0.18
+2 Thr → Ala	130	1807	0.06
+3 Leu → Ala	137	1805	0.07
+4 Gly → Ala	85	1880	0.04
+5 Asn → Ala	62	2320	0.02
Del (×39 to −43)	39	1129	0.03
−34 Arg → Ala	209	1215	0.14
−34 Arg → Asp	520	136	0.79

[a]CHO cells were transfected with expression vectors containing the native preprorenin sequence or the indicted mutations (numbering with respect to the amino terminus of renin) and the secreted prorenin was assayed for active renin and prorenin (active renin following trypsin activation minus active renin) by an AI-generating assay. Methods and details are given in Chu *et al.* (1991).

cell material revealed that it had been cleaved to remove 32 amino acids of the prosegment (Shinagawa *et al.*, 1991). These results indicate that prorenin can be activated by cleavage other than at the site used by the kidney and AtT-20 cells, and that enzymes that do this are present in several cell types. Although there is no evidence that these enzymes have access to prorenin in the placenta or CHO cells, this remains an open possibility.

B. DETERMINANTS FOR PRORENIN PROCESSING IN AtT-20 CELLS

The finding that AtT-20 cells accurately sort prorenin to the regulated secretory pathway and process prorenin to renin using the same cleavage sites employed in the kidney provides a model for understanding the processing *in vivo*. We do not know whether AtT-20 cells use the same enzyme as the kidney, but this system serves as a useful model and it is important to learn about extrarenal as well as renal prorenin processing. Since the kidney and AtT-20 cells cleave prorenin dominantly at the one site, then either the enzyme is the same, or there are features of the prorenin structure that favor proteolytic cleavage at this site. As discussed in Section V,D above, prorenin can be activated by a number of enzymes, including trypsin, cathepsin B, and kallikrein. In most cases the precise site of cleavage has not been determined. However, we have recently shown that cathepsin B uses the same site as the kidney and appears not to degrade renin further under certain conditions (discussed in Section V,D). Trypsin preferentially cleaves prorenin at the correct site, but with time degrades renin further (Carilli et al., 1988a; Heinrickson et al., 1989). Thus, whereas the correct cleavage site appears to be in general more accessible to proteolysis, it is not the only site, and enzymes such as the PPE in the kidney and AtT-20 cells have a substrate specificity that is influenced by more than the dibasic pair Lys-Arg at position -2 and -1 of the prosegment, as there are six additional dibasic pairs in prorenin (Imai et al., 1983).

To address this specificity for the AtT-20 cells and ultimately the renal PPE, we have begun to mutate the amino acids around the cleavage site. Mutated preprorenin expression vectors were transfected into AtT-20 cells and the expressed prorenin was examined for processing to renin. Some preliminary results are shown in Table VI. Mutation of any of the first four amino-terminal amino acids of renin to Ala does not affect prorenin processing. The percentage of active renin ranged from 26 to 34% in the absence of forskolin and 47 to 62% following secretogogue stimulation compared to the value of 24% (unstimulated) and 49% (stimulated) for the native vector. Mutation of the $+5$ Asn to Ala appeared to enhance the efficiency of processing somewhat, with values of 42 and 80% before and after stimulation, respectively. By contrast, mutation of each of the five carboxyl-terminal amino acids of the prosegment decreased the proportion of prorenin converted to renin. This was expected for the dibasic residues of the cleavage site where the -2 Lys $\rightarrow$ Ala and -1 Arg $\rightarrow$ Ala mutations resulted in unstimulated values of 4 and 6%, respectively, and stimulated values of 6 and 18%, respectively. These levels of active renin were near the value of native prorenin (discussed earlier). However, conversion of -4 Pro of the prosegment to Ala decreased the active renin to 12% (from 24%) in basal conditions and to 10% (from 49%) under stimulated conditions. This represents an 80% decrease under stimulated conditions. The conversion of -3 Met and -5 Gln decreased the active renin to 28% and 30%, respectively, of the total as compared to 49% for the native, suggesting some influence of these residues on processing. These results clearly indicate that amino acids outside the normal cleavage site have a significant influence on the processing of prorenin by the AtT-20 cell enzyme. To see if the first four amino acids alone dictate processing we mutated -4 Pro $\rightarrow$ Ala, -15 Asp $\rightarrow$ Pro, and -13 Ala $\rightarrow$ Lys, creating the sequence Pro-Met-Lys-Arg-Leu in the $-15/-12$ region (Fig. 10). This mutated prorenin

TABLE VI

Effect of Mutations of the Amino Acids
Surrounding Prorenin Cleavage Site on
Processing of Prorenin in AtT-20 Cells[a]

Mutation	Active renin (%)	
	Unstimulated	Plus forskolin
Native	24	49
−5 Gln	18	30
−4 Pro	12	10
−3 Met	20	28
−2 Lys	4	6
−1 Arg	6	18
+1 Leu	33	56
+2 Thr	34	47
+3 Leu	26	62
+4 Gly	26	54
+5 Asn	42	80

[a] Vectors that express human preprorenin containing either native sequences or the indicated mutations (all to Ala) were transfected into cultured AtT-20 cells and the renin and prorenin released under basal conditions or in response to 10^{-5} M forskolin were measured by techniques described in Chu *et al.* (1990, 1991). Data taken from Chu *et al.* (1991). Amino acid numbering made with respect to the amino terminus of renin.

Native −43   Leu – Pro – Thr – Asp – Thr – Thr – Thr – Phe – Lys

Native −34   Arg – Ile – Phe – Leu – Lys – Arg – Met – Pro – Ser

Native −25   Ile – Arg – Glu – Ser – Leu – Lys – Glu – Arg – Gly

Native −16   Val – Asp – Met – Ala – Arg – Leu – Gly – Pro – Glu
Mutant   Pro – Met – Lys – Arg

Native −7   Trp – Ser – Gln – Pro – Met – Lys – Arg – RENIN
Mutant   Ala

FIG. 10. Amino acid sequence of the native prorenin prosegment with the dibasic residues shaded, and of a mutated (mutant) prorenin tested for prosegment cleavage in AtT-20 cells.

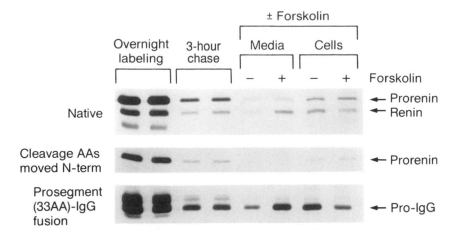

FIG. 11.   Effect of movement of the first four prorenin cleavage sequences 11 amino acids toward
the amino terminus and of inserting the first 33 amino acids of the prorenin prosegment onto an
immunoglobulin on processing and sorting. The prosegment amino acids were mutated as shown in
Fig. 10 [cleavage amino acids (AAs) moved toward NH_2 terminus] and the codons of the signal
peptide and first 33 amino acids of the prorenin prosegment were linked to immunoglobulin gene
sequences in expression vectors. Plasmids containing these and a native preprorenin expression
vector were transfected into AtT-20 cells and the expressed products assayed by pulse–chase studies
for unstimulated and forskolin (10^{-5} M)-stimulated release of the products and for processing of the
preproteins as described in Chu et al. (1990). Results taken from Chu et al. (1991).

was not converted to active renin by the AtT-20 cell enzyme more than the -4
Pro $\rightarrow$ Ala mutation (Fig. 11). These results indicate that this structure alone,
while necessary, is insufficient to confer site-specific processing.

These mutations should be especially useful for testing the renal prorenin
processing enzyme described in Section V,D, and we are currently in the process
of doing this. However, more information is needed to define better the substrate
specificity of the AtT-20 cell enzyme. In our model for prorenin the dibasic
cleavage site is buried in the cleft of prorenin containing the active site. It appears
that prorenin must undergo some conformational change for this site to be accessible to the processing enzyme. In fact, the amino acids around the cleavage
site that influence the processing could conceivably be as important for this
change as for recognition by the processing enzyme. In this context, we found
that the putative renal PPE does not cleave prorenin more efficiently when
prorenin has been previously activated by low temperature or low pH, implying
that the postulated conformational changes may not be rate limiting (Shinagawa
et al., 1990).

However, we have preliminary data that suggest that the conformation of the

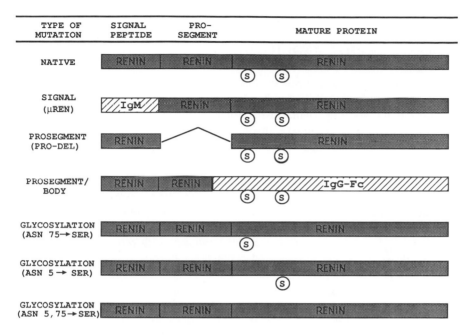

TYPE OF MUTATION	SIGNAL PEPTIDE	PRO-SEGMENT	MATURE PROTEIN
NATIVE	RENIN	RENIN	RENIN
SIGNAL (μREN)	IgM	RENIN	RENIN
PROSEGMENT (PRO-DEL)	RENIN		RENIN
PROSEGMENT/ BODY	RENIN	RENIN	IgG-Fc
GLYCOSYLATION (ASN 75→SER)	RENIN	RENIN	RENIN
GLYCOSYLATION (ASN 5→SER)	RENIN	RENIN	RENIN
GLYCOSYLATION (ASN 5,75→SER)	RENIN	RENIN	RENIN

FIG. 12.   Mutations in preprorenin used to examine the role of various structures in cellular sorting. S, Sugar; IgG, γ-globulin; IgM, immunoglobulin M.

prosegment in prorenin may contribute to the inability of cellular enzymes to cleave at the dibasic residues of the prosegment located upstream from the natural cleavage site (Chu *et al.*, 1991). When the prosegment amino acids −43 to −11 were linked to an immunoglobulin (Fig. 12) and the resulting expression vector transfected into AtT-20 cells, the fusion protein was sorted to the regulated secretory pathway and the prosegment was cleaved (Fig. 11). The precise site of cleavage has not been defined but must be at one of the alternative sites, since the protein encoded by this vector does not contain the natural cleavage site.

C.   STRUCTURES IN PRORENIN AND RENIN INVOLVED IN INTRACELLULAR SORTING AND TRAFFICKING

It is clear that structures on proteins sort them to the regulated pathway and that many proteins (immunoglobulin and angiotensinogen, for example) are not sorted to the regulated pathway in cells that contain such a pathway (Moore *et al.*, 1983). Sorting is dominant, since hybrid molecules containing both regulated and constitutive secretory protein sequences are sorted to the regulated secretory pathway (Moore and Kelly, 1986; Stoller and Shields, 1989). Further, evidence for a sorting receptor for growth hormone and prolactin has been presented (Chung *et al.*, 1989). However, the precise sequences involved in the sorting have not been defined for any protein. Thus, the defining of such sequences for renin is of general interest and important for the specific case of prorenin and renin. In addition to knowing which sequences are responsible for the sorting, it is also important to learn whether any structures on prorenin, renin, or other proteins quantitatively affect their regulated release. The renin–angiotensin system, like other systems (insulin, corticotropin, parathyroid hormone), functions best if it can be completely shut off in some circumstances and markedly stimulated in other situations. The systems in culture that have been studied contain both regulated and constitutive secretory pathways, with mostly unprocessed hormone being released under suppressed conditions, with some basal release of processed hormone. Does this imply that there is low-level secretion of the protein from the regulated pathway with processing occurring only in the granules of the regulated pathway, or is some renin formed in the constitutive pathway? Our recent studies with renal cells suggest that there is predominantly prorenin in the clear vesicles of the constitutive secretory pathway whereas there is a predominance of renin in the dense secretory granules thought to be components of the regulated pathway (Macauley *et al.*, 1991). This finding may imply that all renin release comes from the regulated pathway, which should be considered in interpreting any quantitative effects on release/processing of mutating various prorenin/renin structures.

To identify structures important for sorting that may affect the pathway quantitatively, we prepared a series of mutated prorenin/renin and hybrid protein expression vectors to determine the sequences on renin and prorenin that may be involved in such sorting. These vectors were transfected into AtT-20 cells and the prorenin and renin released were measured under basal conditions and in response to secretogogues. We also performed pulse–chase studies and analyzed the effect of the secretogogue on secreted and intracellular prorenin and renin.

1. Role of the Signal Peptide Sequence

Figure 12 shows the various mutated and hybrid molecules examined and Fig. 13 shows the measurements of secreted prorenin and renin. Replacement of the signal peptide "pre" sequence of prorenin by that of an immunoglobulin (μRen) did not affect prorenin-to-renin processing or secretogogue-stimulated renin release, implying that sorting sequences are not on this portion of the molecule.

2. Role of Glycosylation

Mutation of the two amino acids that are glycosylated resulted in significant changes in the pattern of prorenin and renin release. As shown in Fig. 13, substitution of either one or both of the glycosylated amino acid residues resulted

FOLD INDUCTION

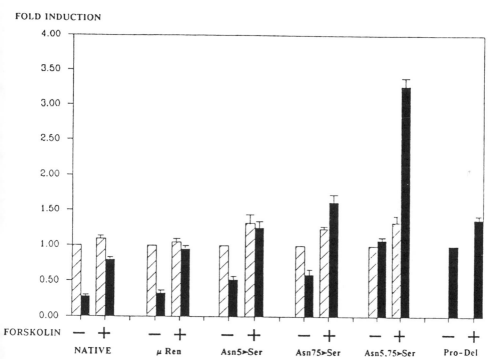

FIG. 13.   Effect of some of the mutations shown in Fig. 12 on basal and stimulated secretion of prorenin (striped bars) and renin (solid bars). Vectors expressing the various preprorenins were transfected into AtT-20 cells and the products were assayed for basal and 10^{-5} M forskolin-stimulated release of renin and prorenin. Reprinted from Chu et al. (1991). The results (mean of 6–12 independent transfections ± SEM) are normalized to the level of prorenin secreted in the absence of secretagogue except for the Pro-Del mutation, where they are normalized to the level of active renin released in the secretagogue-unstimulated state.

in an enhanced release of renin relative to prorenin under both basal and stimulated conditions. The effect of the double mutation was greater than with either of the single mutations. These results imply that the glycosylation in some way restricts the processing of prorenin to renin. This could provide an enhanced ability to suppress the RAS. This would occur at the expense of less renin release under stimulated conditions, but the capacity to activate the system is substantial, and the limiting factor may not be the overall ability of renin-producing tissues to produce renin, but suppressing the system in times of volume overload. This could be important for the overall control of renin release, since it may be advantageous to minimize the release of renin under unstimulated conditions where the RAS needs to be relatively inactive. Thus, the glycosylation of prorenin may ultimately be important for the overall control of the RAS. Paul et al.

TABLE VII
*Total Renin Plus Prorenin Released Expressed
from Various Vectors[a]*

Vector	Prorenin plus renin[b]	
	Unstimulated	Plus forskolin
Native	258.0 ± 13.3	375.1 ± 16.7
Asn-5 → Ser	152.5 ± 7.1	253.5 ± 11.2
Asn-75 → Ser	166.0 ± 20.3	291.0 ± 34.6
Asn-5, 75 → Ser	163.3 ± 31.7	355.8 ± 64.6
Pro-Del	28.6 ± 6.0	36.5 ± 23.3

[a]The experiment was performed as described in Fig. 14. Data taken from Chu *et al.* (1990).
[b]Nanograms AI per milliliter per hour, ± SEM.

(1988) found that blockade of glycosylation led to decreased intracellular transit time for prorenin in transfected AtT-20 cells. Our data do not address this issue and they did not address prorenin activation.

Evidence has been presented that the sugar residues of prorenin may acquire 6-phosphomannose moieties that are known to sort proteins to lysosomes (Faust *et al.*, 1987; Nakayama *et al.*, 1990). In frog oocytes and mouse L cells, prorenin appears to go to lysosomes, where it is destroyed (Faust *et al.*, 1987; Nakayama *et al.*, 1990). There is no evidence that a substantial level of such sugar-mediated enhanced destruction occurs in the AtT-20 cells. As Table VII shows, the total prorenin plus renin produced in AtT-20 cells is similar to the native and double glycosylation site-mutated prorenin and is only slightly decreased (33 and 22%, respectively) for the single mutations. As we have discussed, the dense secretory granules appear to be modified lysosomes and might have 6-phosphomannose receptors (Shinagawa *et al.*, 1991). However, our data do not support a role for these receptors in sorting to the regulated pathway, since such sorting if anything is more, not less, efficient with the deglycosylated prorenin.

3. Role of the Body of Renin

Figure 13 also shows the results when the prorenin prosegment is deleted. In this case there is a constitutive release of active renin and no detectable release of prorenin. These results suggest that renin is capable of being released through the constitutive pathway. Total renin release with the prosegment-deleted renin increases in response to the secretogogue by about 1.3-fold (Fig. 13 and Table I), suggesting that at least some of this renin has entered the regulated secretory pathway. The stimulation(1.3-fold) of renin in this case would be expected to be less than that of native renin, since the primary product of the constitutive secretory pathway with the native vector is prorenin, whereas it is renin with the

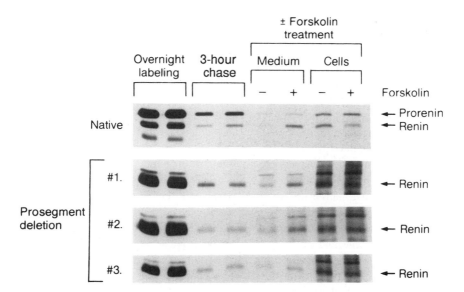

FIG. 14.   Effect of forskolin on secretion of renin expressed from a vector lacking prorenin prosegment amino acids. Vectors expressing the native and prosegment-deleted preprorenins were transfected into AtT-20 cells and pulse–chase experiments performed as described in Chu *et al.* (1990). The results from three separate experiments with the prosegment-deleted vectors are shown.

prosegment-deleted renin. A more meaningful measurement of the efficiency of sorting to the regulated pathway with the prosegment-deleted prorenin is to compare the effect of the secretogogue on total prorenin plus renin release with the native expression vector vs renin release with the prosegment-deleted vector. The value of 1.3-fold with the prosegment-deleted vector compares favorably with the value of 1.5-fold with the native vector (Table VII). This result further suggests that most of the prorenin-to-renin conversion occurs in the regulated pathway and that this conversion per se is not required for sorting of the protein to the regulated pathway. These results also imply that the body of the renin contains sequences that sort it to the regulated pathway. They do not demonstrate that the efficiency of sorting with the prosegment-deleted prorenin is as high as with native prorenin.

That the body of renin contains sequences that sort it to the regulated pathway is also suggested by pulse–chase studies. As shown in Fig. 14, following the pulse and chase, and then secretogogue stimulation, there is a clear increase in the release of renin into the medium and a corresponding decrease in intracellular renin. Shown for comparison are results with the native preprorenin expression vector. In this case both prorenin and renin are present. The secretogogue increased the release of renin and decreased the intracellular renin, but did not affect the cellular or secreted quantities of prorenin. Inspection of the gels suggests that the fold stimulation of renin release is greater with the native protein. However, this conclusion cannot be made with certainty, because the background of constitutive pathway renin complicates interpretation of the results.

The total renin produced with the prosegment-deleted vector was around 10% that of total renin plus prorenin expressed from the native vector (Table VII). Thus, the prosegment-deleted renin is less efficiently expressed than the native proteins. Nagahama *et al.* (1989) also found that the production of prosegment-deleted renin in CHO cells was only 5% that of prorenin. These workers further found that renin mRNA levels from the prosegment-deleted vector were about half those from a native preprorenin expression vector, implying that the decreased renin production is not due to decreased mRNA levels. Instead, it appears more likely that either the translation of the mRNA is impaired or else the product is unstable and degraded by the cell. This could be due to defective folding of the renin and raises the issue of whether the prorenin prosegment plays a role in the folding of the molecule. It is also possible that processed renin is less stable in the vesicles in which it resides prior to its entry into the dense secretory vesicles of the regulated secretory pathway. These possibilities are currently being investigated.

4. Role of the Prosegment

We have begun to examine a series of hybrid vectors containing preprorenin gene sequences linked to those of constitutively released proteins such as immunoglobulin. Figure 12 shows the results of one such experiment, in which the vector contains the sequence of the preprorenin signal peptide and the first 33 amino acids of the prorenin prosegment. Surprisingly, pulse–chase studies demonstrated that this protein was cleaved and that there was an increased release of the product with a decrease in its cellular content in response to the secretogogue. We have not yet characterized this product, but the results suggest that the prosegment amino acids do contain a sorting sequence. This result needs to be followed up by using several other constitutively released proteins and with different prosegment constructs, but if confirmed, the results would be the first demonstration of a small peptide fragment capable of directing sorting.

The findings that there are sorting sequences on both the prosegment and on the body of renin would initially suggest that there are two different structures, each of which could direct sorting. This finding may also explain why other investigators looking at different proteins failed to find deletions that block sorting. Perhaps proteins evolved in this manner to make sorting more efficient, and there is some precedent of this with other proteins for other types of sorting. For example, at least two different sorting sequences were found on the glucocorticoid receptor (Picard and Yamamoto, 1987).

A second possible interpretation of these results is suggested by an examination of the prorenin model and the renin structure. As discussed earlier, the distal amino acids of the prorenin prosegment occupy the same position on the body of the renin fitting into a hydrophobic groove that is occupied by the amino terminus of mature renin. Thus, it is conceivable that in prorenin the prosegment amino acids direct sorting, whereas with renin, the amino-terminal amino acids in the same position of the prorenin prosegment replace the function ordinarily carried out by the prosegment. This possibility is currently being tested.

VIII. Summary

This article describes investigations of several aspects of the molecular biology of the human renin gene and the three-dimensional structure of renin and its precursor, prorenin. Because of the importance of the RAS in hypertension, heart failure, renal failure, and possibly other disorders such as atherosclerosis, it is critical to understand the detailed control of this system. This control involves regulation at the transcriptional level, folding of prorenin, sorting of prorenin to a regulated pathway where it is proteolytically cleaved to renin and released in response to secretogogues, constitutive release of uncleaved prorenin, and nonproteolytic activation of prorenin. Currently there is great interest not only in the control of renin in the kidney, the sole source of circulating renin, but also at extrarenal sites where RAS activity may regulate cardiovascular functions.

The renin gene was found to be expressed significantly in the renal juxtaglomerular cells and several other cell types. Most tissue culture cells did not express the gene; exceptions were cultured SK-LMS-1 cells and cAMP-stimulated human lung fibroblasts. Cultured human uterine–placental cells expressed the human renin gene at levels higher than in other cell types assessed. Renin mRNA had the same start site in the placental cells as the kidney and was regulated by calcium ionophores and cAMP. Thus, these cells provide primary nontransformed human cells to study the homologous human promoter.

Transfected renin promoters showed cell type-specific expression and cAMP responsiveness in these cells in constructs containing as few as 102 bp of 5'-flanking DNA. DNA upstream from this appears to contain an inhibitory element(s) that may have some tissue specificity in its distribution. The cAMP response is not due to cAMP induction of a transcription factor that secondarily affects the renin promoter. A novel element may be involved, since the promoter does not contain a CRE element that mediates many cAMP responses, and the cells do not appear to respond to another known cAMP-responsive transcription factor, AP-2. Studies with transfected vectors expressing a mutant cAMP-responsive protein kinase A regulatory subunit suggest that cAMP is not responsible for basal renin promoter activity in the placental cells. By contrast, cAMP induces in essence gene activation in WI26VA4 transformed human lung fibroblasts in which renin mRNA levels increase by up to 150-fold in response to forskolin. Thus, cAMP may activate renin gene expression under certain circumstances and tissue-specific renin gene expression may be directed by more than one mechanism.

A model system in which a preprorenin expression vector is expressed in cultured AtT-20 cells demonstrates features of regulated renin release seen in the kidney and possibly extrarenal tissues. These cells release prorenin constitutively and sort it to the regulated secretory pathway where it is processed to renin and released in response to regulatory stimuli. The prorenin processing enzyme in these cells (and in the kidney) selectively cleaves prorenin after one of seven dibasic residue pairs. Studies with mutated expression vectors reveal that amino acids upstream from the cleavage site are critical for recognition by the prorenin processing enzyme, particularly a proline at position −4 of the prosegment.

A putative prorenin processing enzyme was purified from human kidney. This enzyme accurately cleaved prorenin to renin, did not further degrade renin, and had an acidic pH optimum consistent with its possible function in granules of the regulated secretory pathway. Microsequencing of the purified material and inhibitor studies reveal that the enzyme is the thiol protease, cathepsin B. These findings, plus those that indicate that cathepsin B is localized in the dense secretory granules of renal juxtaglomerular cells, suggest that cathepsin B is the renal prorenin processing enzyme.

Recombinant prorenin was produced in CHO cells and this protein and renin derived from it were purified. The prorenin was identical to renal inactive renin, confirming its identity as prorenin. When prorenin was infused into monkeys for 40 minutes such that plasma levels were raised by about threefold, prorenin did not elicit renin-like effects, implying that if circulation-derived prorenin is active, it must be present at higher concentrations and/or for longer times.

Crystals were prepared from the purified renin and used for X-ray diffraction analyses to obtain electron density maps of renin. These were used to solve the renin three-dimensional structure with the aid of a model for human renin based on the three-dimensional structure of porcine pepsinogen and the primary structure of renin. This model agreed with the determined structure in the core of renin, but not in certain surface loops that probably are critical for the differences in substrate specificity between renin and pepsin. Thus, these modeling techniques are probably not refined enough to predict detailed three-dimensional structures in certain critical portions of the molecule, a point that is relevant for the use of structure for drug design. The structures of pepsin, pepsinogen, and renin were nonetheless used to obtain a crude model for the prorenin three-dimensional structure that could be experimentally tested. This model predicts interactions between the prorenin prosegment and renin in regions around the active site of the molecule and through a β strand near the prosegment amino terminus that fits into a hydrophobic groove on the back of the molecule.

Mutated preprorenin expression vectors were employed to define the role of various prosegment amino acids in inactivating prorenin. Mutation of individual amino acids around the active site did not activate prorenin, except for the +1 Leu → Ala mutation that increased prorenin activity by three- to fourfold. Mutation of −34 Arg → Ala or Asp increased prorenin activity from 4 to 17 and 39%, respectively, implying a repulsive effect between the −34 Asp and a positive group(s) in the body of renin. These results and others imply that the amino terminal amino acids of the prosegment that are part of the β strand are critical for the stabilization of inactive prorenin.

Mutated preprorenin expression vectors transfected into the AtT-20 cells were also used to examine the structures on prorenin and renin that effect their intracellular fates. The results show that both the body of renin and the prorenin prosegment, but not the signal peptide sequence, appear to contain sorting sequences. The prorenin sorting sequence is contained in the first 33 amino acids of the prosegment amino terminus; this is the first demonstration of a sorting sequence localized to a small peptide fragment. The prorenin prosegment may also play a role in folding and/or stability of prorenin and renin, because prosegment-deleted renin was produced only about 10% as efficiently as native prorenin. Removal of the glycosylation residues results in an increased efficiency of delivering renin through the regulated pathway. We speculate that the sugar residues, by impairing sorting, may provide a mechanism to help suppress renin release under conditions that inactivate the system.

ACKNOWLEDGMENT

This work was supported by NIH rants 1 RO1 HL35706, 1 P50 HL44404, and R01 AM30254, and by a Grant-in-Aid from the American Heart Association, California Chapter.

REFERENCES

Acker, G. M., Galen, F. X., Devaux, C., Foote, S., Papernik, E., Pesty, A., Ménard, J., and Corvol, P. (1982). *J. Clin. Endocrinol. Metab.* **55,** 902–909.

Admiraal, P. J. J., Derkx, F. H. M., Jan Danser, A. H., Pieterman, H., and Schalekamp, M. A. D. H. (1990). *Hypertension* **15,** 44–55.

Akahane, K., Umeyama, H., Nakagawa, S., Moriguchi, I., Hirose, S., Iizuka, K., and Murakami, K. (1985). *Hypertension* **7,** 3–12.

Barrett, G. L., Morgan, T. O., Smith, M., Alcorn, D., and Aldred, P. (1989). *Hypertension* **14,** 385–395.

Baxter, J. D., Perloff, D., Hsueh, W., and Biglieri, E. G. (1987). *In* "Endocrinology and Metabolism" (P. Felig, J. D. Baxter, A. E. Broadus, and L. A. Frohman, eds.), 2nd ed., pp. 693–798. McGraw-Hill, New York.

Baxter, J. D., James, M. N. G., Chu, W. N., Duncan, K., Haidar, M. A., Carilli, C. T., and Reudelhuber, T. L. (1989). *Yale J. Biol. Med.* **62,** 493–501.

Beato, M. (1989). *Cell (Cambridge, Mass.)* **56,** 335–344.

Blaine, E. H., Nelson, B. J., Seymour, A. A., Schorn, T. W., Sweet, C. S., Slater, E. E., Nussberger, J., and Boger, J. (1985). *Hypertension* **7**, Suppl. I, I-66–I-71.

Blundell, T., Sibanda, B. L., and Pearl, L. (1983). *Nature (London)* **304**, 273–275.

Boger, R. S., Glassman, H. N., Cavanaugh, J. H., Schmitz, P. J., Lamm, J., Moyse, D., Cohen, A., Kleinert, H. D., and Luther, R. R. (1990). *Hypertension* **15**, 835–839.

Brunner, H. R., Waeber, B., and Nussberger, J. (1989). *Am. J. Med.* **87**, Suppl. 6B, 15S–18S.

Burt, D. W., Nakamura, N., Kelley, P., and Dzau, V. J. (1989). *J. Biol. Chem.* **264**, 7357–7362.

Campbell, D. J. (1989). *Am. J. Hypertens.* **2**, 266–275.

Carilli, C. T., Vigne, J.-L., Wallace, L. C., Smith, L. M., Wong, M. A., Lewicki, J. A., and Baxter, J. D. (1988a). *Hypertension* **11**, 713–716.

Carilli, C. T., Wallace, L. C., Smith, L. M., Wong, M. A., and Lewicki, J. A. (1988b). *J. Chromatog.* **444**, 203–208.

Cherry, J. R., Johnson, T. R., Dollard, C., Shuster, J. R., and Denis, C. L. (1989). *Cell (Cambridge, Mass.)* **56**, 409–419.

Chidgey, M. A. J., and Harrison, T. M. (1990). *Eur. J. Biochem.* **190**, 139–144.

Chobanian, A. V., Haudenschild, C. C., Nickerson, C., and Drago, R. (1990). *Hypertension* **15**, 327–331.

Chu, W. N., Baxter, J. D., and Reudelhuber, T. L. (1990). *Mol. Endocrinol.* **4**, 1905–1913.

Chu, W. N., Baxter, J. D., and Reudelhuber, T. L. (1991). Submitted for publication.

Chung, K.-N., Walter, P., Aponte, G. W., and Moore, H.-P. H. (1989). *Science* **243**, 192–197.

Churchill, P. C. (1985). *Am. J. Physiol.* **249**, F175–F184.

Clegg, C. H., Correll, L. A., Cadd, G. G., and McKnight, G. S. (1987). *J. Biol. Chem.* **262**, 13111–13119.

Corvol, P., Chauveau, D., Jeunemaitre, X., and Ménard, J. (1990). *Hypertension* **16**, 1–11.

Derkx, F. H. M., Wenting, G. J., Man In'T Veld, A. J., Verhoeven, R. P., and Schalekamp, M. A. D. H. (1978). *Clin. Sci. Mol. Med.* **54**, 529–538.

Deschepper, C. F., Mellon, S. H., Cumin, F., Baxter, J. D., and Ganong, W. F. (1986). *Proc. Natl. Acad. Sci. U.S.A.* **83**, 7552–7556.

Doi, Y., Atarashi, K., Franco-Saenz, R., and Mulrow, P. J. (1984). *Hypertension* **6**, Suppl. I, I-124–I-129.

Duncan, K. G., Haidar, M. A., Baxter, J. D., and Reudelhuber, T. L. (1990). *Proc. Natl. Acad. Sci. U.S.A.* **87**, 7588–7592.

Duncan, K. G., Haidar, M. A., Pearce, D., Reudelhuber, T. L., and Baxter, J. D. (1991). In preparation.

Dzau, V. J. (1986). *Hypertension* **8**, 553–559.

Dzau, V. J., Brody, T., Ellison, K. E., Pratt, R. E., and Ingelfinger, J. R. (1987a). *Hypertension* **9**, Suppl. III, III-36–III-41.

Dzau, V. J., Ellison, K. E., Brody, T., Ingelfinger, J., and Pratt, R. E. (1987b). *Endorinology (Baltimore)* **120**, 2334–2338.

Ehlers, M. R. W., and Riordan, J. F. (1989). *Biochemistry* **28**, 5311–5318.

Ekker, M., Tronik, D., and Rougeon, F. (1989). *Proc. Natl. Acad. Sci. U.S.A.* **86**, 5155–5158.

Erdos, E. G. (1990). *Hypertension* **16**, 363–370.

Everett, A. D., Carey, R. M., Chevalier, R. L., Peach, M. J., Gomez, R. A., and Geary, K. M. (1990). *J. Clin. Invest.* **86**, 169–175.

Evin, G., Galen, F.-X., Carlson, W. D., Handschumacher, M., Novotný, J., Bouhnik, J., Ménard, J., Corvol, P., and Haber, E. (1988). *Biochemistry* **27**, 156–164.

Faust, P. L., Chirgwin, J. M., and Kornfeld, S. (1987). *J. Cell Biol.* **105**, 1947–1955.

Feig, P. U., and Rutan, G. H. (1989). *Ann. Intern. Med.* **111**, 451–453.

Ferrario, C. M., Barnes, K. L., Block, C. H., Brosnihan, K. B., Diz, D. I., Khosla, M. C., and Santos, R. A. S. (1990). *Hypertension* **15**, Suppl. 1, I-13–I-19.

Fisch, T. M., Prywes, R., and Roeder, R. G. (1987). *Mol. Cell. Biol.* **7**, 3490–3502.

Fritz, L. C., Arfsten, A. E., Dzau, V. J., Atlas, S. A., Baxter, J. D., Fiddes, J. C., Shine, J., Cofer, C. L., Kushner, P., and Ponte, P. A. (1986). *Proc. Natl. Acad. Sci. U.S.A.* **83,** 4114–4118.

Fritz, L. C., Haider, M. A., Arfsten, A. E., Schilling, J. W., Carilli, C., Shine, J., Baxter, J. D., and Reudelhuber, T. L. (1987). *J. Biol. Chem.* **262,** 12409–12412.

Frohlich, E. D., Iwata, T., and Sasaki, O. (1989). *Am. J. Med.* **87,** Suppl. 6B, 19S–23S.

Fukamizu, A., Nishi, K., Nishimatsu, S., Miyazaki, H., Hirose, S., and Murakami, K. (1986). *Gene* **49,** 139–145.

Fukamizu, A., Nishi, K., Cho, T., Saitoh, M., Nakayama, K., Ohkubo, H., Nakanishi, S., and Murakami, K. (1988). *J. Mol. Biol.* **201,** 443–450.

Gavras, H. (1988). *N. Engl. J. Med.* **319,** 1541–1543.

Geistefer, A. A. T., Peach, M. J., and Owens, G. (1988). *Circ. Res.* **62,** 749–756.

Glorioso, N., Troffa, G., Tonolo, G., Manunta, P., Mandeddu, P., Melis, M. G., Pazzola, A., Pala, F., Maioli, M., Mameli, P., Salardi, S., and Bianchi, G. (1989). *Am. J. Hypertens.* **2,** 920–923.

Gomez, R. A., Lynch, K. R., Chevalier, R. L., Everett, A. D., Johns, D. W., Wilfong, N., Peach, M. J., and Carey, R. M. (1988). *Am. J. Physiol.* **254,** F900–F906.

Hano, T., Shiotani, M., Baba, A., Ura, M., Nakamura, Y., Tomobuchi, Y., Nishio, I., and Masuyama, Y. (1990). *Am. J. Hypertens.* **3,** 206S–209S.

Hardman, J. A., Hort, Y. J., Catanzaro, D. F., Tellam, J. T., Baxter, J. D., Morris, B. J., and Shine, J. (1984). *DNA* **3,** 457–468.

Harrison, T. M., Chidgey, M. A. J., Brammar, W. J., and Adams, G. J. (1989). *Proteins: Struct., Func., Genet.* **5,** 259–265.

Hatsuzawa, K., Hosaka, M., Tsutomo, N., Masahiro, N., Shoda, A., Murakami, K., and Nakayama, K. (1990). *J. Biol. Chem.* **265,** 22075–22078.

Heinrickson, R. L., Hui, J., Zarcher-Neily, H., and Poorman, R. A. (1989). *Am. J. Hypertens.* **2,** 367–390.

Higashimori, K., Mizuno, K., Shigeo, N., Boehm, F. H., Marcotte, P. A., Egan, D. A., Holleman, W. H., Heusser, C., Poisner, A. M., and Inagami, T. (1989). *J. Biol. Chem.* **246,** 14662–14667.

Hirata, Y., Tomita, M., Fujita, T., and Ikeda, M. (1986). *Hypertension* **8,** 883–889.

Hjorth, A. L., Khanna, N. C., and Firtel, R. A. (1989). *Genes Dev.* **3,** 747–759.

Hobart, P. M., Fugliano, M., O'Conner, B. A., Shaefer, I. M., and Chingwin, J. M. (1984). *Proc. Natl. Acad. Sci. U.S.A.* **81,** 5020–5030.

Hoeffler, J. P., Meyer, T. E., Yun, Y., Jameson, J. L., and Habener, J. F. (1988). *Science,* **242,** 1430–1433.

Hollenberg, N. K. (1985). *Am. J. Med.* **79,** Suppl. 3C, 1–2.

Holm, I., Ollo, R., Panthier, J. J., and Rougeon, F. (1984). *EMBO J.* **3,** 557–562.

Hsueh, W. A., and Baxter, J. D. (1991). Prorenin. *Hypertension* **17,** 469–479.

Hsueh, W. A., Do, Y. S., Shinagawa, T., Tam, H., Ponte, P. A., Baxter, J. D., Shine, J., and Fritz, L. (1986). *Hypertension* **8,** Suppl. II, II-78–II-83.

Imagawa, M., Chiu, R., and Karin, M. (1987). *Cell (Cambridge, Mass.)* **51,** 251–260.

Imai, T., Miyazaki, H., Hirose, S., Hori, H., Hayashi, T., Kageyama, R., Ohkubo, H., Nakanishi, S., and Murikami, K. (1983). *Proc. Natl. Acad. Sci. U.S.A.* **80,** 7405–7409.

Inagami, T., and Murakami, K. (1980). *Biomed. Res.* **1,** 456–475.

Inagami, T., Okamoto, H., Ohtsuki, K., Shimamoto, K., Chao, J., and Margolius, H. S. (1982). *J. Clin. Endocrinol. Metab.* **55,** 619–627.

Inagami, T., Mizuno, K., Naruse, M., Nakamaru, M., Naruse, K., Hoffman, L. H., and McKenzie, J. C. (1989). *Am. J. Hypertens.* **2,** 311–319.

Itoh, H., Nakao, K., Yamada, T., Morii, N., Shiono, S., Sugawara, A., Saito, Y., Mukoyama, M., Arai, H., and Imura, H. (1988). *Hypertension* **11,** Suppl. I, I-57–I-61.

Itoh, S., Carretero, O. A., and Murray, R. D. (1985). *Kidney Int.* **27**, 762–767.

Itskovitz, J., Sealey, J. E., Glorioso, N., and Rosenwaks, Z. (1987). *Proc. Natl. Acad. Sci. U.S.A.* **84**, 7285–7289

Iwao, H., Fukui, K., Kim, S., Nakayama, K., Ohkubo, H., Nakanishi, S., and Abe, Y. (1988). *Am. J. Physiol.* **18**, E129–E136.

James, M. N. G., and Sielecki, A. R. (1986). *Nature (London)* **319**, 33–38.

James, M. N. G., Russel, R., Sielecki, A. R., Chu, W. N., Baxter, J. D., and Reudelhuber, T. L. (1991). In preparation.

Jin, M., Wilhelm, M., Lang, R. E., Unger, T., Lindpaintner, K., and Ganten, D. (1988). *Am. J. Med.* **84**, Suppl. 3A, 28–37.

Jones, C. A., Sigmund, C. D., McGowan, R. A., Kane-Haas, C. M., and Gross, K. W. (1990). *Mol. Endocrinol.* **4**, 375–383.

Kifor, I., and Dzau, V. J. (1987). *Circ. Res.* **60**, 422–428.

Kim, S., Iwao, H., Nakamura, N., Ikemoto, F., Yamamoto, K., Mizuhira, V., and Yokofujita, J. (1987). *Am. J. Physiol.* **253**, E621–E628.

Kleinert, H. D., Martin, D., Chekal, M. A., Kadam, J., Luly, J. R., Plattner, J. J., Perun, T. J., and Luther, R. R. (1987). *Hypertension* **11**, 613–619.

Kotchen, T. A., and Guthrie, G. P., Jr. (1988). *Am. J. Cardiol.* **62**, 41G–46G.

Kotchen, T. A., Guthrie, G. P., Jr., Cottrill, C. M., McKean, H. E., and Kotchen, J. M. (1982). *J. Clin. Endocrinol. Metab.* **54**, 808–814.

Lancet (1987). **2**, 311–313.

Lancet (1990). **2**, 718–719.

Laragh, J. H. (1989). *Am. J. Med.* **87**, Suppl. 6B, 2S–14S.

Lees, K. R., MacFadyen, R. J., and Reid, J. L. (1990). *Am. J. Hypertens.* **3**, 266S–272S.

Lenz, T., and Sealey, J. E. (1990). In "Hypertension: Pathophysiology, Diagnosis and Management" (J. H. Laragh and B. M. Brenner, eds.), pp. 1319–1328. Raven Press, New York.

Lenz, T., Sealey, J. E., Lappe, R. W., Carilli, C., Oshiro, G. T., Baxter, J. D., and Laragh, J. H. (1990). *Am. J. Hypertens.* **3**, 257–261.

Lever, A. K. (1989). *Am. J. Hypertens.* **2**, 276–285.

Luetscher, J. A., Kraemer, F. B., Wilson, D. M., Schwartz, H. C., and Bryer-Ash, M. (1985). *N. Engl. J. Med.* **312**, 1412–1417.

Lumbers, E. R. (1971). *Enzymologia* **40**, 329–336.

Macauley, L., Do, Y. S., Koss, M., Anderson, P., Baxter, J. D., and Hsueh, W. A. (1991). Submitted for publication.

Makrides, S. C., Mulinari, R., Zannis, V. I., and Gavras, H. (1988). *Hypertension* **12**, 405–410.

Matsumura, Y., Kawazoe, S., Ichihara, T., Shinyama, H., Kageyama, M., and Morimoto, S. (1988). *Am. J. Physiol.* **255**, F614–F620.

Mento, P. F., Holt, W. F., Murphy, W. R., and Wilkes, B. M. (1989). *Hypertension* **13**, 741–748.

Miller, C. C. J., Carter, A. T., Brooks, J. I., Lovell-Badge, R. H., and Brammer, W. J. (1989). *Nucleic Acids Res.* **17**, 3117–3128.

Miyazaki, H., Fukamizu, A., Hirosi, S., Hayachi, T., Hori, H., Ohkubo, H., Nakanishi, S., and Murikami, K. (1984). *Proc. Natl. Acad. Sci. U.S.A.* **81**, 5999–6003.

Moffett, R. B., McGowan, R. A., and Gross, K. W. (1986). *Hypertension* **8**, 874–882.

Moore, H. P. H., and Kelly, R. B. (1986). *Nature (London)* **321**, 443–446.

Moore, H. P. H., Walker, M. D., Lee, F., and Kelly, R. B. (1983). *Cell (Cambridge, Mass.)* **35**, 531–538.

Morice, A. H., Lowry, R., Brown, M. J., and Higenbottam, T. (1987). *Lancet* **2**, 1116–1118.

Mullins, J. J., Burt, D. W., Windass, J. D., McTurk, P., George, H., and Brammar, W. J. (1982). *EMBO J.* **1**, 1461–1466.

Mullins, J. J., Sigmund, C. D., Kane-Haas, C., and Gross, K. W. (1989). *EMBO J.* **8**, 4065–4072.

Mullins, J. J., Peters, J., and Ganten, D. (1990). *Nature (London)* **344**, 541–544.

Nagahama, M., Nakayama, K., Hori, H., and Murakami, K. (1989). *FEBS Lett.* **259**, 202–204.

Nakamura, N., Soubrier, F., Ménard, J., Panthier, J.-J., Rougeon, F., and Corvol, P. (1985). *Hypertension* **7**, 855–859.

Nakamura, N., Burt, D. W., Paul, M., and Dzau, V. J. (1989). *Proc. Natl. Acad. Sci. U.S.A.* **86**, 56–59.

Nakayama, K., Nagahama, M., Kim, W.-S., Hatsuzawa, K., Hashiba, K., and Murakami, K. (1989). *FEBS Lett.* **257**, 89–92.

Nakayama, K., Hatsuzawa, K., Kim, W.-S., Hashiba, K., Yoshino, T., Hori, H., and Murakami, K. (1990). *Eur. J. Biochem.* **191**, 281–285.

Nishizuki, Y. (1983). *Trends Biochem. Sci.* **8**, 13–16.

Ogihara, T., Mikami, H., Katahira, K., and Otsuka, A. (1991). *Am. J. Hypertens.* **4**, 46S–51S.

Ohkubo, H., Kawakami, H., Kakehi, Y., Takumi, T., Arai, H., Yokota, Y., Iwai, M., Tanabi, Y., Masu, M., Hata, J., Iwao, H., Okamoto, H., Yokoyama, M., Nomura, T., Katsuki, M., and Nakanishi, S. (1990). *Proc. Natl. Acad. Sci. U.S.A.* **87**, 5153–5157.

Oliver, J. A., and Sciacca, R. A. (1984). *J. Clin. Invest.* **74**, 1247–1251.

Panthier, J.-J., Dreyfus, M., Tronik-Le Roux, D., and Rougeon, F. (1984). *Proc. Natl. Acad. Sci. U.S.A.* **81**, 5489–5493.

Park, C. S., Sigmon, D. H., Han, D. S., Honeyman, T. W., and Fray, J. C. S. (1986). *Am. J. Physiol.* **86**, R531–R536.

Paul, M., Nakamura, M., Pratt, R. E., and Dzau, V. J. (1988). *J. Hypertens.* **6**, Suppl. 4, S487–S489.

Picard, D., and Yamamoto, K. R. (1987). *EMBO J.* **6**, 3333–3340.

Piccini, N., Knopf, J. L., and Gross, K. L. (1982). *Cell (Cambridge, Mass.)* **30**, 205–213.

Pinet, F., Corvol, M. T., Dench, F., Bourguignon, J., Feunteun, J., Ménard, J., and Corvol, P. (1985). *Proc. Natl. Acad. Sci. U.S.A.* **82**, 8503–8507.

Pinet, F., Mizrahi, J., Laboulandine, I., Ménard, J., and Corvol, P. (1987). *J. Clin. Invest.* **80**, 724–731.

Pinet, F., Corvol, M.-T., Bourguignon, J., and Corvol, P. (1988). *J. Clin. Endocrinol. Metab.* **67**(6), 1211–1220.

Poisner, A. M., and Poisner, R. (1987). *In* "In Vitro Methods for Studying Secretion" (A. M. Poisner and W. U. Trifaro, eds.), pp. 155–169. Elsevier, New York.

Poisner, A. M., Wood, G. W., Poisner, R., and Inagami, T. (1981). *Endocrinology (Baltimore)* **109**, 1150–1155.

Poorman, R. A., Palermo, D. P., Post, L. E., Murakami, K., Kinner, J. H., Smith, C. W., Reardon, I., and Heinrickson, R. L. (1986). *Proteins* **1**, 139–145.

Pratt, R. E., Flynn, J. A., Hobart, P. M., Paul, M., and Dzau, V. J. (1988). *J. Biol. Chem.* **263**, 3137–3141.

Rosenthal, J., Thurnreiter, M., Plaschke, M., Geyer, M., Reiter, W., and Dahlheim, H. (1990). *Hypertension* **15**, 848–853.

Rougeon, F., Chambraud, B., Foote, S., Panthier, J.-J., Nageotte, R., and Corvol, P. (1981). *Proc. Natl. Acad. Sci. U.S.A.* **78**, 6367–6371.

Rubattu, S., Marion, D. N., Peterson, M., and Sealey, J. E. (1989). *Endocrinology (Baltimore)* **125**, 1533–1539.

Schule, R., Rangarajan, P., Kliewer, S., Ransone, L. J., Bolado, J., Yang, N., Verma, I. M., and Evans, R. M. (1990). *Cell (Cambridge, Mass.)* **62**, 1217–1226.

Sealey, J. E., and Laragh, J. H. (1990). *In* "Hypertension: Pathophysiology, Diagnosis and Management" (J. H. Laragh and B. M. Brenner, eds.), pp. 1287–1318. Raven Press, New York.

Sealey, J. E., and Rubattu, S. P. (1989). *Am. J. Hypertens.* **2**, 358–366.

Sealey, J. E., Atlas, S. A., and Laragh, J. H. (1980). *Endocr. Rev.* **1**, 365–391.

Seo, M. S., Fukamizu, A., Nomura, T., Yokoyama, M., Katsuki, M., and Murakami, K. (1990). *J. Cardiovasc. Pharmacol.* **16,** Suppl. 4, S8–S10.

Shaw, K. J., Do, Y. S., Anderson, P. W., Shinagawa, T., Dubeau, L., and Hsueh, W. A. (1989). *J. Clin. Invest.* **83,** 2085–2092.

Sheng, M., Dougan, S. T., McFadden, G., and Greenberg, M. E. (1988). *Mol. Cell. Biol.* **8,** 2787–2796.

Shier, D. N., Kusano, E., Stoner, G. D., Franco-Saenz, R., and Mulrow, P. J. (1989). *Endocrinology (Baltimore)* **125,** 486–491.

Shinagawa, T., Do, Y. S., Baxter, J. D., Carilli, C., Schilling, J., and Hsueh, W. A. (1990). *Proc. Natl. Acad. Sci. U.S.A.* **87,** 1927–1931.

Shinagawa, T., Do, Y. S., Baxter, J. D., and Hsueh, W. A. (1991). In preparation.

Sibanda, B. L., Blundell, T., Hobart, P. M., Fogliano, M., Bindra, J. S., Dominy, B. W., and Chirgwin, J. M. (1984). *FEBS Lett.* **174,** 1–10.

Sielecki, A. R., Hayakawa, K., Fujinaga, M., Murphy, M. E. P., Fraser, M., Muir, A. K., Carilli, C. T., Lewicki, J. A., Baxter, J. D., and James, M. N. G. (1989). *Science* **243,** 1346–1351.

Sigmund, C. D., Jones, C. A., Mullins, J. J., Kim, U., and Gross, K. W. (1990). *Proc. Natl. Acad. Sci. U.S.A.* **87,** 7993–7997.

Skeggs, L. T., Jr., and Dorer, F. E. (1989). *Am. J. Hypertens.* **2,** 768–779.

Skinner, S. L., Thatcher, R. L., Whitworth, J. A., and Horowitz, J. D. (1986). *Lancet* **1,** 995–997.

Sogawa, K., Fuji-Kuriwama, Y., Mizukami, Y., Ichihara, Y., and Takahashi, K. (1983). *J. Biol. Chem.* **258,** 5306–5311.

Sola, C., Tronik, D., Dreyfus, M., Babinet, C., and Rougeon, F. (1989). *Oncogene Res.* **5,** 149–153.

Soubrier, F., Panthier, J.-J., Houot, A.-M., Rougeon, F., and Corvol, P. (1986). *Gene* **41,** 85–92.

Stoller, T. J., and Shields, D. (1989). *J. Cell Biol.* **108,** 1647–1655.

Swales, J. D., and Samani, N. J. (1990). *Am. J. Hypertens.* **3,** 890–892.

Taugner, R., Hackenthal, E., Helmcher, U., Ganten, D., Kugler, P., Marin-Grey, M., Norbiling, R., Unger, T., Lowckwald, I., and Keilbach, R. (1982). *Klin. Wochenschr.* **60,** 1218–1222.

Taugner, R., Whalley, S., Angermuller, S., Buhrle, C. P., and Hackenthal, E. (1985a). *Cell Tissue Res.* **239,** 575–587.

Taugner, R., Buhrle, C. P., Nobiling, R., and Kirschke, H. (1985b). *Biochemistry* **83,** 103–108.

Taugner, R., Murakami, K., and Kim, S. J. (1986). *Biochemistry* **85,** 107–109.

Toffelmire, E. B., Slater, K., Corvol, P., Ménard, J., and Schambelan, M. (1989). *J. Clin. Invest.* **83,** 679–687.

Tronik, D., Dreyfus, M., Babinet, C., and Rougeon, F. (1987). *EMBO J.* **6,** 983–987.

Tronik, D., Ekker, M., and Rougeon, F. (1988). *Gene* **69,** 71–80.

Unger, T., and Gohlke, P. (1990). *Am. J. Cardiol.* **65,** 3-I–10-I.

Verburg, K. M., Kleinert, H. D., Kadam, J. R. C., Chekal, M. A., Mento, P. F., and Wilkes, B. M. (1989). *Hypertension* **13,** 262–272.

Ville, D. B. (1979). *In* "Placenta-A Neglected Experimental Animal" (P. Beaconsfield and C. Villee, eds.), pp. 74–86. Pergamon, Oxford.

Wang, P. H., Do, Y. S., Macauley, L., Tatsuo, S., Anderson, P. W., Baxter, J. D., and Hsueh, W. A. (1991). *J. Biol. Chem.* (in press).

Weih, F., Stewart, A. F., Boshart, M., Nitsch, D., and Schutz, G. (1990). *Genes Dev.* **4,** 1437–1449.

Welch, W. J., Ott, C. E., Guthrie, G. P., Jr., and Kotchen, T. A. (1983). *Endocrinology (Baltimore)* **113,** 2086–2091.

Williams, G. H. (1988). *N. Engl. J. Med.* **319,** 1517–1525.

Wilson, D. M., and Luetscher, J. A. (1990). *N. Engl. J. Med.* **323,** 1101–1106.

Wood, J. M., Stanton, J. L., and Hofbauer, K. G. (1987). *J. Enzyme Inhib.* **1,** 169–185.

Yamaguchi, T., Naito, Z., Stoner, G. D., Franco-Saenz, R., and Mulrow, P. J. (1990). *Hypertension* **16,** 635–641.

Yamamoto, K. K., Gonzalez, G. A., Biggs, W. H., III, and Montminy, M. R. (1988). *Nature (London)* **334,** 494–498.

Yamauchi, T., Nagahama, M., Watanabe, T., Ishizuka, Y., Hori, H., and Murakami, K. (1990). *J. Biochem. (Tokyo)* **107,** 27–31.

Yang-Yen, H.-F., Chambard, J.-C., Sun, Y.-L., Smeal, T., Schmidt, T. J., Drouin, J., and Karin, M. (1990). *Cell (Cambridge, Mass.)* **62,** 1205–1215.

DISCUSSION

S. McKnight. I wondered if prorenin might be phsophorylated. I was studying the site where the cleavage of prorenin can be seen, and its sequence contains a Lys-Arg-X-Thr-Leu, which might be a good site for the cyclic AMP-dependent protein kinase. Perhaps phosphorylation could somehow be involved in the regulation of the specific processing of prorenin.

J. Baxter. That is an interesting suggestion. We do not have any evidence that it is phosphorylated, but have never really studied that.

A. Means. Do synthetic peptides made against the prosegment inhibit the activity of the prorenin processing enzyme?

J. Baxter. We have not examined this possibility.

B. O'Malley. I would like to clarify what you have said. Do you maintain that sorting is perhaps a product of two different structural regions in the molecule? Are they synergistic together relative to either one acting alone?

J. Baxter. We do not know whether both of these sorting sequences are functional in sorting with intact prorenin. The magnitude of sorting with the prosegment immunoglobulin fusion protein is about the same that we obtain with native prorenin, but we have not done it enough times to say whether the fold is exactly the same as with the native form. In addition, we do not know which sequences function within intact prorenin. We are currently testing two models. One is that there are two sorting sequences that may or may not function synergistically. The other is that the sorting sequences are on the prosegment and that after renin undergoes this large conformational change following removal of the prosegment, the amino-terminal sequences of renin replace those of the prosegment as sorting sequences, since they occupy the same position previously held by the prosegment.

B. O'Malley. It seems to me that if you have two different regions required for sorting, they should either act synergistically at one step or work at different steps in sorting. Otherwise, why hold that duplication?

J. Baxter. We do not have enough information to answer your question. This situation could be analogous to the case with the steroid receptors that appear to have more than one nuclear localization signal, but I do not think we know whether those function synergistically. This issue may be a hard one to answer because of the relatively modest changes in secretion that we get.

B. O'Malley. Does cold-temperature activation of processing play any role in exacerbating frostbite physiology in humans?

J. Baxter. I do not know, but this is a most interesting suggestion. Prorenin has 1–8% the activity of renin at 37°C, but has almost 100% renin activity at 0°C. Thus, a considerable proportion of the prorenin might be active in a cold extremity, and this could conceivably lead to enhanced angiotensin generation.

B. O'Malley. Could you summarize what is known about genetic defects in any of the steps within this pathway?

J. Baxter. The most interesting general clinical issue with respect to renin is with essential hypertension. When the plasma renin levels in people with essential hypertension are examined, one finds that about 70% of these subjects have normal plasma renin levels, 5 to 20% have very suppressed renin levels, and another few percent, perhaps up to 5%, have elevated renin levels. There may be functional differences in these groups as well. When prehypertensives are examined, one finds that their plasma renin levels vary much more than those of control subjects. Whether these differences in renin gene expression are due to intrinsic differences in the gene or are secondary to other factors that affect expression of the gene is unknown, but it is an issue that we ultimately would like to address.

I probably should also have mentioned with respect to the processing of prorenin to renin that the only disease of which I am aware that may result in abnormal protein processing is diabetes mellitus. In very early diabetes, sometimes even before there is micro-albuminuria, there may be a relative increase in the plasma levels of prorenin versus renin. With time and with further renal damage, hyporeninemic hypoaldosteronism can develop. Some of these patients have normal to even high levels of plasma prorenin. It is controversial whether all of the prorenin comes from the kidney, but it may be of renal origin. Thus there may be a metabolic derangement that prevents the processing of prorenin to renin. There are restriction fragment length polymorphism (RFLP) studies in a strain of hypertensive rats which suggest linkage of an RFLP in the renin gene to the hypertension. Equivalent studies in man have not been done as it is difficult to obtain meaningful data due to difficulties in defining the phenotype and in analyzing large families.

P. Kelly. You mentioned the role of the carbohydrate moieties in the processing of renin, but you also mentioned that you had to deglycosylate the renin produced in CHO cells for it to be active.

J. Baxter. I am sorry I misled you. The deglycosylation, under most conditions where we have tested it, has no detectable effect on the activity of renin. We removed the sugar moieties from the CHO cell-synthesized renin to facilitate its crystallization. Several different glycosylated forms are secreted by these cells as is detected by isoelectric focusing studies, and such heterogeneous mixtures are not good for crystallization. Thus, we attempted to obtain more homogeneous material for crystallization by treating the renin preparations with endoglycosidase F. Of interest, however, is that in spite of this treatment, which resulted in a molecular weight change as based on sizing gels, the crystal structure analysis revealed residual sugar moieties on both of the glycosylation sites.

P. Kelly. What is known about the renin receptor or other sites of action?

J. Baxter. The renin "receptor" is a speculated entity predominantly addressed by Jean Sealey. Whereas renin ordinarily acts as an enzyme, the finding of prorenin, but not renin, in some tissues has led to the suggestion that prorenin or renin might have actions unrelated to its catalytic activity and that these molecules may bind to specific cellular sites. These could either mediate some noncatalytic event or else be activated. However, to date, evidence for such a "receptor" has not been published.

M. New. John, I would like to remind you that there is one other natural hyperreninemic state which is found in the fetus and in the newborn. There is marked hyperreninemia; its function is not known.

Cellular and Molecular Analysis of Pancreatic Islet Cell Lineage and Differentiation

GLADYS TEITELMAN[1]

Department of Neurology, Cornell University Medical College, New York, New York 10021

I. Introduction

The mature pancreas is formed by exocrine tissue (acinar cells and ducts) and endocrine cells, which are clustered into the islets of Langerhans. Islets are randomly distributed within the exocrine tissue lying in close proximity to elements of the vascular system. There are four main islet cell types, each defined by the hormone they produce and by the distinct characteristics of their respective storage granules. The α cells contain glucagon (Glu), β cells contain insulin (IN), and δ and PP cells contain somatostatin (Som) and pancreatic polypeptide, respectively.

Because of the importance of insulin in regulating carbohydrate metabolism, the endocrine pancreas has been the focus of active research for many years and its hormonal products were among the first to be isolated and characterized (Steiner, 1983; Habener, 1987). More recently, DNA complementary to the mRNA of human and rat insulin has been isolated, cloned, and sequenced, and the chromosomal location of the human gene has been determined. Similarly, the DNA sequences that code for Glu, Som, and PP have been identified (Habener, 1987; Steiner *et al.*, 1989). These studies have revealed that pancreatic hormones, like most other regulatory peptides, are initially synthesized as longer precursor molecules. These prohormones apparently contain the information required for correct intracellular sorting and transport (Hand and Oliver, 1981). Several elegant studies (i.e., Steiner, 1983; Rothman and Orci, 1990) have established a correlation between maturation of the insulin molecule from its precursor substances and a pathway of subcellular transport of the hormone and its ultimate transformation, within secretory granules, into its mature form.

Great efforts are also underway to understand the molecular mechanisms that control cell-specific expression of pancreatic hormone gene. It is generally agreed that gene expression is regulated by an interaction between certain nuclear proteins and specific nucleotide sequences located outside the coding regions of

---

[1] Present address: Department of Anatomy and Cell Biology, SUNY, Health Science Center at Brooklyn, Brooklyn, New York 11203.

the genes. The focus of molecular analysis of hormone gene expression currently lies in the identification of those signals and the elucidation of their mode of action (Hwung *et al.*, 1989; Whelan *et al.*, 1989; see also Habener, 1987).

The availability of immunological and molecular probes to identify the various hormones has also provided new tools to examine the mechanisms that govern the differentiation of pancreatic islet cells. From the perspective of a developmental biologist, the fundamental questions regarding the endocrine pancreas revolve around the role of genetic and epigenetic signals on differentiation of islet cells and on stability and/or plasticity of the differentiated phenotype. This article will selectively examine recent advances on these issues and will identify areas which remain to be explored.

II. Methods of Analysis

Most of the studies that will be discussed relied on immunohistochemical techniques that were used to visualize one or more antigens in the same tissue sections. In other instances, immunohistochemistry was combined with autoradiography. Therefore, in order to facilitate understanding of the different experimental designs, a brief description of these techniques will be presented.

A.   SINGLE-LABEL IMMUNOHISTOCHEMISTRY

Adult mice were perfused through the heart with fixative solution. The pancreas was removed and postfixed for 1 hour in the same solution. Embryos were fixed by immersion in fixative. Tissues were sectioned in a Cryostat microtome and the sections mounted onto microscope slides. The tissue sections were sequentially incubated with 0.25% Triton X-100, a blocking solution of goat serum in buffer to prevent nonspecific binding of antibodies to tissue proteins, and overnight in a solution of specific antibodies against a pancreatic hormone diluted in buffer. The following day the bound antibody was visualized by labeling with an avidin–biotin–peroxidase complex (ABC) and subsequent demonstration of the peroxidase product by incubation of slides in 3,3'-diaminobenzidine (DAB) and H_2O_2 (Teitelman *et al.*, 1981).

B.   DOUBLE-LABEL IMMUNOHISTOCHEMISTRY

Sections were incubated with a highly diluted primary antibody and subsequently, with donkey anti-rabbit IgG labeled with [125]I. A second primary antibody layer was applied and visualized with an avidin–biotin–peroxidase complex and DAB (Fig. 1). The sections were then dipped in photographic emulsion (Ilford L-4, Polysciences, Inc., Warrington, PA) and incubated in the dark for 2 to 3 weeks. Sections were then developed, fixed, and coverslipped for

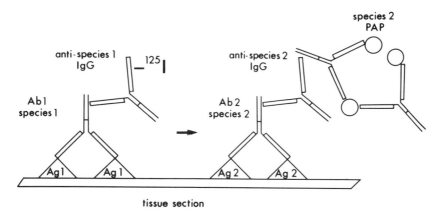

FIG. 1.   Schematic representation of an immunohistochemical procedure followed to label two antigens in the same tissue section. First (Ag 1) and second (Ag 2) primary antibodies were visualized by the presence, over the tissue section, of silver grains and DAB precipitate, respectively.

microscope examination. With this double-label procedure, the presence of silver grains indicated the location of the first primary antisera and, therefore, of the first tissue antigen of interest. The second primary antisera was visualized by the brown reaction product of the DAB precipitate (Fig. 1; Pickel *et al.*, 1985).

C.   COMBINED IMMUNOCYTOCHEMISTRY AND AUTORADIOGRAPHIC VISUALIZATION OF [^{3}H]THYMIDINE INCORPORATION

In order to determine the labeling index of pancreatic islet cells, animals were injected with [^{3}H]thymidine ([^{3}H]TdR), perfused, and processed for single-label immunocytochemistry. After the DAB step, sections were dipped in photographic emulsion, stored in the dark for 2 to 3 weeks, developed, and fixed as indicated above. The presence of silver grains over the nuclei indicated that those cells had incorporated the isotope into DNA. Cells that also expressed the pancreatic hormone of interest contained DAB reaction product in the cytoplasm (Rothman *et al.*, 1980).

III.   Animal Models

A.   TRANSGENIC MICE

The generation of transgenic animals represents a new experimental approach to examine the mechanisms that regulate gene expression. The basic strategy to produce transgenic animals involves the introduction of a gene into the pronucleus of fertilized eggs that are then transferred into the oviducts of pseudopregnant females. The injected DNA is integrated into the mouse genome and is normally segregated in the progeny. Founder mice generated in this way are mated and give rise to stable pedigrees expressing the transgene. Several recent reviews describe the gene transfer technology and its applications to the study of the regulation of cell-specific gene expression (Palmiter and Brinster, 1986; Hogan *et al.*, 1986; Hanahan, 1989; Gordon, 1989). Studies described in this article examined tissues of several lines of transgenic mice that harbor a hybrid insulin gene. RIP–Tag2 mice carry a transgene consisting of the 5′-flanking region of the rat insulin II gene linked to the sequence encoding the simian virus 40 (SV40) large T antigen (Tag). RIR–Tag2 mice harbor a gene with the insulin promoter inverted with respect to Tag. These mice express Tag specifically in β cells of the pancreas and develop insulinomas as a consequence of oncogene expression. In addition, a line of transgenic mice containing a hybrid insulin–human placental lactogen (PLA) gene was analyzed (Alpert *et al.*, 1988a). RIP–Tag and RIR–Tag mice were generated and first described by Hanahan (1985) and Efrat and Hanahan (1987).

B. OBESE MICE

Obese mice have been a remarkably useful model for the study of obesity and diabetes. Mice of the C57BL/6J *ob/ob* strain are mildly hyperglycemic and markedly hyperinsulinemic. The primary defect in the *ob/ob* mice appears to be related to a deficiency in the hypothalamic satiety centers. The pancreatic islets of *ob* mice are characterized by β cell hypertrophy–hyperplasia and a concomitant reduction in the population of α, δ, and PP cells. The diabetic condition in *ob* mice disappears with time and animals eventually lose weight and develop improved glucose tolerance (Herberg and Coleman, 1977; Like, 1977; Shafrir, 1990).

C. DIABETIC MICE

Mice of the C57BL/Ls *db/db* have been extensively investigated as a model of insulin-resistant diabetes. These mice are obese and normoglycemic up to the age of 1 month, when they develop hyperglycemia and hyperinsulinemia that peak between 9 to 12 weeks. Between 3 and 6 months the insulinemia decreases due to β cell failure, the mice lose weight, become severely diabetic, and then die between 8 and 10 months of age (Herberg and Coleman, 1977; Like, 1977; Shafrir, 1990). Obese and diabetic mice were provided by Jackson Laboratories (Bar Harbor, ME).

D. CD-1 MICE

This is a line of white outbred mice that was used for studies of normal pancreatic development and to characterize the phenotype of the different islet cell types in control adult mice. They provided observations that served as a comparison with the results obtained from the different animal models mentioned above.

IV. Role of Genetic and Epigenetic Factors on Pancreatic Islet Cell Differentiation

A. MODELS OF ISLET CELL DIFFERENTIATION

In mammals, the pancreas develops by fusion of dorsal and ventral primordia, which appear as evaginations of the gut. This occurs in embryos of about 20 somites, corresponding to day 10 of development (E-10) in mouse and E-11 in rat (Altman and Dittmer, 1983). The two primitive glands grow independently, forming both endocrine and exocrine tissue, and finally merge during midgestation (Wessels and Evans, 1968; Pictet and Rutter, 1972). Each primordium retains its own duct, which empties the products of the exocrine tissue into the gut lumen. It is generally believed that, in embryos, the pancreatic duct contains undifferentiated precursor cells that migrate away from the duct to originate islet and acinar cells (Pictet and Rutter, 1972). Endocrine cell differentiation occurs as soon as the pancreatic cell precursors emerge from the pancreatic duct.

The timetable of appearance of the different hormones during development was determined using immunochemical, ultrastructural, and histological techniques. In mouse and rat, glucagon cells first differentiated in 20-somite embryos, concomitant with the formation of the pancreatic primordia (Wessels and Evans, 1968; Pictet and Rutter, 1972; Rall et al., 1973; Alpert et al., 1988a; Reddy and Elliot, 1988). At this stage, cells containing glucagon (Glu$^+$ cells) were seen within the mucosa of the gut and clustered in the mesenchyme next to the gut cavity, thus forming the first pancreatic islets (Fig. 2). The time of initiation of glucagon expression in cells of the pancreatic primordia coincided with appearance of secretory granules characteristic of α cells (Pictet and Rutter, 1972), suggesting a correlation between biochemical and ultrastructural differentiation. Cells containing insulin immunoreactivity (IR) were first seen at E-12 in mouse (Alpert et al., 1988a; Reddy and Elliot, 1988) and E-12.5 in rat (Yoshinari and Daikoku, 1982; Reddy and Elliot, 1988), the same time the hormone was detected by radioimmunoassay (Pictet and Rutter, 1972; Rall et al., 1973). The other two pancreatic hormones, somatostatin (Som) and pancreatic polypeptide (PP), appeared in mouse at days E-16 and postnatal day 1, re-

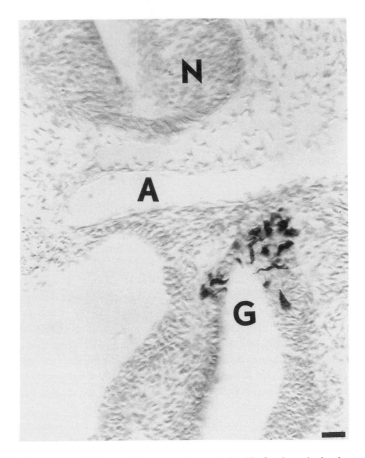

FIG. 2.   Appearance of glucagon in mouse embryos at day 10 of embryonic development (cross section). Note the presence of cells containing immunoreactive glucagon in the primordium of the dorsal pancreas. N, Neural tube; A, aorta; G, gut. Bar: 15 μm.

spectively (McIntosh *et al.*, 1977; Reddy and Elliot, 1988; Alpert *et al.*, 1988b).

While these studies revealed a sequence of appearance of pancreatic islet cell types, they did not provide insight into the mechanisms that lead to their selective differentiation. Conceivably, each cell type could arise from a separate line of precursor cells; in this case, differentiation would be determined solely by intrinsic (genetic) signals with little or no influence from the environments (Fig. 3, model A). Alternatively, all the cells may arise from a common, multipotential precursor that gradually produces phenotypically distinct cell types. In this model, differentiation would result from complex interactions between the cells

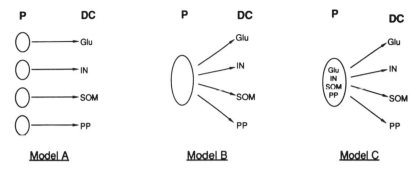

FIG. 3. Hypothetical models of islet cell differentiation. Model A: Each cell type arises from a unipotential precursor (P). The differentiation of these cells is guided by genetic cues and is independent from environmental signals. Model B ("activation model"): A multipotential precursor cell gives rise to all four islet cell types of the pancreas. According to this model, progenitor cells are undifferentiated and expression of cell-specific traits is induced by epigenetic signals. Model C ("activation–repression model"): this model proposes that an initial event occurs in islet progenitor cells that activates the synthesis of all four pancreatic hormones. During maturation, cells retain only one hormone and inhibit expression of the others. Although other models are also possible, for simplicity only these three possibilities are discussed. DC, Differential cells.

and tissue-derived factors. Furthermore, in this latter model two different modes of differentiation can be envisioned.

In the "activation model" precursors are undifferentiated multipotential cells that are induced by their environment to give rise to a variety of cell types, each expressing a unique phenotype. In pancreas, according to this model, different factors would promote the appearance of each of the hormones and the differentiation of the four islet cell types (Fig. 3, model B).

In the "activation–repression model" progenitor cells are initially induced to express traits specific for most (or all) of the cell types they will originate. However, during the course of maturation, these primitive islet cells retain the expression of only one marker and inhibit the expression of others (Fig. 3, model C).

Different approaches have been used in other systems to ascertain the role of genetic and environmental factors in differentiation. Precursor cells of the vertebrate central nervous system (CNS) and peripheral nervous system (PNS) have been labeled *in vitro* with dyes (Jacobson, 1986; Bronner-Fraser and Fraser, 1988; Serbedzija *et al.*, 1990), isotopes (Weston, 1970), or by retroviral-mediated gene transfer (Rossant, 1986; Sanes *et al.*, 1990; Turner *et al.*, 1990), and the progeny that originated from each individual cell have been determined. *In vitro*, clonal analysis of cultured progenitor cells was used to examine the phenotypic diversity of cells that descended from a common precursor (Sieber-Blum *et al.*, 1988; Sieber-Blum, 1989). Finally, grafting experiments were performed in avian embryos to examine the differentiation of cells of the peripheral nervous system from identified precursor cells (Le Douarin, 1982). From these and related studies a consensus has begun to emerge: in contrast to the predeterministic pattern that characterizes invertebrate development (Davidson, 1990), in vertebrate embryos, the environment guides the differentiation of multipotential precursors cells.

It is likely that differentiation of pancreatic islet cells is also regulated by environmental influences. Over 20 years ago, classic studies (Rutter *et al.*, 1964, 1978; Pictet and Rutter, 1972) clearly described the basic features of the initial stages of cytodifferentiation of pancreatic epithelia into endocrine cells and demonstrated that morphogenesis and differentiation of pancreatic epithelia into endocrine cells were dependent on a factor secreted by mesenchymal cells. Several years later, this mesenchymal factor was identified as a member of the insulin-like growth factor family (Burgess and Nicola, 1983).

The sequence of appearance of pancreatic hormones and the events that lead to maturation of the islet cell phenotypes may also be regulated by environmental cues. Consistent with this proposition was the observation that the sequence of differentiation of some islet cell types *in vivo* differed from that observed *in vitro*. During development *in vivo*, Som[+] cells appeared before PP[+] cells. However, when the dorsal pancreatic primordium was removed from E-11 embryos and allowed to differentiate *in vitro*, the appearance of PP[+] cells preceded that of Som[+] cells (Teitelman *et al.*, 1987a). Differences in the cellular milieu may regulate not only the sequence of appearance of islet cell types but also their relative proportion in various regions of the pancreas. It has been reported that, in adults, PP cells are more abundant in the ventral lobe than in the dorsal lobe of the gland (Orci, 1982). Since these two regions arise from the ventral and dorsal primordia, respectively, it has been proposed that differences in the distribution of PP cells were due to the origin of this cell types from the ventral primordium (Rahier *et al.*, 1979). However, when the dorsal pancreas was removed from E-11 mouse embryos and maintained *in vitro* in isolation from the ventral primordium, PP cells continued to differentiate (Teitelman *et al.*, 1987a), suggesting that differences in the numbers of PP cells, rather than being determined by site of origin, may be controlled by local tissue factors.

Very little is known, however, of the mechanisms that regulate islet cell differentiation. The basic strategy used to examine this crucial question, similar to that followed in other developmental systems (i.e., Anderson, 1989; Stockdale, 1989), involves the characterization of (1) the cell lineage relationships of pancreatic islet cells, (2) the phenotypes of cells at different levels in the hierarchy of differentiation, (3) the (presumed epigenetic) signals that regulate these processes, (4) the mechanisms of transmission of the chemical signals to the cell nucleus, and (5) nuclear factors that regulate proper gene expression during development.

Recently, the cell lineage relationships of pancreatic islet cells have been examined. Because of the internal location of the pancreas, it is difficult to label individual cells to follow their progeny *in vivo*. Therefore, noninvasive approaches were devised to examine islet cell differentiation. The development of immunocytochemical techniques to determine whether embryonic pancreatic cells contain more than one hormone provided a valuable tool to identify the phenotype of the progenitor cells and study their cell lineage relationships. The assumption was made that if islet cells coexpressed two or more hormones during development, these mixed cells probably represented precursors that would differentiate into mature cells containing a single hormone.

When sections of embryonic mouse pancreas were processed for the dual localization of insulin and glucagon, cells coexpressing both hormones were detected at all stages of development. At E-12 all the cells that contained insulin were Glu$^+$. The converse was not true; i.e., cells were detected that were Glu$^+$ but did not contain immunoreactive insulin (Table I). The number of double-labeled cells decreased during development. Thus, by E-15, a significant number of cells contained only insulin or only glucagon. These single-labeled cells were intermingled in the pancreatic primordia with mixed phenotype cells. At birth and in adults only a small percentage of islet cells were both IN$^+$ and Glu$^+$ (Alpert *et al.*, 1988a). These results suggested that islet cells that contained either IN or Glu originated from precursors that coexpressed both hormones. As predicted by the activation–inhibition model (Fig. 3, model B), during maturation

TABLE I

Coexpression of Insulin and Glucagon during Development[a]

Age	Proportion of IN$^+$ cells producing glucagon	Proportion of Glu$^+$ cells producing insulin
E-12	1.00	0.11 ± 0.007
E-15	0.73 ± 0.09	0.31 ± 0.06
E-16	0.17 ± 0.10	0.15 ± 0.03
P-1	0.09 ± 0.05	0.10 ± 0.04
Adult	0.02 ± 0.006	0.09 ± 0.04

[a]Three pancreata were processed for each developmental stage. The proportion of cells containing both IN and Glu was determined in entire pancreata at E-10 to E-12 and in at least 10 endocrine cell clusters from E-14 to P-1. The number of cells containing both hormones was expressed as the mean fraction (± standard error) of the cells immunoreactive for one particular hormone. Reprinted from Alpert *et al.* (1988a).

TABLE II

Coexpression of Insulin in Developing δ and PP Cells[a]

Proportion of Som+ cells IN+ at E-17	Proportion of PP+ cells IN+ at P-1
0.18 ± 0.05	0.20 ± 0.11

[a]At the time of onset of PP and somatostatin (Som) expression, CD-1 pancreatic sections were processed for the dual localization of one of the hormones with insulin. Calculations as in Table I. Reprinted from Alpert *et al.* (1988a).

most of the cells retained the expression of only one hormone and inhibited the appearance of the other.

Islet cell precursors expressed other hormones in addition to Glu and IN (Alpert *et al.*, 1988a). Thus, at E-17, 17% of the IN+ cells also contained somatostatin while at P-1 approximately 20% of IN+ cells contained PP (Table II). After birth, very few cells coexpressed insulin and either somatostatin or PP (Table II). The evaluation of the different single and mixed phenotypes expressed by islet cells at various times of development allowed the formulation of a more comprehensive model of islet cell lineage and differentiation (Alpert *et al.*, 1988a). In this model, precursor cells expressing several hormones undergo a multistep process of maturation that is characterized by a gradual segregation of each phenotype to a different cell type. This model also predicts that cells at intermediate stages of maturation would express a chimeric phenotype (Fig. 4).

Islet cell precursors are not unique in their ability to coexpress markers specific for each of the cell types they will originate. Sympathetic neurons and adrenal chromaffin cells arise from common bipotential precursors (Anderson, 1989) of neural crest origin (Le Douarin, 1988) that initially express traits characteristic of endocrine and neural cells (Vogel and Weston, 1990; G. Teitelman, M. J. Evinger, and M. Erlich, unpublished findings, 1990). In pituitary gland, growth hormone (GH)-secreting somatotropes and prolactin (PRL)-synthesizing lactotropes arise from precursors that contain both hormones (Neil *et al.*, 1987; Behringer *et al.*, 1988). Presumably, in these systems, differentiation entails a choice among the various cell-specific traits expressed by precursor cells, and involves, as in the pancreas, retention of some properties and inhibition of others.

B. FACTORS CONTROLLING THE TIMETABLE OF ACTIVATION OF PANCREATIC-SPECIFIC GENES

Following the identification of a developmental pathway for pancreatic islet cells, an understanding of the mechanisms that lead to their differentiation in-

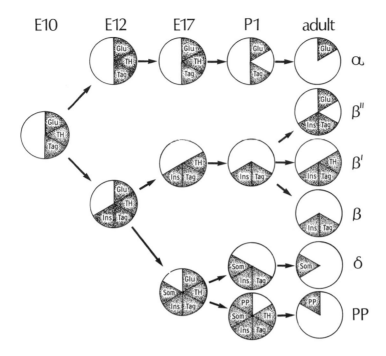

FIG. 4. Schematic diagram of the pancreatic cell lineage. Shown is a proposed pathway for the formation of pancreatic endocrine cells in mouse embryos. The model is consistent with the colocalization of endocrine markers in the same cell during development. It also shows the timetable of appearance and distribution of a hybrid insulin gene in transgenic (RIP1–Tag2) mice (Sections IV,B and C). These mice harbor a transgene consisting of the regulatory information flanking the rat insulin II gene linked to the sequence encoding the simian virus 40 (SV40) large T antigen (Tag). At E-10, cells containing glucagon, TH, and Tag appear in the pancreatic duct. These cells could give rise to the other population of cells as illustrated. For clarity the figure illustrates only the time when each cell type first appears. It should be noted that many of the mixed cell types are present throughout prenatal development and some persist in adults. Cells containing both somatostatin and TH and cells containing PP and TH appear transiently during development. Furthermore, cells expressing both somatostatin and PP have been identified, suggesting the possible conversion of a Som+ cell into a PP+ cell, as indicated. Appearance of IN+ Glu+ cells and of IN+ TH+ cells in adults is discussed in Sections V,A and C, respectively. Glu, Glucagon; TH, tyrosine hydroxylase; Ins, insulin; Som, somatostatin; PP, pancreatic polypeptide; E, embryonic day; P, postnatal day. Modified from Alpert et al. (1988a).

volves the characterization of the signals that regulate the timetable of appearance of the hormones in precursor cells and the segregation of these markers into independent cell types. It is reasonable to assume that differentiation of islet cells is guided, as in other systems, by tissue-derived soluble factors and by specific histiotypical associations. However, a prerequisite for this analysis,

which has not yet been accomplished, is the development of techniques for
(1) the identification of islet cell precursors and (2) their isolation and growth *in
vitro*. In addition, future advances in this critical area of research depend on the
establishment of appropriate tissue culture conditions that will allow the analysis
of the phenotype of embryonic islet cells maintained in serum-deprived media
and in the absence of other cell types.

In contrast to the poorly understood cell biological processes underlying islet
cell differentiation, the application of recombinant DNA technology has greatly
accelerated the gathering of information concerning the molecular mechanisms
that regulate cell-specific expression of hormone genes. It is generally believed
that cell-specific expression is regulated at the transcription level by trans-acting
proteins that bind with short DNA sequences located upstream from the transcription initiation site (Darnell, 1982; Habener, 1987). The generation of transgenic mice provided a novel approach to identify nucleotide sequences that are

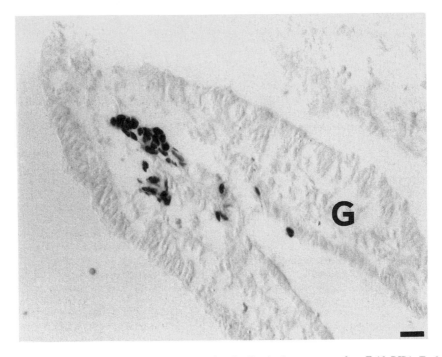

FIG. 5.   Appearance of cells immunoreactive for Tag in the pancreas of an E-10 RIP1–Tag2
transgenic mouse embryo. Cells producing Tag were localized in embryonic pancreatic sections by
immunostaining with antibodies specific for the oncoprotein, followed by treatment with a horse-
radish peroxidase-conjugated secondary antibody and incubation with DAB. Thus Tag immunoreactivity is visualized by the presence of a black DAB precipitate. Immunoreactive cells are localized
around the gut (G). Bar: 30 μm.

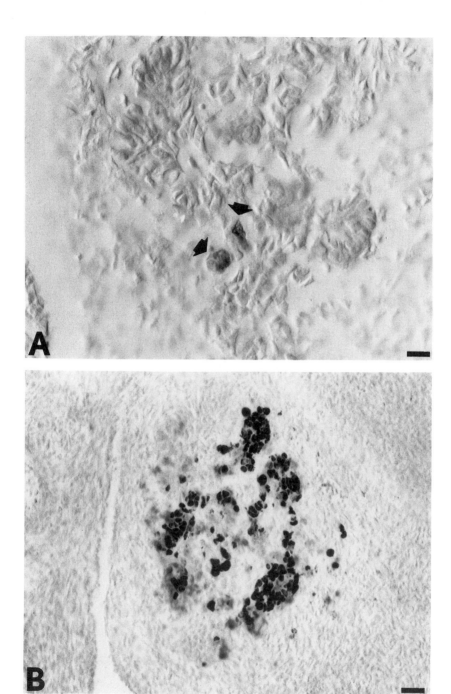

FIG. 6. Comparison of the number of IN⁺ and Tag⁺ cells in E-12 RIP–Tag2 pancreatic sections. Sections of RIP–Tag2 pancreas were processed for the immunocytochemical localization of either insulin or Tag. (A) Only a few cells immunoreactive for insulin are present in the E-12 pancreas. The arrows indicate the presence of a cell exhibiting insulin immunoreactivity. Bar: 20 μm. (B) At E-12, considerably more cells immunostain for Tag. Bar: 40 μm.

required for correct expression of islet-specific genes during development. Transgenic embryos from a line of mice (RIP–Tag2) that contained a hybrid insulin gene composed of the insulin promoter linked to the sequences encoding for an oncogene (Tag) were examined using immunocytochemical techniques to localize expression of the transgene (Alpert et al., 1988a).

Abundant cells containing immunoreactive Tag were first observed in transgenic embryos at E-10 of development as the pancreatic diverticulum was forming (Fig. 5). The appearance of Tag at this time preceded by 2 days the immunodetection of insulin (Fig. 6). Analysis of other lines of transgenic mice harboring hybrid genes that also contained the insulin promoter region indicated that this premature expression of the reporter molecule Tag was not due to the position of integration of the transgene into the host chromosome or to the presence of the SV40 early region. Taken together, these finding suggested that there were additional regulatory elements outside the promoter/enhancer regions included in the transgenes that control the time of initiation of insulin synthesis during development. Conceivably, negative regulatory elements located upstream (Laimins et al., 1986) and/or within the noncoding sequences of the endogenous insulin gene may be involved in the control of proper temporal expression. This proposition implies the existence of DNA sequences that function as "developmental clocks." This is a testable hypothesis because, if correct, transgenic lines with hybrid gene constructs containing additional portions of the insulin gene should express the fusion gene at the correct time in development.

C. CELL-SPECIFIC EXPRESSION OF A HYBRID INSULIN GENE IN PANCREAS DURING DEVELOPMENT

While the previous section examines whether the hybrid insulin gene contains the information required to initiate its expression at the same time in development as the endogenous insulin gene, this section describes studies that were carried out to determine whether the regulatory sequences of the transgene were sufficient for correct cell-specific regulation of expression in pancreas during embryogenesis. When the distribution of Tag$^+$ cells was examined in transgenic embryos of the RIP–Tag2 lineage, it was found that, at E-10, when the pancreas first differentiated, the oncoprotein was present in a significant number of cells in the primordia (Alpert et al., 1988a). These stained cells were located in groups throughout the gland and appeared along the pancreatic duct. By E-12 considerably more cells containing Tag were localized in groups along the ducts and in newly forming islets. The number of Tag cells increased throughout development such that by E-17 the majority of islet cells were Tag$^+$ (Fig. 6). Tag was not detected in acinar tissue. To determine the identity of the Tag$^+$ cells, sections of pancreas removed from transgenic mice from day 12 of embryonic development to adult were processed for dual immunocytochemical localization of Tag and

TABLE III

Coexpression of Tag in Cells Producing Endocrine Hormones in RIP–Tag2 Mice[a]

Day of development	Glu+ Tag+	In+ Tag+	Som+ Tag+	PP+ Tag+
E-10	1.0			
E-11	1.0			
E-12	1.0			
E-14	0.97 ± 0.02	1.0		
E-15	0.93 ± 0.04	1.0		
E-17	0.38 ± 0.07	1.0	0.36 ± 0.07	
P-1	0.35 ± 0.06	1.0	0.29 ± 0.08	0.25 ± 0.05
P-60	0.02 ± 0.01	1.0	0.08 ± 0.03	0.04 ± 0.03

[a]Three pancreata were processed for each developmental stage. The proportion of cells containing both Tag and the indicated hormone was determined in entire pancrata at E-10 to E-12, and in at least 10 endocrine cell clusters from E-14 tp P-1. The number of cells containing both Tag and the indicated endocrine hormone was expressed as the mean fraction (± standard error) of the cells immunoreactive for the hormone. From Alpert *et al.* (1988a).

insulin. Other sections were processed for immunolocalization of the oncogene and either glucagon, somatostatin, or PP. Throughout pre- and postnatal development, all the cells that contained the transgene also expressed insulin. In addition, some Tag+ cells contained either Som, Glu, or PP although their numbers decreased during development. The proportion of Tag+ Som+, Tag+ Glu+, and Tag+ PP+ cells was similar in all stages examined to that of cells that contained insulin and other pancreatic hormones (Table III). These studies demonstrated, therefore, that the information contained in the hybrid insulin gene of RIP1–Tag2 mice specifies transient expression of the pancreas in a fashion that reflects the expression of the endogenous gene (Fig. 4).

D.   NEUROBLASTS OF THE PERIPHERAL AND CENTRAL NERVOUS SYSTEM EXPRESS A HYBRID INSULIN GENE

Although it is generally believed that insulin gene expression is restricted to islet cells, the hybrid insulin gene was expressed not only by endocrine cells of pancreas but also by precursor cells of the peripheral and central nervous system of transgenic embryos.

It is now well established that in the CNS, proliferation of neuronal precursors occurs when these cells are located in the germinal or ventricular layer of the neural tube (Jacobson, 1978). When these cells become postmitotic, they migrate into the mantle layer and differentiate into neurons. In general, in brain, neurogenesis and differentiation occur during early and midgestation, while during late embryogenesis proliferative cells of the ventricular layer give rise to glial cells (Jacobson, 1978). Our immunocytochemical studies indicated that, in RIP–

Tag2 mice, the hybrid insulin gene was transiently expressed in neuronal precursors of the central and peripheral nervous system (Alpert *et al.*, 1988a,b). Thus, in the brain of E-10 embryos, neuroblasts located in the ventral region of the germinal (proliferating) layer of the fore-, mid-, and hindbrain and spinal cord contained Tag (Fig. 7). However, the expression of Tag in neuroblasts of the CNS was transient since cells of the germinal layer did not contain the oncogene after midgestation (S. Alpert, D. Hanahan, and G. Teitelman, unpublished results, 1988). Consequently, neurons of newborn and adult brain did not express Tag (G. Teitelman and T. Milner, unpublished results, 1991). These observations indicate, therefore, that in transgenic embryos, CNS neurons that were born during early development originated from precursors that expressed a hybrid

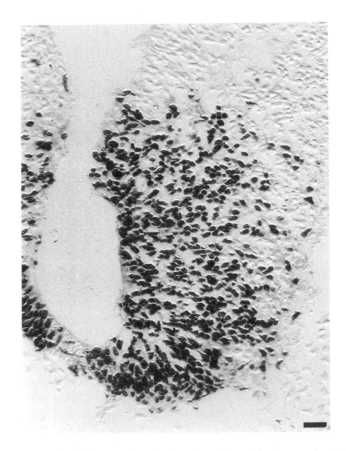

FIG. 7.   Cells containing Tag are present in the embryonic central nervous system (CNS). Photomicrograph illustrates the presence of Tag-containing cells in the forebrain of E-10 RIP1–Tag embryos. Immunostained cells are found in the germinal layer of the neural tube. Bar: 45 μm.

insulin gene. A similar pattern of expression by neuronal precursors was observed in the peripheral nervous system. Thus, precursor cells of neural crest origin contained Tag during their ventral migration (Fig. 8) but did not express the hybrid insulin gene after they aggregated and differentiated, forming the sympathetic ganglia (Alpert *et al.*, 1988a,b).

Neuroblasts of the CNS and PNS, however, did not contain IN immunoreactivity, suggesting that these cells did not express the endogenous insulin gene. It is possible that, in RIP–Tag2 embryos, the fusion gene lacks regulatory elements(s) that prevent endogenous insulin gene expression in neuroblasts. There are many examples of ectopic expression of fusion genes in transgenic animals. In brain, a transgene containing the mouse methallothionein-1 promoter linked to either rat or human growth hormone was found in neurons that did not contain endogenous methallothionein or growth hormone (Russo *et al.*, 1988). Since neurons expressing the methallothionein–growth hormone fusion gene had no developmental or functional relationship it was postulated that ectopic expression

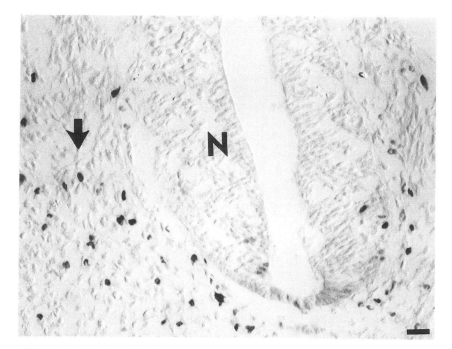

FIG. 8.   Localization of Tag in histological sections of E-11 RIP–Tag embryos. Immunoreactive (Tag+) cells, located alongside the neural tube (N), presumably are neural crest cells in their ventral pathway of migration to form the sympathetic ganglia. Arrow indicates the dorsoventral axis. Some Tag+ cells are also found within the neural tube. Bar: 40 μm.

was due to properties of the chimeric sequence (Russo *et al.*, 1988). In contrast, expression of the hybrid gene in RIP–Tag2 embryos occurred in cells that had in common the fact that they were neuroblasts, that is, they had not yet differentiated into neurons. Moreover, transient expression of Tag in a large number of neuroblasts of transgenic embryos harboring an insulin–Tag fusion gene was not due to the presence of the oncogene but, rather, it was determined by the regulatory sequences of the transgene. Thus, transgenic mice harboring a hybrid gene containing the 5′-flanking region of the rat preproglucagon gene linked to the DNA sequences encoding Tag, as expected, expressed the hybrid gene in α cells of the pancreas and in precursors and mature glucagon neurons of brain (Efrat *et al.*, 1988). Therefore, the distribution of neuronal precursors and neurons expressing the oncogene in brains of Glu–Tag mice was different from the distribution of Tag$^+$ cells in brains of mice harboring the hybrid insulin gene. These observations suggest the possibility that regulatory sequences similar to those that control transgene expression in RIP–Tag mice could be activated in subsets of neuronal precursors of nontransgenic mice during the early stages of neuronal differentiation; if so, this step could indicate the initial stage of commitment of those cells to a specific fate. It remains to be determined whether or not the same sequences that inhibit endogenous insulin expression in neuroblasts also regulate proper temporal expression of the hormone in pancreas.

The studies with transgenic animals, therefore, have provided new insight into the mechanisms that regulate insulin expression during development and have also documented similarities between islet cells and neurons. These observations and the fact that, as described below, islet cell precursors express both neuronal and endocrine traits when they first differentiate, raise the possibility that neurons and pancreatic islet cells may originate from precursors that display a strikingly similar repertoire of active genes. Section V,E below will examine whether this similarity is an indication of a lineage relationship.

V. Phenotypic Plasticity of Adult Islet Cells

A. ISLET CELL PRECURSORS AND SENESCENT CELLS EXPRESS A NEURONAL ENZYME

The expression of the hybrid insulin gene by neuroblasts of the peripheral and central nervous system provided yet another example of the ability of embryonic cells reported to express some properties transiently during development (Kosik *et al.*, 1988; Wolozin *et al.*, 1988; Vacher and Tilghman, 1990). Over 10 years ago it was reported that cells of the murine embryonic gut, believed to be neuronal progenitor cells, expressed tyrosine hydroxylase (TH, tyrosine monooxygenase), a neuronal enzyme, in an impermanent fashion (Cochard *et al.*, 1978; Teitelman *et al.*, 1978). Similarly, in pancreas, some cells of the pancreatic duct transiently expressed TH during development (Teitelman and Lee, 1987;

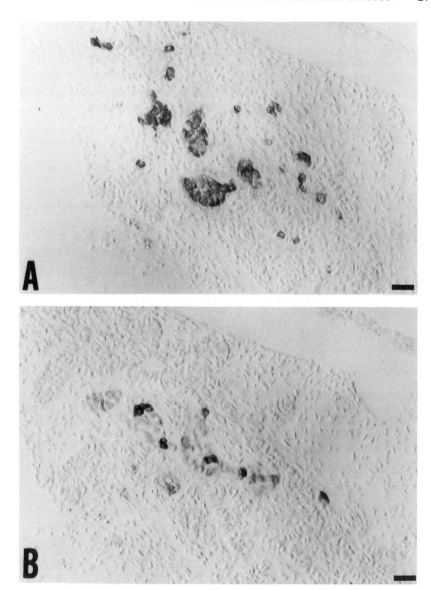

FIG. 9.   Distribution of TH+ cells in embryonic pancreas. Immunocytochemical localization of glucagon (A) and TH (B) in consecutive sections of day-14 mouse pancreas. Note the similar distribution of both types of cells along the pancreatic duct, suggesting that they belong to the same islet. In contrast, clusters of glucagon cells located at a distance from duct lack cells containing TH (not shown). Bar: 45 μm. From Teitelman and Lee (1987).

Fig. 9). It has generally been assumed that islet cell precursors present in the embryonic pancreatic duct migrate into the surrounding connective tissue stroma where they differentiate into the various islet cell types (Picter and Rutter, 1972). Taken together, these observations raised the possibility that TH may be a marker of islet cell precursors and that, during migration from the pancreatic epithelium to the stroma, the cells inhibit expression of the enzyme while they initiate synthesis of the pancreatic hormones.

However, in contrast to the previously proposed site of differentiation of pancreatic islet cells in the connective tissue (Pictet and Rutter, 1972), it was recently found that progenitor cells expressed not only TH but also pancreatic hormones when they were still located in the pancreatic duct (Teitelman and Lee, 1987). Thus, double-label immunohistochemical studies revealed that at E-10, islet cell precursors located in the duct contained TH and Glu (Alpert *et al.*, 1988a). Two days later, when IN first appeared, IN[+] cells present in the duct contained TH and Glu (Fig. 4; Teitelman and Lee, 1987; Alpert *et al.*, 1988a). In RIP–Tag embryos, these progenitor cells expressed the oncogene in addition to the hormones and the neuronal enzyme (Fig. 4; Alpert *et al.*, 1988a). These observations indicated that islet cell precursors present in the pancreatic duct expressed both endocrine (i.e., IN and Glu) and neuronal traits (TH) when they first differentiated (Fig. 4). Moreover, these findings also suggested that the appearance of those initial differentiated traits may be induced by cues released by other cells of the pancreatic duct or, as proposed by Dudek and Lawrence (1988), by cells from neighboring tissues.

Islet precursors underwent a second step in the pathway of differentiation when they left the pancreatic duct and migrated into the surrounding stroma. At

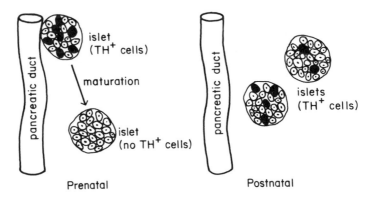

FIG. 10.   Schematic diagram illustrates the distribution of TH[+] cells in pancreas of pre- and postnatal mice. In embryos, cells containing TH are found in the epithelium of the pancreatic duct or in islets emerging from the duct. Presumably, these TH[+] cells are precursors that inhibit the expression of the neuronal enzyme during maturation. Mature islets, which do not contact the duct, lack TH cells. After birth, all the islets contain a few TH cells.

this stage developing islet cells suppressed the expression of the neuronal enzyme but retained their endocrine properties (Fig. 10). The embryonic endocrine cells gradually matured and, in adults, most of them contained only one hormone (Figs. 4 and 10).

The distribution of cells containing TH in pancreas changed after birth. One of the main differences involved the absence of TH cells from the pancreatic duct of postnatal and adult mice. In addition, differences were also found between islets of embryonic and postnatal mice: i.e., while in embryos the distribution of cells expressing the neuronal enzyme was restricted to islets found in apposition or still connected to the duct, all islets of pups and adult mice contain TH$^+$ cells (Fig. 10).

These intriguing results raised the question of whether TH observed in adult islets occurred in precursor cells or, alternatively, whether TH was reexpressed by a mature islet cell. Immunocytochemical studies of adult mouse pancreas revealed that TH colocalized with insulin in β cells while other islet cell types did not contain the enzyme (Teitelman et al., 1988). To ascertain whether these TH–insulin cells were precursor cells that proliferated and increased in number, thereby affecting islet growth, two different types of nontumorigenic β cell hyperplasia were examined. In pregnant and mutant obese (ob/ob) mice, the increase in β cell number was transient (Helleström and Sweene, 1985; Shafrir, 1990).

In those two cases of transient islet growth the observed increase in β cell number was accompanied by a corresponding increase in the percentage of TH–insulin cells (Teitelman et al., 1988; Fig. 11). Conceivably, the relatively rare TH–insulin cells present in all pancreatic islets of 3-month-old mutant mice proliferated, thereby giving rise to the large number of TH–insulin cells observed in islets of older mice. To test this, mutant and control mice were injected with [^{3}H]thymidine ([^{3}H]TdR) and the pancreas examined with immunohistochemical techniques to localize TH or insulin followed by autoradiographic localization of the isotope. The isotope was localized in the nucleus of immunostained insulin cells (Fig. 12A). Since the time elapsed between injection of the isotope and sacrifice of the animal was so brief, it is unlikely that these proliferating cells were undifferentiated precursors that initiated insulin synthesis after completion of the mitotic cycle. Rather, these cells probably contained insulin prior to incorporation of the isotope into DNA. Since most cells that go through the S phase of the cell cycle divide, it is likely that β cells containing the isotope were indeed proliferating. At all sites examined, islets of mutant mice contained significantly more proliferating IN$^+$ cells per islet than control groups, which accounted for the increase in islet size. In contrast to IN "only" cells, cells containing TH did not incorporate [^{3}H]TdR (Fig. 12B). The fact that all the cells that expressed TH also contained insulin demonstrated that the subset of β cells that contain the catecholamine enzyme (IN$^+$ TH$^+$) did not divide. Analysis of hyperplastic islets of pregnant mice further confirmed that

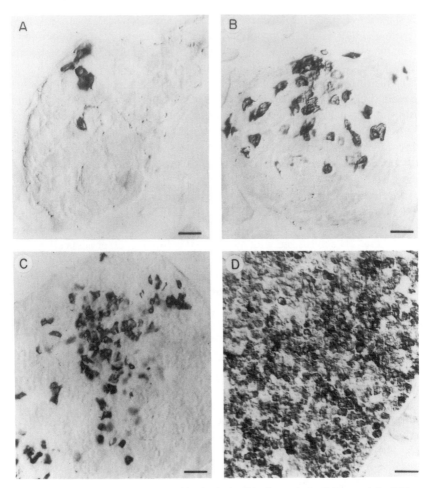

FIG. 11.   Increase in the number of TH cells during islet hyperplasia: Cells containing TH were
localized in pancreatic sections by immunostaining with antibodies specific for tyrosine hydroxylase.
(A) Cells containing TH in the pancreas of CD-1 normal mouse. Bar: 30 μm. (B) Cells containing
TH in the pancreas of a pregnant mouse. Bar = 30 μm. (C) Pancreatic sections from a 5-month-old
mutant obese mouse. Note the large number of TH cells present. Bar = 75 μm. (D) Hyperplastic
pancreatic islet of a 9-week-old RIP–Tag2 mouse. Most of the cells contain TH. Bar = 75 μm. From
Teitelman *et al.* (1988).

islet growth was due to proliferation of insulin-only cells, whereas β cells containing TH were mitotically quiescent. These unequivocal findings, therefore,
demonstrated that the IN^+ TH^+ cells observed in adult islets were not precursor
cells. Instead, we concluded that these cells were a previously unrecognized
subpopulation of mature β cells characterized by their failure to proliferate.

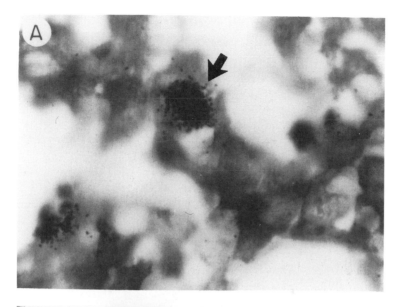

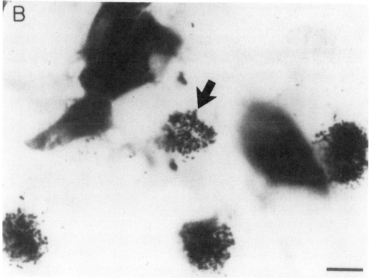

FIG. 12.   TH cells in hyperplastic islets do not divide. Mutant obese mice were injected with [³H]thymidine, and the pancreata were removed 1 hour later. Sections were immunostained and processed for autoradiography. (A) Section immunostained with specific antisera against insulin. All cells labeled with [³H]thymidine (arrow) also contained the peptide. Bar = 6 μm. (B) Another section stained with TH antibodies demonstrates that TH cells have not incorporated [³H]thymidine (arrow) into their nuclei. Bar = 6 μm. From Teitelman *et al.* (1988).

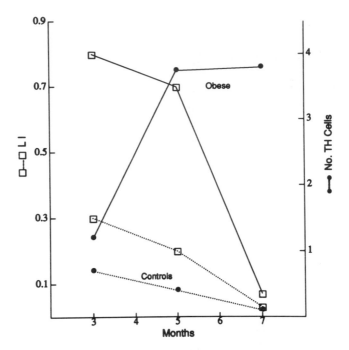

FIG. 13. Temporal variation in islet cell types in pancreas of obese mice. The proportion of labeled cells was determined in at least 10 islets per pancreas with at least three pancreata processed from each time point. The number of labeled cells in an islet was expressed as a percentage of the total number of cells in that islet. The average of these percentages is referred to as labeling index (LI). The labeling index was determined in slides of pancreas stained with insulin antisera and processed for autoradiographic visualization of [³H]thymidine. The percentage of TH cells was determined in at least 10 islets per pancreas. Three pancreata were processed for each time point. The number of stained cells in an islet is expressed as a percentage of the total number of cells for the islet. From Teitelman *et al.* (1988).

Analysis of the temporal variation in islet cell populations revealed that the increase in the percentage of proliferating IN⁺ cells was transient and it was followed by an increase in the percentage of nondividing IN⁺ TH⁺ cells (Fig. 13). The fact that large numbers of nondividing TH–insulin cells accumulated following active proliferation raised the possibility that, in adult mice, β cells reexpressed TH only after they exhausted their mitotic potential (Fig. 14). It has been suggested that β cells have a limited proliferative capacity (Logothetopoulos, 1972; Helleström and Sweene, 1985) and nonproliferating TH–insulin cells may, therefore, represent cells that have become senescent and are destined to be eliminated (Teitelman *et al.*, 1988). The finding that TH was expressed by immature cells of embryos and by senescent cells of adult islets

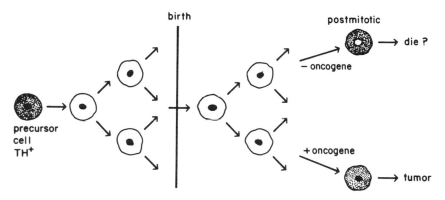

FIG. 14.   The presence of TH is a useful marker of different stages of the life cycle of pancreatic
β cells. Pancreatic β cells originate from precursors present in the·pancreatic duct that express
endocrine properties and the neuronal enzyme (dark cytoplasm). During maturation, the cells retain
the endocrine traits but inhibit TH expression. The IN⁺ cells (clear cytoplasm) proliferate, thereby
increasing islet size. This model postulates that IN⁺ cells have a limited mitotic potential. When β
cells exhaust their ability to proliferate, they become mitotically quiescent cells and reexpress TH.
IN⁺ cells expressing an oncogene undergo the same "aging" process as the nontumorigenic counterparts and become immortal tumor cells after the reappearance of TH. As in embryonic precursors, the
presence of the neuronal enzyme in neoplastic β cells is also transient, since fully developed tumors
rarely contain TH⁺ cells. Proliferating cells have dark nuclei; nonproliferating cells have clear nuclei.

suggests that the enzyme could become an extremely useful marker to study the
life cycle of β cells *in vivo* and *in vitro*.

B.   IN THE PRESENCE OF AN ONCOPROTEIN, SENESCENT β CELLS BECOME IMMORTAL TUMOR CELLS

The third case of islet cell growth examined was that occurring in transgenic
RIP1–Tag2 mice. In this line of mice the β cells of the islets expressed an
oncoprotein (Tag) that resulted in formation of hyperplastic islets and β cell
tumors in adult mice (Hanahan, 1985). When the rate of cell proliferation was
determined using a sequential immunological–autoradiographic procedure (described in the previous section), it was found that, in the prencoplastic period, the
number of proliferating IN⁺ cells, i.e., β cells that incorporated [³H]thymidine
into their nuclei, increased gradually with time. In older animals, the rate of cell
proliferation had increased dramatically and several islets had developed into
large tumors in which most IN⁺ cells were labeled with isotope (Table IV).
The number of TH–insulin cells also increased during the preneoplastic period
(Fig. 11 and Table IV). In contrast to the nontransgenic models discussed above,
however, TH–insulin cells proliferated in RIP–Tag mice. The percentage of

TABLE IV

Insulin-Producing Cell Types in Pancreatic Islets of RIP–Tag2 Transgenic Mice[a]

Age (Weeks)	Genotype	[³H]Thymidine	TH	TH + [³H]thymidine	Islet type
4	RIP–Tag2	3.0 ± 0.2	9.27 ± 1.35	0.4 ± 0.08	A
	Control	1.0 ± 0.2	2.80 ± 0.36	0	
6	RIP–Tag2	5.3 ± 0.5	8.28 ± 1.38	0.7 ± 0.1	A
	Control	1.4 ± 0.14	2.78 ± 0.24	0	
9.5	RIP–Tag2	4.9 ± 0.5	32.1 ± 4.7	0.6 ± 0.1	B
		14.0 ± 1.0	50.7 ± 5.2	8.8 ± 0.4	C
		22.4 ± 1.1	0	0	D
	Control	0.43 ± 0.03	1.5 ± 0.19	0	

[a]Proportions of cells containing [³H]thymidine (third column) in at least 10 islets per pancreas with at least 2 pancreata processed for each time point. The number of labeled cells in an islet was expressed as a percentage of the total number of cells in that islet. The average of these percentages is referred to as the labeling index (LI). The percentage of TH cells (fourth column) was determined in at least 10 islets per pancreas. Three pancreata were processed for each time point. The number of stained cells in an islet is expressed as a percentage of proliferating TH cells for that islet. The percentage of proliferating TH cells (fifth column) was determined in slides stained with TH antisera and then processed for visualization of the isotope. Those values were calculated as a ratio of the number of dividing TH cells over the total number of TH cells. Unstained cells were considered insulin cells. Reprinted from Alpert *et al.* (1988a).

dividing TH–insulin cells was initially low but increased significantly during tumor formation, suggesting that TH^+ IN^+ cells containing an oncogene escaped the usual senescent pathway and instead gave rise to neoplastic β cells (Teitelman *et al.*, 1988; Fig. 13). However, fully developed tumors formed by actively proliferating β cells contained few or no TH cells (Table IV). Thus, the patterns of TH expression in β cells containing an oncogene identified a stage in the preneoplastic period that preceded the transformation of the cells into tumor cells. During that stage critical secondary events may occur that will lead to the formation of immortal tumor cells (Hanahan, 1989). Neoplastic β cells, however, no longer expressed the neuronal enzyme.

TABLE V

Labelling Index of Glucagon Cells[a]

	2 months		5 months	
db/db	0.1	0.02	0.15	0.01
control	0.08	0.003	0.07	0.005

[a]Control and *db/db* mice, 2 and 5 months old, were injected with tritiated thymidine(H-tdR) which labels proliferating cells. The animals were sacrificed one hour later and the pancreas was processed for immunocytochemical staining followed by autoradiography.

C.   MATURE β CELLS REEXPRESS OTHER EMBRYONIC ANTIGENS

The observation that mature β cells reexpressed TH following intensive proliferation induced by either hyperglycemia, hormones, or oncogenes provided
the first indication that the phenotype of adult islet cells is labile. This raised the
possibility that mature β cells may acquire other traits, in addition to TH, that
characterize islet precursor cells of embryos. Since in the proposed model of
pancreatic islet cell differentiation progenitor cells were able to synthesize the
four pancreatic hormones (Fig. 4), the question was raised as to whether insulin-
containing cells of adults could reexpress not only the catecholamine enzyme but
also other pancreatic hormones. In normal animals a subset of insulin cells also
contained glucagon (Alpert *et al.*, 1988a). Studies of islet cell composition in
pancreas of adult diabetics (C57B/16Ks *db*/*db*) indicated that this subset of IN[+]
Glu[+] cells increases during development of the disease (G. Teitelman and M.
Moustakos, unpublished observation). Those preliminary findings suggested that
one of the possible sources of these chimeric cells were β cells that had reiniti-
ated the expression of glucagon.

In the initial phase of those studies the number of different islet cell types in
control and *db* mice was determined. This analysis confirmed earlier observations (Gapp *et al.*, 1983; Helleström and Sweene, 1985) that the percentage of
glucagon cells in islets of mutant *db*/*db* mice increased gradually during

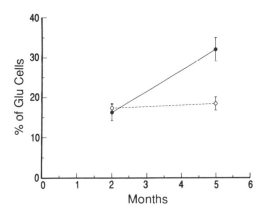

FIG. 15.   Temporal increase in the percentage of Glu cells in *db* mice. Each pancreas was serially
sectioned and one of every five sections was collected and immunostained for glucagon. Islets for
analysis were chosen randomly and at least 20 counted for each animal. Three animals of each type
(age/phenotype) were analyzed. In each islet, the total number of cells was counted using a gridded
reticule eyepiece at ×400 magnification and Nomarski optics. The results are expressed as a percentage of that islet cell type. (●), *db*/*db*; (○), control.

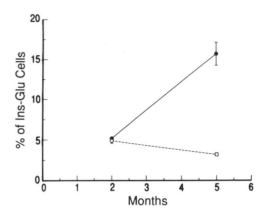

FIG. 16. Temporal increase in the percentage of mixed insulin–glucagon cells in *db* mice. Pancreata from 2- and 5-month-old *db/db* (●) and control (○) mice were processed for double immunocytochemical localization of glucagon and insulin in the same tissue section as described in Pickel *et al.* (1985) and Fig. 1. The percentage of cells containing both hormones was determined as described in Table I.

postnatal development, reaching a level that was significantly higher than in controls by 5 months of age (Fig. 15). Further studies indicated that the increase in the number of Glu[+] cells was not due to selective cell proliferation, since α cells in *db* mice had a very low level of cell division (Table V), or to shrinkage of the islets from β cell death, since, at the stages examined, islets of mutant and control mice contained similar numbers of IN[+] cells (G. Teitelman, unpublished observations). Instead, the increase in the percentage of glucagon cells in *db* mice appeared to be due to an increased proportion of the subset of β cells that contained glucagon in addition to insulin (Fig. 16). These results, therefore, strongly suggested that β cells were able to reinitiate expression of traits that, like TH, are characteristic of precursor cells and that reappearance of each of these embryonal antigens was elicited by specific and perhaps independent cues. The expression of multiple pancreatic hormones by cloned cell lines derived from islet cell tumors (Phillipe *et al.*, 1987; Madsen *et al.*, 1986) also may reflect a regression, during tumorigenesis, of neoplastic cells to a pluripotent phenotype characteristic of embryonal islet cells.

D.   ISLET CELLS *in Vitro* FROM NEURITES

The surprising phenotypic versatility of mature pancreatic endocrine cells is also exemplified by their ability to form neurite-like extensions *in vitro*

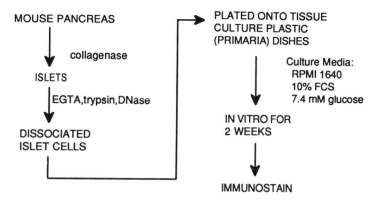

FIG. 17.   Flow diagram of the method used to examine adult islet cells in dissociated cell culture. FCS, Fetal calf serum.

(Teitelman, 1990). It has been found that when islets isolated from adult white outbred mice (CD-1) were dissociated into single cells, placed in culture, and immunostained with insulin antibodies after several days *in vitro*, approximately 20% of the IN[+] cells extended long stained processes (Figs. 17 and 18). Although other islet cell types, i.e., α, δ, PP, and TH cells were also occasionally seen, none of these cells extended processes. However, the low number of non-β cells found in these cultures limited our ability to draw definite conclusions regarding their neurite-forming capacities. The cellular elongations produced by dissociated β cells *in vitro* contained, in addition to insulin, an intermediate filament protein specific to neurites (Teitelman, 1990; Fig. 19). It should also be pointed out that, although the mammalian pancreas contains intrinsic parasympathetic ganglia (Le Douarin, 1982), the IN[+] cells forming processes in culture are not these neurons since autonomic neurons do not contain insulin (Fontaine *et al.*, 1977; Le Douarin, 1982).

In contrast to cultures of dissociated islet cells, islets that were explanted without dissociation did not extend processes. This suggested that neurite extension was facilitated by disruption of the normal histiotypical associations among cells of pancreatic islets and not by factors present in culture (Teitelman, 1990).

E.   PROPOSED MODEL OF ISLET CELL DIFFERENTIATION

It is now well established that islet cells and neurons of adults share many functional properties and express a large number of common traits (Le Douarin, 1988). Peripheral and/or central neurons contain peptides similar or identical to pancreatic glucagon (Tager *et al.*, 1980; Han *et al.*, 1986; Efrat *et al.*, 1988) somatostatin and PP (Reichlin, 1983; Palkovitz, 1984; Tatemoto *et al.*, 1982).

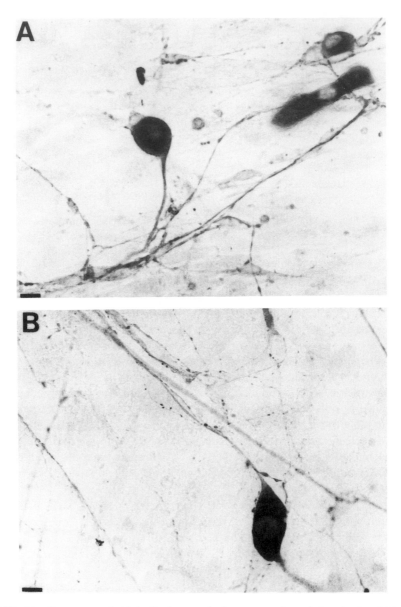

FIG. 18.   Some pancreatic islet cells of mouse display a neuronal morphology *in vitro*. Islet cells in culture were immunolabeled with an antiserum to insulin. The DAB reaction product decorates both perikarya- and neurite-like processes. Photomicrographs illustrate two examples of neurite-bearing IN⁺ cells. These cells are surrounded by numerous unstained cells. Bar: (A) 10 μm; (B) 7 μm. From Teitelman (1990).

Conversely, islet cells not only extend neurites and express TH but also display other neuronal traits such as neuron-specific enolase (Polak *et al.*, 1984), the cell-surface antigens PGP 9.5 (Thompson *et al.*, 1983), A2B5, and receptors for tetanus toxin (Eisenbarth *et al.*, 1982), the epinephrine-synthesizing enzyme phenylethanolamine *N*-methyltransferase (PNMT; Teitelman and Evinger, 1989), the enzyme glutamic acid decarboxylase (GAD; Baekkeskov *et al.*, 1990) and its transmitter product γ-aminobutyric acid (GABA; Garry *et al.*, 1986). This enormous repertoire of shared traits by these two cell types highlights a long-standing controversy regarding the embryological origin of pancreatic endocrine cells. The classical cell lineage theory holds that, during development of vertebrates, endoderm gives rise to gut, mesoderm to connective tissue, and ectoderm to skin and nervous tissue. Pearse (1976, 1977) initially proposed that pancreatic islet cells originate from the neural crest. However, work in birds (Fontaine *et al.*, 1977; Andrew, 1984; Le Douarin, 1982) and mammals (Pictet and Rutter, 1972; Pictet *et al.*, 1976) indicated an alternative possibility, namely, that although islet cells express a neuronal phenotype they originate from endodermal precursors (Le Douarin, 1988). Should the latter be true, islet cells and neurons will join the growing list of known cases of convergent differentiation; the most notable examples are connective tissue derivatives which, in the head, originate from neuroectoderm, and elsewhere in the body from mesoderm (Le Douarin, 1982).

While many of the neuronal properties of islet cells are probably required for their physiological activities, it is difficult to account for the presence of other neuronal traits solely on functional grounds. Thus, while islet cells express the catecholamine biosynthetic enzymes TH and PNMT, they do not synthesize endogenous catecholamines (Teitelman and Evinger, 1989). Similarly, islet cells do not require neurites to function properly. It is unlikely, however, that the appearance of these properties is due to the ability of islet cells to adopt a variety of unrelated phenotypes, since islet cells do not express other "inappropriate traits." Rather, it is possible that activation of pancreatic peptide hormone genes occurs concurrently with that of a set of neuronal-specific genes. The striking functional, morphological, and biochemical homology displayed by islet cells and neurons (Pearse, 1976; Habener, 1987; Alpert *et al.*, 1988a; Efrat *et al.*, 1988) lends support to this proposition.

An evaluation of the results presented here suggests a model of pancreatic islet cell differentiation. According to this model (Fig. 20) the first stage of differentiation involves the activation of a set of "neuroendocrine genes" that will enable the cells to express all four hormones in addition to neuronal traits. Endocrine and exocrine precursor cells probably diverge by this stage, since islet cells do not express specific exocrine markers (Pictet and Rutter, 1972; Teitelman *et al.*, 1987b). Once progenitor cells are committed to an endocrine pathway, their range of phenotypic traits becomes progressively more restricted until they

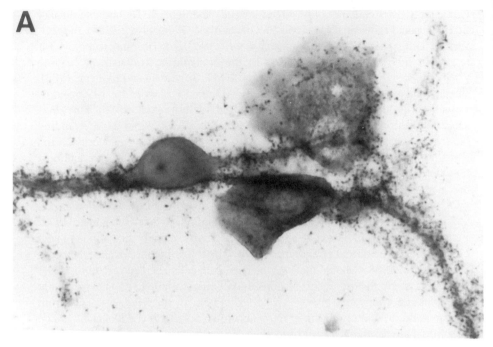

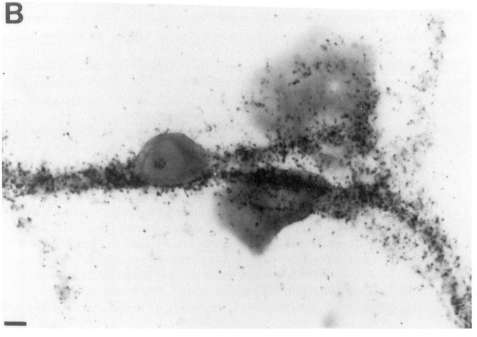

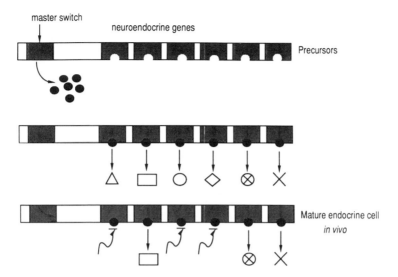

FIG. 20. Proposed model of islet cell differentiation. This hypothetical model proposes that the initial stage of differentiation of pancreatic islet cell precursors involves the activation of a large set of "neuroendocrine" genes. The result of this event is the appearance, in embryonic endocrine cells, of many neuronal and endocrine traits. This model also proposes that, during maturation, the expression of some traits is inhibited. Consistent with findings reported here, the model also postulates that this inhibition of expression can be overcome by a variety of epigenetic cues.

retain traits characteristic of only one cell type. However, this model also proposes that inhibition of those evanescent traits is not permanent since it can be released by specific environmental cues such as hormones, hyperglycemia, and disruption of cellular contacts. The identity of the signals that regulate phenotypic expression of islet cells of embryos and adults is largely unknown. Once the mechanisms that modulate islet cell expression are characterized it will be possible to induce experimentally alterations in phenotype. Conceivably, these changes in cell properties may be caused by and/or be correlated to perturbations

←

FIG. 19. Processes of "neuronal-like" cells contain neurofilament. Double immunocytochemical localization of insulin (visualized with DAB) and neurofilament (visualized with ^{125}I-labeled IgG) antisera in process-bearing islet cells. (A) and (B) are photomicrographs of the same field taken at different planes of focus: in (A) the focus is on the cells while in (B) the focus is on the overlying silver grains. The photographs illustrate a double-labeled "neuronal-like" cell in which the perikarya contains insulin while the neurite-like extension contains insulin and neurofilament. Two other cells, labeled only with insulin, display the typical morphology of β cells in culture. A network of doubly labeled processes, present to the right of these cells, is located at a different plane of focus. Bar: 4 μm. From Teitelman (1990).

of metabolic processes. If so, identification of the signals that regulate the differentiation of pancreatic islet cells and the stability of the differentiated phenotype may provide clues to the mechanisms that lead to β cell dysfunction.

ACKNOWLEDGMENTS

The author wishes to thank Drs. L. Iacovitti, M. Ehrlich, and D. Hanahan for providing helpful suggestions and criticism. This work was performed while the author was a member of the Laboratory of Neurobiology, Cornell University Medical College, New York, New York and was partially supported by a grant from NIDDK.

REFERENCES

Alpert, S., Hanahan, D., and Teitelman, G. (1988a). *Cell (Cambridge, Mass.)* **53**, 295–308.
Alpert, S., Hanahan, D., Moustakos, M., and Teitelman, G. (1988b). *Soc. Neurosci. Abstr.* **39.3.**
Altman, P. L., and Dittmer, D. S. (1983). *In* "Biology Data Book, 2nd Edition" Vol. 1, pp. 178–180. Federation of American Societies for Experimental Biology, Bethesda, Maryland.
Anderson, D. J. (1989). *Soc. Dev. Biol.* **47**, 17–36.
Andrew, A. (1984). *In* "Proceedings of the Fifth E. K. Fernstrom Symposium" (S. Falkner, R. Hakanson, and F. Sundler, eds.), pp. 91–111. Elsevier; Amrsterdam.
Baekkeskov, S., Aanstoot, H. J., Christgau, S., Reetz, A., Solimena, M., Cascalho, M., Folli, F., Richter-Olsesen, H., and De Camilli, P. (1990). *Nature (London)* **347**, 151–156.
Behringer, R. R., Mathews, L. S., Palmiter, R. D., and Brinster, R. L. (1988). *Genes Dev.* **2**, 453–461.
Bronner-Fraser, M., and Fraser, S. E. (1988). *Nature (London)* **335**, 161–164.
Burgess, A., and Nicola, N. (1983). "Growth Factors and Stem Cells." Academic Press, New York.
Cochard, P., Goldstein, M., and Black, I. B. (1978). *Proc. Natl. Acad. Sci. U.S.A.* **75**, 2986–2990.
Darnell, J. E. (1982). *Nature (London)* **297**, 365–371.
Davidson, E. (1990). *Development (Cambridge, U.K.)* **180**, 365–389.
Dudek, R. W., and Lawrence, I. E., Jr. (1988). *Diabetes* **37** (70), 892–900.
Efrat, S., and Hanahan, D. (1987). *Mol. Cell. Biol.* **7**, 192–198.
Efrat, S., Teitelman, G., Anwar, M., Ruggiero, D., and Hanahan, D. (1988). *Neuron* **1**, 605–613.
Eisenbarth, G. S., Shimizu, K., Bowring, M. A., and Well, S. (1982). *Proc. Natl. Acad. Sci. U.S.A.* **79**, 5066–5070.
Fontaine, J., Le Lievre, C., and Le Douarin, N. M. (1977). *Gen. Comp. Endocrinol.* **33**, 394–404.
Gapp, D. A., Leiter, E. H., Coleman, D. L., and Schwizer, R. W. (1983). *Diabetologia* **25**, 439–443.
Garry, D. J., Sorenson, L., Elde, R. P., Maley, B. E., and Madsen, A. (1986). *Diabetes* **35**, 1090–1096.
Gordon, J. W. (1989). *Int. Rev. Cytol.* **115**, 171–229.
Habener, J. F. (1987). "Molecular Cloning of Hormone Genes." Humana Press, Clinton, New Jersey.
Han, V. K. M., Hynes, M. A., Jin, C., Towle, A. C., Landes, J. M., and Lund, P. K. (1986). *J. Neurosci. Res.* **16**, 97–107.
Hanahan, D. (1985). *Nature (London)* **315**, 115–122.
Hanahan, D. (1989). *Science* **246**, 1265–1275.
Hand, A. R., and Oliver, C., eds. (1981). "Basic Mechanisms of Cellular Secretion." Academic Press, New York.
Helleström, C., and Sweene, I. (1985). *In* "The Diabetic Pancreas" (B. W. Volk and E. R. Arquilla, eds.), pp. 53–80. Plenum, New York.

Herberg, L., and Coleman, D. L. (1977). *Metab. Clin. Exp.* **26,** 59–99.

Hogan, B., Constantini, F., and Lacy, E. (1986). "Manipulating the Mouse Embryo." Cold Spring Harbor Lab., Cold Spring Harbor, New York.

Hwung, Y. P., Crowe, D. T., Peyton, M., Tsai, S. Y., and Tsai, M. J. (1989). *In* "Current Communications in Molecular Biology: Perspectives on the Molecular Biology and Immunology at the Pancreatic β Cell" (D. Hanahan, H. O. McDevitt, and G. F. Cahill, eds.), Cold Spring Harbor Lab., Cold Spring Harbor, New York.

Jacobson, M. (1978). "Developmental Neurobiology," 2nd ed. Plenum, New York.

Jacobson, M. (1986). *Annu. Rev. Neurosci.* **8,** 71–102.

Kosik, K. S., Orecchio, L. D., Bakalis, S., and Neve, R. L. (1988). *Neuron* **2,** 1389–1397.

Laimins, L., Holmgren-Konig, M., and Khoury, G. (1986). *Proc. Natl. Acad. Sci. U.S.A.* **83,** 3151–3155.

Le Douarin, N. M. (1982). "The Neural Crest." Cambridge Univ. Press, London.

Le Douarin, N. M. (1988). *Cell (Cambridge, Mass.)* **53,** 169–171.

Like, A. A. (1977). *In* "The Diabetic Pancreas" (B. W. Volk and K. F. Wellmann, eds.), pp. 381–423. Plenum, New York.

Logothetopoulos, J. (1972). *In* "Handbook of Physiology" (R. Q. Grup and E. B. Astwood, eds.), Vol. 1, Sect 7, pp. 67–76. Am. Physiol. Soc., Washington, D.C.

Madsen, O. D., Larsson, L. I., Rehfeld, J. F., Schwartz, T. W., Lernmark, A., Labrecque, A. D., and Steiner, D. F. (1986). *J. Cell Biol.* **103,** 2025–2034.

McIntosh, R., Pictet, R. L., Kaplan, S. L., and Grumbach, M. M. (1977). *Endocrinology* **101,** 825–829.

Neil, J. D., Smith, P. F., Luque, E. H., Munoz de Toro, Nagy, G., and Nulchahey, J. J. (1987). *Recent Prog. Horm. Res.* **43,** 175–225.

Orci, L. (1982). *Diabetes* **31,** 538–545.

Palkovitz, M. (1984). *Neuroendocr. Perspect.* **3,** 1–69.

Palmiter, R. D., and Brinster, R. L. (1986). *Annu. Rev. Genet.* **20,** 465–499.

Pearse, A. G. E. (1976). *Nature (London)* **262,** 92–94.

Pearse, A. G. E. (1977). *Med. Biol.* **55,** 115–125.

Phillippe, J., Chick, W. L., and Habener, J. F. (1987). *J. Clin. Invest.* **79,** 531–358.

Pickel, V. M., Chan, J., and Milner, T. A. (1985). *J. Histochem. Cytochem.* **34,** 707–718.

Pictet, R. L., and Rutter, W. J. (1972). *In* "Handbook of Physiology" (D. F. Steiner and M. Frenkel, eds.), Sect. 7, Vol. 1, pp. 25–66. *Am. Physiol. Soc.,* Washington, D.C.

Pictet, R. L., Rall, L. B., Phelps, P., and Rutter, W. J. (1976). *Science* **191,** 191–193.

Polak, J. M., Bloom, S. R., and Marango, P. J. (1984). *In* "Evolution and Tumor Pathology of the Neuroendocrine System" (S. Falkmer, R. Hakanson, and F. Sundler, eds.), p. 433. Elsevier, Amrsterdam.

Rahier, J., Wallon, J., Gepts, W., and Haot, J. (1979). *Cell Tissue Res.* **200,** 359–366.

Rall, J. B., Pictet, R. L., Williams, R. H., and Rutter, W. J. (1973). *Proc. Natl. Acad. Sci. U.S.A.* **70,** 3478–3482.

Reddy, S., and Elliot, R. B. (1988). *Experientia* **44,** 1–90.

Reichlin, S. (1983). *In* "Brain Peptides" (D. T. Krieger, M. J. Brownstein, and J. B. Martin, eds.), pp. 711–752. Wiley, New York.

Rossant, J. (1986). *Trends Genet.* **2,** 302–303.

Rothman, J. E., and Orci, L. (1990). *FASEB J.* **4,** 1460–1468.

Rothman, T. P., Specht, L. A., Gershon, M. D., Joh, T. H., Teitelman, G., Pickel, V. M., and Reis, D. J. (1980). *Proc. Natl. Acad. Sci. U.S.A.* **77,** 6221–6225.

Russo, A. F., Crenshaw, E. B., III, Lira, S. A., Simmons, D. M., Swanson, L. W., and Rosenfeld, M. G. (1988). *Neuron* **1,** 311–320.

Rutter, W. J., Wessels, N. K., and Grobstein, C. (1964). *Natl. Cancer. Inst. Monogr.* **13,** 52–65.

Rutter, W. J., Pictet, R. L., Harding, J. D., Ghirgwin, J. M., MacDonald, R. J., and Przyblyla, A.

E. (1978). *In* "Molecular Control of Proliferation and Differentiation" (J. Papaconstantinou and W. J. Rutter, eds.). Academic Press, New York.

Sanes, J. R., Rubestein, J. L. R., and Nicholas, J. F. (1990). *EMBO J.* **5,** 3133–3142.

Serbedzija, G. N., Frazer, S. C., and Bronner-Fraser, M. (1990). *Development (Cambridge, U.K.)* **108,** 605–612.

Shafrir, E. (1990). *In* "Diabetes Mellitus: Theory and Practice" (H. Rifkin and D. Porte, eds.), 4th ed., pp. 229–340. Elsevier, New York.

Sieber-Blum, M. (1989). *Science* **243,** 1608–1611.

Sieber-Blum, M., Kumar, S. R., and Riley, D. A. (1988). *Brain Res.* **476,** 69–83.

Steiner, D. F. (1983). *Harvey Lect.* **78,** 191–229.

Steiner, D. F., Bell, G. I., and Tager, H. S. (1989). *In* "Endocrinology" (L. J. DeGroot, ed.), p. 1263. Saunders, Philadelphia, Pennsylvania.

Stockdale, F. E. (1989). *Symp. Soc. Dev. Biol.* **47.**

Tager, H., Hohenboken, M., Markese, J., and Dinerstein, R. J. (1980). *Proc. Natl. Acad. Sci. U.S.A.* **77,** 6229–6233.

Tatemoto, K., Carlquist, M., and Mutt, V. (1982). *Nature (London)* **296,** 659–660.

Teitelman, G., and Lee, J. K. (1987). *Dev. Biol.* **121,** 454–466.

Teitelman, G., Joh, T. H., and Reis, D. J. (1978). *Brain Res.* **158,** 229–234.

Teitelman, G., Joh, T. H., and Reis, D. J. (1981). *Proc. Natl. Acad. Sci. U.S.A.* **78,** 5225–5229.

Teitelman, G., Lee, J. K., and Reis, D. J. (1987a). *Dev. Biol.* **120,** 425–433.

Teitelman, G., Lee, J. K., and Alpert, S. (1987b). *Cell Tissue Res.* **250,** 435–439.

Teitelman, G., Alpert, S., and Hanahan, D. (1988). *Cell (Cambridge, Mass.)* **52,** 97–105.

Teitelman, G. (1990). *Dev. Biol.* **142,** 368–379.

Teitelman, G., and Evinger, M. J. (1989). "Current Communications in Molecular Biology. The Pancreatic Cell: Development, Cell and Molecular Biology and Immunology. Cold Spring Harbor Lab., Cold Spring Harbor, New York.

Thompson, R. J., Dorani, J. F., Dhillon, A. P., and Rode, J. (1983). *Brain Res.* **278,** 224–228.

Turner, D. L., Snyder, E. Y., and Cepko, C. L. (1990). *Neuron* **4,** 833–845.

Vacher, J., and Tilghman, S. M. (1990). *Science* **250,** 1732–1735.

Vogel, K. S., and Weston, J. (1990). *Dev. Biol.* **139,** 1–12.

Wessels, N. K., and Evans, J. (1968). *Dev. Biol.* **17,** 413–446.

Weston, J. A. (1970). *Adv. Morphog.* **8,** 41–114.

Whelan, J., Poon, D., Weil, P. A., and Stein, R. (1989). *Mol. Cell. Biol.* **9,** 3253–3259.

Wolozin, B., Scicutella, A., and Davies, P. (1988). *Proc. Natl. Acad. Sci. U.S.A.* **85,** 6202–6206.

Yoshinari, M., and Diakoku, S. (1982). *Anat. Embryol.* **165,** 63–70.

DISCUSSION

P. Epstein. In many models in which β cells are destroyed an increase in the number of glucagon-staining cells is seen, and it is usually thought that this is related to a reduction in insulin activity depressing glucagon synthesis and perhaps α cell proliferation. In the *db/db* model is there elevated or reduced insulin?

G. Teitelman. The levels of insulin depend on the age of the animals. The levels of insulin increase after birth and peak at around 2 to 3 months. Between 3 and 6 months the levels of circulating insulin decrease, reaching subnormal levels. After that time the animals become severely diabetic. My studies were performed during the initial stages of the disease, prior to the time the animals suffered β cell destruction. The role of insulin levels in the regulation of α cell number is not clear. In *db/db* mice there is a large increase in α cell number concomitant with the presence of high levels of insulin. However, islets of *db/db* mice contain many α cells even after the levels of insulin are reduced by modification of the diet (Leiter *et al.*, *J. Nutr.* **113:** 184–190, 1983).

P. Epstein.   Are serum insulin levels elevated at or before the time you see more glucagon cells?

G. Teitelman.   The levels of insulin are elevated before and during the time the number of glucagon cells increase, which occurs between 1 and 2 months after birth. There are, however, differences between individual animals, and I have recently found islets of younger *db* mice containing more glucagon cells than in controls. It should be stressed that these animals are hyperinsulinemic from the beginning.

P. Epstein.   So insulin levels are high when you see the increase in double-labeled cells?

G. Teitelman.   Yes, the levels of insulin are high when the percentage of cells that contain both glucagon and insulin increase above control levels.

P. Epstein.   So this would suggest that this is a totally different phenomenon than the one I mentioned in which low levels of insulin led to increases in glucagon staining.

G. Teitelman.   You are actually suggesting that a reduction in insulin levels may induce glucagon cells to divide. I am not aware of studies performed to evaluate the role of hypoinsulinemia on α cell proliferation.

P. Epstein.   In some cases in which β cells are destroyed increased glucagon staining is seen.

G. Teitelman.   I agree. In instances in which there is a destruction of β cells by toxins such as streptozotocin or alloxan, there is a reduction in the size of the islets and an apparent increase in the percentage of non-β cell types. It has not been determined whether this increase is due to cell proliferation and/or whether there is an increase in the total number of non-β cells per pancreas.

P. Epstein.   True, but are you saying that this invalidates the increase in the amount of glucagon?

G. Teitelman.   The increase could be more apparent than real. Unless there is an evaluation of the labeling index of α cells and of the number of Glu⁺ cells in experimental and control animals, the observations could be misleading.

P. Epstein.   So you are saying that under these circumstances you do not actually have an absolute increase in the number?

G. Teitelman.   What I am saying is that the increase may be apparent. I cannot make more conclusive statements without further analysis of each particular case.

P. Epstein.   Have you seen this phenomenon in any other model?

G. Teitelman.   Yes, in animals injected with streptozotocin there is an increase in the number of glucagon cells per islet. In this case it is likely that the increase in α cell number is due to a reduction in the number of β cells.

P. Epstein.   Why did you select the *db* model to look for the mechanisms that lead to the increase in the number of α cells?

G. Teitelman.   Because there were reports in the literature (Coleman, D. L., and Hummel, K. P., *Diabetologia* **9**, 287, 1973) indicating that there was an increase in the percentage of α cells in islets of *db* mice during development of diabetes.

A. E. Boyd III.   Have you had an opportunity to study any of the model systems that people use to study insulin secretion, specifically the HIT cell hamster insulin tumor, or the RIN cells to see how they compare with these various markers?

G. Teitelman.   No, I have not examined those cell lines. I am also hesitant to study cell types from other species without first examining their development *in vivo* since they may differ from mouse in the types of molecules they express. For instance, while islet cells in mouse express the catecholamine enzyme tyrosine hydroxylase, which is a very good marker for examining different stages of the β cell cycle, islet cells of rat do not contain TH. It is not known whether islet cells of rat and/or hamster express other molecules transiently during development and senescence.

A. Powers.   Is it possible that the cells that become the islets migrate prior to removing the neural tube?

G. Teitelman.   Yes, it is possible.

A. Powers.   If you follow the sympathetic or the parasympathetic neurons that wind up in the islets, at what time of development do they migrate out of the neural crest?

G. Teitelman.   In mammals the neural crest cells that give rise to the sympathetic nervous system migrate between day 9 to day 12 of development [Erickson *et al.*, *Dev. Biol.* **134:** 1120, 1989; Serbedzija *et al.*, *Development (Cambridge, U.K.)* **108:** 605, 1990]. It is generally believed that neural crest cell precursors that populate the gut and pancreas, where they differentiate into parasympathetic neurons, also migrate during that time period.

A. Powers.   When you remove the neural crest and the islets still develop, are the neurons that wind up in your islets still present as well?

G. Teitelman.   Studies by Le Douarin and co-workers (see Le Douarin, *The Neuroblast,* Cambridge University Press, 1982) demonstrated that when the neural crest is removed, the pancreas does not contain parasympathetic neurons. In addition, studies of Pictet *et al.* (*Science* **191,** 191–193, 1976) demonstrated that embryonic tissues that were not populated by the neural crest contained islet cells. These are the cells that in adults extend neuritic processes when they are maintained *in vitro*.

A. Means.   I think that you mentioned that you and Doug Hanahan have also expressed T antigen in transgenic mice using the 5'-flanking region of the glucagon gene. Could you briefly summarize your results?

G. Teitelman.   In that study (Efrat *et al.*, *Neuron* **1,** 605, 1988) the expression of a hybrid gene containing the 5'-region of the preproglucagon gene linked to *Tag* was followed during development. This study demonstrated that the regulatory region present in the hybrid gene coded for proper temporal and tissue-specific expression. The oncogene (*Tag*) was first seen in the pancreas at day 10 of development, the same day glucagon first appeared. Throughout development and in adults, expression of *Tag* was restricted to α cells. The hybrid gene was also expressed in embryos by proliferating neuroblasts of the hindbrain and in adults by the glucagon neurons of the brainstem. Although in the pancreas expression of the transgene resulted in development of cell tumors, expression of *Tag* in proliferating cells of the brain and later in neurons did not result in tumor formation. Similarly, transgenic animals that harbor a hybrid insulin–*Tag* gene and express the oncogene in proliferating neuroblasts do not develop tumors of the central or peripheral nervous system (S. Alpert, D. Hanahan, and G. Teitelman, unpublished results).

J. Kirkland.   Neonates with hypoglycemia secondary to hyperinsulinemia may improve with medical treatment, such as with diazoxide or Sandostatin. Surgical removal of the pancreas may be required for those who do not respond to medical treatment. Have you had an opportunity to examine pancreatic tissue from these neonates with these immunohistochemical methods?

G. Teitelman.   Not yet. That type of tissue is difficult to obtain in a good state of preservation, which is a requirement for immunohistochemical analysis.

R. Hazelwood.   What evidence do you have in your adult mouse studies, in which you found both insulin and glucagon in the same cell, that this is not merely cytoplasmic exchange via gap junctions?

G. Teitelman.   That is a very good question. To demonstrate that cells are synthesizing both hormones will require additional studies. For instance, ultrastructural analysis of cells to ascertain whether they contain typical insulin and glucagon storage vesicles and studies to test whether insulin and glucagon mRNAs are localized to the same islet cells.

P. Epstein.   If I am not mistaken, I think RIN cells, at least some classes of RIN cells, are known to switch hormone type in culture. Are you aware of this?

G. Teitelman.   Yes, I am aware that studies of Phillipe *et al.* (*J. Clin. Invest.* **79,** 351, 1987) reported phenotypic changes of RIN cells. In addition, α cell tumors developed by transgenic mice harboring a hybrid glucagon–*Tag* gene occasionally synthesize insulin (Efrat, Teitelman, and Hanahan, unpublished results).

P. Epstein.   Do they ever express both insulin and glucagon?

G. Teitelman.   No, they express either glucagon or insulin.

P. Epstein.   It is a curious situation that the cell would have both insulin and glucagon because the hormones oppose one another, and normally the secretory stimuli are totally opposite. I could imagine a very confused cell in a physiologically very messed-up situation.

G. Teitelman. That is correct. Those cells appear only during development and in adults in extreme physiological conditions.

A. E. Boyd III. Paul Robertson has a paper in press in *Diabetes* in which a hamster insulin tumor insulinoma is discussed. These HIT cells produce both insulin and a little bit of glucagon and cosecrete these hormones. Clinically, we occasionally see patients who have tumors that produce both glucagon and insulin, but it is rare that we see human tumors that produce both hormones.

R. Hazelwood. About 10 years ago Lelio Orci told me that he had found in many cases dual localization of hormones in the same cell. He said that it was not an uncommon finding to him. Are you suggesting that dual secretion will occur/does occur in your cells?

G. Teitelman. Yes, in preliminary experiments I found that the percentage of PP and somatostatin cells also increase in *db* mice with time. Double-label experiments indicated that many of these PP and δ cells also contain glucagon, suggesting that pancreatic islet cells of that mutant strain of mice contain several hormones. Whether islet cells of *db/db* mice secrete several hormones remains to be determined.

R. Hazelwood. The circulation of PP in diabetics is extremely high. Levels sometimes go up anywhere from 400 to 1000% over the normal circulation levels, yet we have never been able to identify PP as being important in any metabolic process dealing with carbohydrate metabolism. It seems to be more of a marker than anything else.

We do a lot of comparative work in our laboratory and we work a lot with birds. Birds have been said to be "diabetic normally," they just don't know it! A very strong similarity can be seen on comparing the biochemical and physiological parameters of type II human diabetics with those of a normal bird. Frequently, we employ an animal model with a 99% reduction in pancreas volume to study hormogenesis, but the small piece that is left is the splenic lobe, which is extremely difficult to get out of most birds. The exception to that is the duck. So you really cannot make a bird diabetic. You remove 99% of the pancreas and it is no worse off metabolically than if it had its entire pancreas. If we study that piece of splenic lobe we find that 1–3 days after operation the glucagon levels are very high in the plasma and the insulin levels are very low. Within 4 or 5 days, they return to normal and the insulin/glucagon molar ratios are normal. Then we continue our studies for perhaps 10 more days and find the insulin/glucagon molar ratios have now shifted strongly catabolic, the glucagon cell being expressed more and more in secretory action than that of the β cell. Perhaps this relates to the first question asked by Dr. Epstein in terms of do you see increases in α cell activity when you reduce the pancreatic volume of β cells. Streptozotocin is one model, but when you reduce the volume by surgical extirpation you have a different scenario. The bird offers a very valuable comparative animal in such research studies, and I think it is one that has been overlooked, unfortunately.

G. Teitelman. I have not examined pancreas after partial pancreatectomy. I would like to stress again, however, that variations in hormone levels do not necessarily imply differences in cell numbers.

RECENT PROGRESS IN HORMONE RESEARCH, VOL. 47

Sulfonylurea Signal Transduction

Aubrey E. Boyd III,* Lydia Aguilar-Bryan,† Joseph Bryan,*
Diana L. Kunze,‡ Larry Moss,*·† Daniel A. Nelson,†
Arun S. Rajan,† Hussein Raef,† Hongding Xiang,†
and Gordon C. Yaney†

*Departments of *Cell Biology, †Medicine, and ‡Molecular Physiology and Biophysics,
Baylor College of Medicine, Houston, Texas 77030*

I. Introduction

A. DEVELOPMENT OF SULFONYLUREAS

During World War II, Janbon and colleagues (1942) found that certain sulfonamide derivatives being used to treat typhoid fever caused hypoglycemic symptoms. Following the close of the war derivatives of these antibiotics that contained the sulfonylurea moiety were first used to treat patients with diabetes. Several early observations pointed to the pancreas as the site of action of these potent drugs. Sulfonylureas did not alter the blood glucose levels of pancreatectomized dogs and, furthermore, were ineffective in young patients who we would now classify as insulin deficient of type I diabetics (Loubatieres, 1944, 1957a,b). After two generations of drug development, sulfonylureas have become the major therapeutic agents used to treat non-insulin-dependent diabetes (NIDDM) (reviewed in Gerich, 1989; Boyd and Huynh, 1990). In the last decade, advances in the molecular and cellular pharmacology of sulfonylureas suggest that unraveling their actions on the β cells of patients with NIDDM will lead to an understanding of the defect in signal transduction characteristic of this disease which affects over 11,000,000 people in the United States.

B. TYPE II DIABETES: THE ROLL OF SULFONYLUREAS

NIDDM (also called type II diabetes) is a heterogeneous disorder characterized by fasting hyperglycemia, excess glucose production by the liver, varying degrees of resistance to both endogenous and exogenous insulin, and unresponsiveness of the β cell to the major physiologic insulin secretagogue, glucose. Most patients with NIDDM are obese. Epidemiologic studies show that the

299

frequency of NIDDM varies around the world from a low incidence in Europe, Japan, and the United Kingdom (1–2%), to moderate frequency in Australia (approximately 3%), to even higher numbers of the population being involved in the United States (6%) (Zimmet, 1983). When one examines the prevalence of this disease among different ethnic groups in the United States, the variations are striking. In some native American Indian tribes like the Pima, NIDDM exists in epidemic proportions with about 40% of adults over the age of 20 being affected (Bennett et al., 1976). Mexican Americans and black Americans have a higher prevalence (about 10%) than do whites (about 5%) living in the United States (Harris, 1990). Acculturation and modernization of societies, where individuals who previously lived as hunter/gatherers or who performed heavy physical labor and now have much less physical activity, have been associated with this striking increase in the prevalence of NIDDM (Zimmett et al., 1990).

Studies by Perley and Kipnis (1966) and Seltzer et al. (1967) first indicated that patients with NIDDM manifested a characteristic defect in glucose-stimulated insulin secretion. Porte and colleagues (see Brunzell et al., 1976), Palmer et al. (1976), Halter and Porte (1979), Ward et al. (1984a,b), and others (Reaven and Miller, 1968; Deckert et al., 1972; Bogardus et al., 1984) have characterized insulin secretion in patients with NIDDM. In patients with mild fasting hyperglycemia (fasting blood glucose levels over 120 mg/dl) the acute release of insulin after administration of intravenous glucose is absent. In contrast, the immediate release of insulin with other secretagogues like the β-adrenergic agonist, isoproterenol, amino acids, or the sulfonylureas is intact. The potentiation by glucose of insulin secretion by nonglucose secretagogues also is abnormal and there is an impaired steady state relationship between the insulin and glucose levels. Thus, with minimal fasting hyperglycemia there is a marked increase in insulin levels. As the fasting glucose levels rise further the insulin levels do not keep pace and the patients develop relative insulin deficiency. Patients who have mild fasting hyperglycemia (fasting glucose levels between 140 and 200 mg/dl) usually are treated effectively with diet and exercise, but many also require sulfonylureas.

Studies suggest that the hypoglycemic actions of sulfonylureas reside in their ability to inhibit the activity of a class of ion channels, K+ channels, which are also inhibited by adenosine triphosphate (ATP). Glucose, which must be metabolized (thus increasing intracellular ATP) to elicit a secretory signal, inhibits the efflux of K+ through these same channels. The sulfonylureas may improve the blood glucose level in the patient with NIDDM by partially overcoming the abnormality in signal transduction which is unique to the diabetic β cell. Thus, understanding the molecular details of how glucose and sulfonylureas interact in the β cell to signal insulin release should provide crucial information about the etiology of NIDDM.

In this article, we will focus on our studies of sulfonylurea signal transduction.

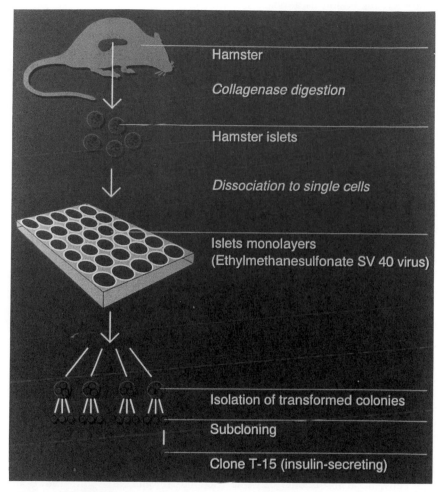

FIG. 1. The derivation of the HIT cell line. [Reproduced with permission from Signal Transduction and Disease, A. E. Boyd, III, M.D., Current Concepts, published by the Upjohn Co. (1991).]

The references are not all inclusive, and focus only on some of those papers that relate directly to our recent work. Several years ago we decided to develop a model system with which we could investigate the molecular mechanisms by which insulin secretion is regulated. These methods have been reviewed extensively (Boyd, 1988; Rajan *et al.*, 1990) and will not be detailed here. After studying the two major insulin-secreting cultured cell lines, we decided to focus on HIT cells, derived from a simian virus 40 (SV40)-transformed hamster insulin-secreting tumor, as a model system. The HIT cell line was developed by Santerre *et al.* (1981) by isolating pancreatic islets from the hamster, dispersing the islets into single cells, transforming the cell isolates with SV40, and cloning out insulin-secreting cell lines (Fig. 1). These clonal cells retain a great deal of differentiated function and respond to all of the secretagogues and inhibitors of insulin secretion that we have tested, except galanin. As expected, the cells do not contain galanin receptors (P. Robertson, personal communication, 1990). The use of clonal cells provides large number of homogeneous cells for biochemical and secretory studies. We have taken advantage of a number of methods to study sulfonylurea signal transduction in the β cell. These include (1) perifusion and static incubations of HIT cells with sulfonylureas to determine dose dependence of various biologic effects of these drugs on the β cell, (2) ligand-binding studies, (3) fluorescent measurements of the free cytosolic calcium level ($[Ca^{2+}]_i$), (4) the patch-clamp technique to study individual ion channels, (5) new methods of protein purification, (6) chemical synthesis of sulfonylurea analogs, and (7) cloning of ion channels and the sulfonylurea receptor from HIT cell cDNA libraries. We will start at the β cell membrane and work inward, defining the steps in signal transduction that lead to the release of insulin elicited by sulfonylureas.

II. Results

A.   IDENTIFICATION OF THE SULFONYLUREA RECEPTOR

High-affinity sulfonylurea receptors have been identified and characterized by our group on HIT cell membranes using two different sulfonylurea ligands: tritiated glyburide ([³H]glyburide; specific activity 1.6 Ci/mmol), the most potent sulfonylurea in clinical use (Gaines *et al.*, 1988), and an analog of glyburide designed and synthesized to be iodinated with $Na^{125}I$ (Aguilar-Bryan *et al.*, 1990). Lazdunski's group has also characterized the sulfonylurea receptor from RIN cells, a rat insulinoma cell line (Schmid-Antomarchi *et al.*, 1987), and Panten and colleagues (1989), using [³H]glyburide, have identified the receptor on mouse pancreatic islet membranes. In all of these studies, the binding affinity of glyburide to the receptor is quite high with K_d values in the low nanomolar range. Using the labeled, high specific activity (>2000 Ci/mmol), iodinated glyburide analog, called iodoglyburide (Fig. 2), Scatchard analysis showed two populations of binding sites. The high-affinity site had a K_d of 0.36 nM and a density of 1.6 pmol/mg of membrane protein (Fig. 3). We also identified a second population of low specific activity binding sites with a K_d of 277 nM and a B_{max} of 100 pmol/mg of protein. Using [³H]glyburide we previously identified only the high-affinity binding site which had a K_d of 0.76 nM and a B_{max} of 1.09 pmol/mg of protein (Gaines *et al.*, 1988). The similarity in binding parameters

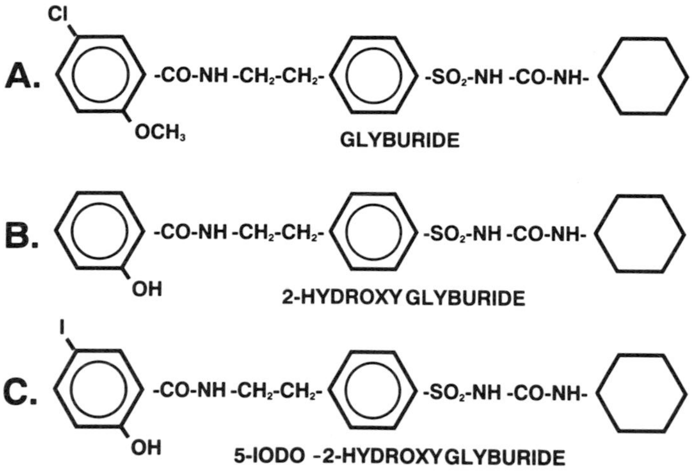

A. Cl— ⬡ —CO-NH-CH₂-CH₂— ⬡ —SO₂-NH-CO-NH— ⬡
OCH₃
GLYBURIDE

B. ⬡ —CO-NH-CH₂-CH₂— ⬡ —SO₂-NH-CO-NH— ⬡
OH
2-HYDROXY GLYBURIDE

C. I— ⬡ —CO-NH-CH₂-CH₂— ⬡ —SO₂-NH-CO-NH— ⬡
OH
5-IODO -2-HYDROXY GLYBURIDE

FIG. 2.   Chemical structures of (A) glyburide, (B) the parent compound 2-hydroxyglyburide, and (C) the new iodinated sulfonylurea, 5-iodo-2-hydroxyglyburide. The iodinated compound was synthesized in four steps as outlined in Aguilar-Bryan *et al.* (1990). Two-dimensional nuclear magnetic resonance was used to confirm the structures of (B) and (C). The iodinated glyburide was labeled with $Na^{125}I$ using chloramine-T and purified on high-performance liquid chromatography (HPLC). [Published with permission from Aguilar-Bryan *et al.* (1990).]

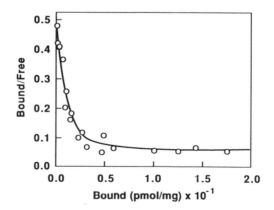

FIG. 3.   Scatchard analysis of the binding of 5^{125}-iodo-2-hydroxyglyburide to HIT cell membranes. HIT cell membranes (200 μg) were incubated with 5^{125}-iodoglyburide at concentrations between 0.01 and 10 nM for 30 min at room temperature. Nonspecific binding was determined with an excess of unlabeled glyburide (1 μM). Binding was terminated by rapid filtration through Whatman GF/F filters and the binding data analyzed according to Scatchard (1949). Modified with permission from Aguilar-Bryan *et al.* (1990).

suggests these two labeled sulfonylureas identify the same protein. We concluded that two or more sulfonylurea-binding proteins are present in HIT cell membranes. To obtain more information about the physiologic relevance of the sulfonylurea receptor, we performed experiments to correlate the binding affinity of iodoglyburide and glyburide with the other biologic processes controlled by these drugs. The free concentration of the drugs was measured in those experiments.

B.   BIOLOGIC ACTIONS OF SULFONYLUREAS IN HIT CELLS

1. $^{86}Rb^+$ Efflux

Sulfonylurea receptor occupancy is thought to inhibit ATP-sensitive K^+ channels, decreasing K^+ efflux, leading to membrane depolarization which opens voltage-gated calcium channels and increases $[Ca^{2+}]_i$. This rise in $[Ca^{2+}]_i$ triggers the exocytosis of insulin. We decided to determine the dose dependence of the biologic effects of glyburide and iodoglyburide on the HIT cell.

The effect of glyburide and iodoglyburide on $^{86}Rb^+$ efflux (a marker for K^+ efflux) was examined in HIT cells prelabeled with $^{86}Rb^+$ and then depleted of ATP by the addition of 2-deoxyglucose and oligomycin. After inhibiting the energy stores the amount of $^{86}Rb^+$ remaining in the HIT cells decreased rapidly, with only about 20% of the tracer still found after 40 min (Fig. 4). The addition

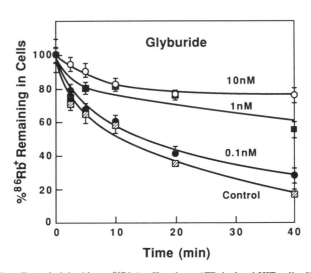

FIG. 4.   The effect of glyburide on $^{86}Rb^+$ efflux from ATP-depleted HIT cells. HIT cells were loaded with $^{86}Rb+$ as described in Hsu *et al.* (1990). The $^{86}Rb+$ remaining in the cells was determined at varying times in control cells unexposed to sulfonylureas or to HIT cells incubated with varying concentrations of glyburide.

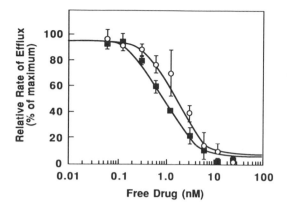

FIG. 5. The dose dependence of increasing concentrations of iodoglyburide (■) and glyburide (○) on $^{86}Rb^+$ efflux. The cells were loaded with $^{86}Rb^+$ as described in Fig. 4 and incubated with varying concentrations of glyburide or iodoglyburide and the $^{86}Rb^+$ remaining at 40 min determined.

of 0.1 nM glyburide to HIT cells had a slight inhibitory effect on $^{86}Rb^+$ efflux. At 1 nM glyburide the inhibition of efflux was approximately half-maximal and at 10 nM, glyburide was even more effective in inhibiting the release of $^{86}Rb^+$ from HIT cells. Similar concentrations of iodoglyburide were effective in inhibiting $^{86}Rb^+$ efflux. The inhibition of $^{86}Rb^+$ efflux and binding and saturation of the receptor by these drugs occur over the same concentration range. The dose dependence of inhibition of $^{86}Rb^+$ efflux for the two sulfonylureas is shown in Fig. 5.

2. Effects on $[Ca^{2+}]_i$ and Insulin Secretion by HIT Cells

In our original studies we did not find a close correlation between the [^{3}H]glyburide concentrations that saturated the sulfonylurea receptor and increased $[Ca^{2+}]_i$. The dose dependence of insulin secretion also was not tightly correlated with sulfonylurea binding to the receptor. It is known that sulfonylureas bind to the albumin which is used to prevent the insulin secreted by the cultured cells from sticking to the Petri dishes or perifusion chambers. From the known concentration of albumin in the insulin secretory buffer and the affinity of binding of albumin to glyburide, we calculated the concentration of free glyburide present under experimental conditions (Gaines *et al.*, 1988). This correction demonstrated an excellent agreement between [^{3}H]glyburide binding, K_d 2.0 nM, and the ED$_{50}$ of insulin secretion of 5.5 nM. Despite this correction for ligand concentration, there still was little correlation between glyburide concentrations required to increase $[Ca^{2+}]_i$, (ED$_{50}$ of 525 nM) and the potency of the drug to stimulate insulin secretion (ED$_{50}$ of 5.5 nM).

Panten *et al.* (1989), using isolated mouse pancreatic islets, showed that at nonsaturating drug concentrations the onset of action of some sulfonylureas on insulin secretion was delayed. The kinetics of the insulin secretory response depended on the lipid solubility of the various sulfonylureas and were particularly delayed with low concentrations of the most hydrophobic drug, glyburide. The same maximal secretory rates were observed with saturating concentrations of all of the sulfonylureas. This work caused us to reexamine sulfonylurea secretory dynamics in the HIT cell.

We confirmed the delay in the insulin secretion and rise in $[Ca^{2+}]_i$ response to low concentrations of glyburide. In contrast, tolbutamide elicited a prompt insulin secretory response and rise in $[Ca^{2+}]_i$ over the entire dose range tested (data not shown). We then tested for the effects of glyburide and iodoglyburide on $[Ca^{2+}]_i$ measuring this parameter over a 10-min period. Figure 6 shows that at high concentrations, such as 100 nM, the effect of glyburide on $[Ca^{2+}]_i$ is quite rapid and a maximal effect occurs within 2 min after addition of the drug to HIT cells. In contrast, the increase in $[Ca^{2+}]_i$ is more delayed at 10 nM, while at 1 and 0.1 nM the $[Ca^{2+}]_i$ rose more slowly still and had not plateaued at 10 min.

At this point, the free glyburide and iodoglyburide concentrations were determined in the following experiments using an ultrafiltration system as described

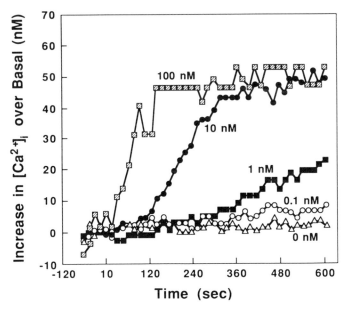

FIG. 6.   The effect of increasing concentratins of glyburide on the $[CA^{2+}]_i$ of HIT cells. The HIT cells were loaded with Fura-2 and $[CA^{2+}]_i$ determined continuously in HIT cells as described in Rajan *et al.* (1990).

by Panten *et al.* (1989). At high drug concentration, the maximal increase in $[Ca^{2+}]_i$ was calculated from the plateau values; while using low concentrations, the 10-min $[Ca^{2+}]_i$ values were used. Both glyburide and iodoglyburide increase $[Ca^{2+}]_i$ over the same free drug concentration range with an ED_{50} of 2.0 nM for the parent compound and 1.5 nM for iodoglyburide. Thus, there is an excellent correlation between the affinity of binding of glyburide and iodoglyburide to HIT cell membranes and the half-maximal increase in $[Ca^{2+}]_i$ elicited by an interaction of these drugs with the HIT cell.

At low concentrations of glyburide the maximal increase in insulin release had not occurred during the 10-min drug stimulation period used in our original experiments. We therefore determined the time to reach maximal insulin secretion. At low concentrations the secretory effects of glyburide and iodoglyburide were complete within a 45-min period. Thus, we studied insulin secretion for a 15-min period after drug addition and a second 30-min period. The data for the 30-min secretory period are shown in Fig. 7. Within the limits of our measurements we were unable to distinguish a difference in the concentrations of glyburide and iodoglyburide which stimulated an increase in insulin secretion. At saturation both drugs stimulated a maximal increase in insulin release of three- to fourfold over the basal level.

Table I compares the ability of glyburide and iodoglyburide to displace ^{125}I-labeled glyburide from HIT cell membranes, and their effects on $^{86}Rb^+$ efflux,

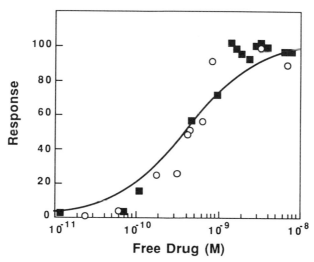

FIG. 7. The dose–response relationships of glyburide (■) and iodoglyburide (○) required to increase insulin secretion. The free iodoglyburide and glyburide concentrations were measured as described in Aguilar-Bryan *et al.* (1990).

TABLE I

Comparison of K_i Values for Displacement of Labeled Iodoglyburide to K_i Values on the Effect of Iodoglyburide and Glyburide on $^{86}Rb^+$ Efflux, ED_{50} Values for $[Ca^{2+}]_i$, and Insulin Secretion

| | K_i (nM) | | ED_{50} (nM) | |
Compound	Binding	$^{86}Rb^+$ efflux	$[Ca^{2+}]_i$	Insulin secretion
Iodoglyburide	0.5	0.7	2.0	0.4
Glyburide	1.0	2.3	1.5	0.4

changes in $[Ca^{2+}]_i$, and insulin secretion. Note the close correlation in the concentrations of free glyburide and iodoglyburide required to alter each of these processes. These data and those published by Schmid-Antomarchi *et al.* (1987) and Panten *et al.* (1989) suggest that the sulfonylurea receptor will either be the ATP-sensitive K^+ channel or a protein which is closely associated with this channel.

The simplest view of the channel configuration is that a single polypeptide contains both the sulfonylurea binding site and the channel activity. However, it is possible that the sulfonylurea receptor is one subunit of a multimeric channel or that the sulfonylurea receptor somehow communicates with the channel. To date there is no evidence to suggest that there is a second messenger link between receptor occupancy and the ATP-sensitive K^+ channel.

C. EFFECTS OF SULFONYLUREAS ON INHIBITION OF K^+ CURRENTS

Although it is difficult in patch-clamp studies to determine accurate K_i values for sulfonylureas on the ATP-sensitive K^+ channels because of run-down, in preliminary studies we have seen that glyburide inactivates the channel in whole-cell patches using low, nanomolar concentrations (Gaines *et al.*, 1987). In patch-clamp studies sulfonylureas are highly specific for the ATP-sensitive K^+ channel in both β cells and cardiac cells (reviewed in Ashcroft and Ashcroft, 1990). In other studies, where the relative potency of various sulfonylureas has been determined, the rank order of inhibition is similar to that which has been seen for binding inhibition of $^{86}Rb^+$ efflux, increased $[Ca^{2+}]_i$, and the initiation of insulin secretion. Glyburide is the most potent, with a K_i of 4 nM for $^{86}Rb^+$ efflux, followed by glipizide (K_i of 6 nM), meglitinide (K_i of 2 μM), and tolbutamide (K_i of 4 μM) (Schmid-Antomarchi *et al.*, 1987; Zunkler *et al.*, 1988; Sturgess *et al.*, 1988; Belles *et al.*, 1987). Dose-dependent inhibition of the channel activity by tolbutamide gives a Hill coefficient of 1, suggesting a one-to-one relationship between the number of sulfonylurea binding sites and the number of ATP-sensitive K^+ channels (Zunkler *et al.*, 1988; Belles *et al.*, 1987).

D. SOLUBILIZATION OF THE SULFONYLUREA RECEPTOR

To test directly the hypothesis that the sulfonylurea receptor is the ATP-sensitive K^+ channel, we devised a purification strategy in the hopes of obtaining an amino acid sequence for the eventual cloning and expression of the receptor.

To do this, it was necessary to solubilize the protein. Most nonionic detergents were capable of solubilizing the receptor and typically yielded approximately 50% of the receptor protein in the soluble fraction. We tested 10 different detergents and determined that digitonin was best suited for purification. As illustrated in Fig. 8, the receptor solubilized in 1% digitonin will bind to ^{125}I-labeled glyburide. The label can be displaced with increasing concentrations of all of the sulfonylureas we have tested. Glyburide, iodoglyburide, and another potent second generation sulfonylurea, glipizide, all displace ^{125}I-labeled glyburide from the solubilized sulfonylurea receptor over the nanomolar range. The first generation sulfonylureas tolbutamide, tolazamide, and chloropropamide displaced the label over the micromolar concentration range. This order of potency is similar to the ability of these drugs to bind to the intact receptor, to stimulate insulin secretion, and to inhibit ^{86}Rb$^+$ efflux (Schmid-Antomarchi *et al.*, 1987; Gaines *et al.*, 1987; Panten *et al.*, 1989).

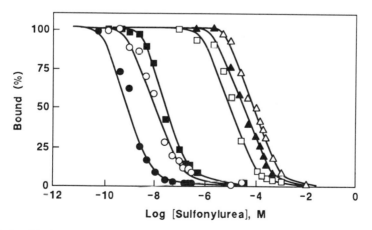

FIG. 8. Displacement of label iodoglyburide from the digitonin-solubilized receptor with six different sulfonylureas. ●, Glyburide; ○, iodoglyburide; ■, glipizide; □, tolazamide; ▲, tolbutamide; △, chlorpropamide.

E. PHOTOAFFINITY LABELING OF
THE SULFONYLUREA RECEPTOR

To follow the sulfonylurea receptor during purification, we developed a method of photolabeling the receptor. We thought this would be possible since Kramer *et al.* (1988) reported that [³H]glyburide photolabeled two β cell membrane proteins of M_r 33,000 and 140,000. The aromatic structure of the glyburide molecule results in the photolabeling of several HIT cell proteins after preincubation of HIT cell membranes with 5^{125}-iodo-2-hydroxyglyburide followed by ultraviolet irradiation. Within 1 min of irradiation of the membranes a 140-kDa protein is detected; at longer time points other, smaller molecular mass proteins, two of M_r 65,000 and 43,000, are labeled (Fig. 9). In the presence of excess unlabeled 5-iodo-2-hydroxyglyburide, the 140-kDa protein is not labeled and the labeling of a 65- and a 43-kDa protein is only partially reduced. The 140-kDa protein appears to be the high-affinity receptor, since it is labeled at nanomolar concentrations of the tracer. The amount of the label in this band is maximal above 22 nM of 5^{125}-iodo-2-hydroxyglyburide, whereas the binding to the lower molecular mass proteins continues to increase well above this concentration. These results suggest that the 140-kDa binding protein is the physiologic receptor

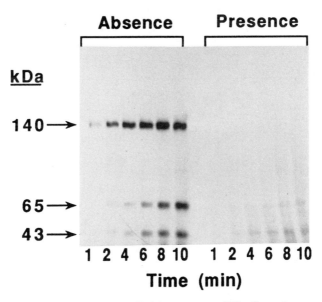

FIG. 9.   Photoaffinity labeling of the sulfonylurea receptor. HIT cell membranes were incubated with 5^{125}-iodo-2-hydroxyglyburide at five different concentrations (4.4–44 nM) in the presence and absence of unlabeled analog (1 μM). The samples were irradiated for 10 min at 4°C, then proteins were separated by SDS-gel electrophoresis and subjected to autoradiography.

and that the other proteins could be the low-affinity binding proteins which, on the basis of the Scatchard analysis, are present in much higher concentration (B_{max} = 100 pmol/mg protein) than are the high-affinity binding sites (B_{max} = 1.6 pmol/mg protein). These low-affinity binding proteins also can be removed by additional washing. The 65-kDa protein appears to be albumin, which is labeled with this procedure.

F. PARTIAL PURIFICATION AND CLONING OF THE 140-kDa PROTEIN

The solubilization receptor was purified by ion-exchange chromatography with DEAE-Sephacel, exclusion chromatography using agarose A1.5M, and hydrophobic interactions using a dodecyl-agarose column as depicted in Fig. 10. The 140-kDa protein was separated further by electrophoresis. The 140-kDa protein was excised from the gel and subjected to digestion with V8 protease. The three most abundant peptides were transferred to immobilon P and then subjected to microsequencing by Dr. Richard Cook of the Howard Hughes Medical Institute

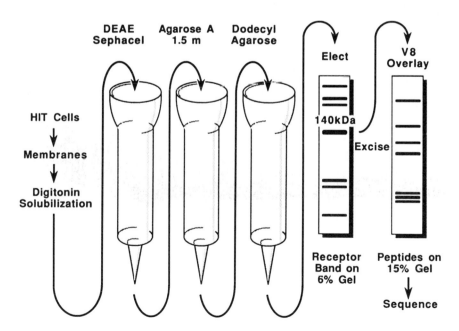

FIG. 10.   Purification scheme. The 140-kDa protein was photolabeled and the protein followed through a three-column scheme: an ion-exchange chromatography on DEAE-Sephacel, exclusion chromatography on an agarose A1.5M column, followed by a hydrophobic interaction using dodecyl-agarose chromatography as described in Aguilar-Bryan et al. (1990).

of the Baylor College of Medicine (Houston, TX). Sequence was obtained on three peptide fragments that were isolated from the 140-kDa protein. This amino acid sequence was used to design degenerate oligonucleotides.

These oligonucleotides have been used to screen λZapII and λgt11 (Hsu *et al.*, 1990) HIT cell cDNA libraries and to carry out polymerase chain reactions (PCRs) in order to clone the cDNA which encodes the receptor. These strategies have yielded a partial cDNA sequence coding for approximately 136 kDa of protein containing two of the three available receptor peptide sequences. Work is in progress to obtain the complete cDNA and the predicted protein sequence.

G. MOLECULAR STUDIES OF CALCIUM CHANNELS

The rise in $[Ca^{2+}]_i$ and insulin secretion triggered by sulfonylureas can be blocked in a reversible manner either by adding various calcium channel blockers to the medium or by chelating extracellular calcium with EGTA. It is now well established that calcium enters β cells through voltage-dependent calcium channels. Two subtypes of voltage-dependent calcium channels have been described in β cells, L type and T type. The predominant current moves through an L-type

```
                            ··············IVS2··············      ··············IVS3··············
 rbskm  CaCh1a  :  ----H--EEMNHISD---VA--II--L-------L---AR--G---------------------IDTFLASSGGLYCLGGGCGNVDPD
 BC3H1  CaCh1a  :  ----H--EEMNHISD---VA--II--L--V---------R--G---------------------IDTFLASSGGLYCLGGGCGNVDPD
 BC3H1  CaCh1b  :  ---H--EEMNHISD---VA--II--L--V---------R--G---------------------ID·············DPD
 rtbrn  C61 (2a):  -----------------------------------------------------------------·················S--
 rtaort RA3 (2) :  ----------------------------------------------------------------//
 rbcar  CaCh2a  :  -----------------------------------------------------------------·············----
 rbcar  CaCh2c  :  -----------------------------------------H--C-A--T--A---V---V--I AIT-VH·············----
 hucar  CaCh2a  : MQBYGQSCLFKIAMNILNMLFTGLFTVEMILKLIAFKPKGYFSDPWNVFDFLIVIGSIIDVILSETN········PAEHTQCSPSMNAE
 hucar  CaCh2c  :  -----------------------------------------H--C-A--T--A---V---V-I AIT-V-·············----
 hucar  CaCh2d  :  -----------------------------------------H--C-A--T--A---V---V-I AIT-V-·············----
 mov    CaCh2a  :  -----------------------------------------------------------------·············S--
 mov    CaCh2d  :  -----------------------------------------H--C-A--T--A---V---V-I AIT-VH············S--
 rtbrn  D55 (3a):  ----E--KM-ND--D----V--V-----V--V------------A--T--S----------A---AD···PSDSENIPLPTATPG-S-
 HIT    CaCh3a  :  ----E--KM-ND--D----V--V-----V--V------------A--T--S----------A---AD···PTESELIPLPTATPG-S-
 HIT    CaCh3b  :  ----E--KM-ND--D----V--V-----V--V------------A--T--S----------A---AD············S-
 HIT    CaCh3d  :  ----E--KM-ND--D----V--V-----V--V------H--T-A--T--A---V--VV-I AIT-V-·············
 mov    CaCh3b  :  ----E--KM-ND--D----V--V-----V--V------------A--T--S----------A---AD············S--
 rtbrn  RB19 (3c): ----E--KM-ND--D----V--V-----V--V------H--T-A--T--A---V--VV-I AIT-V-····PSDSENIPLPTATPG-S-
 rtbrn  A56 (4) :  --KF-A-VAYEN-LVRF-IV--S--SL-CV--VM--GILN--R-A--I---VT-L---T-ILVT-·············
 rtbrn  B10 (5) :  //-SL-C---I---GILN--R-A-D----VT-L---T-IVNF·············
```

FIG. 11.   Shown above is the alignment of the partial amino acid sequence of the α subunit of the L-type calcium channel amplified by polymerase chain reaction (PCR) from various tissues (Perez-Reyes *et al.*, 1990). This section represents approximately one-sixth of the open reading frame for this protein. The variants pertaining to the neuroendocrine channel type in the HIT cell are highlighted. Variant 3a was found to be identical either by PCR or actual retrieval from a λgt11 HIT cell library. All sequences are compared to that of the type 2a human cardiac calcium channel (CaCh2a). ———, Identity; ···, gap; //, limits of available sequence.

channel having a long-lasting current ($t_{1/2}$ of 200–700 msec) that can be blocked by dihydropyridines. Our studies, using the patch-clamp method to characterize calcium current, have shown a remarkable agreement between the amount of L-type current identified, the rise in $[Ca^{2+}]_i$, and the amount of insulin that is secreted by cell depolarization (Keahey *et al.*, 1989). T-type channels, which inactivate more rapidly ($t_{1/2}$ of $\sim$ msec), were more difficult to identify and seen only intermittently. Thus, it would appear, electrophysiologically and pharmacologically, that the predominant calcium channels in HIT cells are L-type channels.

In collaboration with L. Birnbaumer and X. Wei at Baylor, we have been cloning the cDNA encoding the α_1 subunit of the L-type voltage-dependent calcium channels from a λgt11 HIT cell cDNA library. Although L-type calcium channels contain five subunits, the α_1 subunit alone appears to be sufficient to form a conductance pore and to gate this conductance in a voltage-dependent manner (Mikami *et al.*, 1989). This library has been screened with oligonucleotides, the design of which is based on the conserved regions in the amino acid sequence of this subunit across various tissues. Using these oligonucleotides, we have found two types of α_1 subunits. Based on partial sequence, one of the channels falls into the category of a brain L-type calcium channel while the other appears to be cardiac-like. Snutch *et al.* (1990) showed that the brain expresses a heterogeneous family of α_1 subunits containing four calcium channels. Based on partial sequence data, two of these brain channels are related more closely to the cardiac channel and two are related more distantly to skeletal dihydropyridine channels. In addition, the Birnbaumer laboratory (Perez-Reyes *et al.*, 1990) has used the polymerase chain reaction and primers, again from conserved regions, to examine the diversity of the α_1 subunit in a variety of cells and tissues.

Perez-Reyes *et al.* (1990) have identified, in the ovary and HIT cells, "neuroendocrine" calcium channels. These channels bear similarity to the brain channel and are thought to be expressed in both brain and endocrine tissue. Full sequence data on each of these channels is not yet available. Their sequence data suggest that part of the diversity of calcium channel α_1 subunits will result from alternative exons of equal size which are spliced to produce proteins of the same length. Alternatively, exon-skipping events can produce channels of different lengths. Three variants have been identified for the neuroendocrine subtype that differ both in the sequence of the third membrane-spanning region of the fourth repeat unit (IVS3) and in the size of the linker between this and the fourth membrane-spanning segment (IVS4). Limited sequence data on this region of our cloned α_1 subunit confirm the hypothesis of an alternatively spliced transcript concerning the neuroendocrine α_1 subunit. Additionally, the open reading frame 3' of the last repeat unit, as well as the 3' untranslated region, may be shorter than the sequence encoding the α_1 subunit from either cardiac or skeletal muscle.

III. Summary

In the pancreatic β cells the proximal step in sulfonylurea signal transduction is the binding of these clinically important drugs to high-affinity receptors in the β cell membrane. Using HIT cells as a model system, we have established an extremely close correlation between the affinity of binding of glyburide and its analog, iodoglyburide, and the activation of various steps in stimulus–secretion coupling—inhibition of $^{86}Rb+$ efflux, increase in $[Ca^{2+}]_i$ resulting from gating of voltage-gated calcium channels by cell depolarization, and the exocytosis of insulin. Two different L-type channel cDNAs have been identified in an HIT cell library, one neuroendocrine in type and one more cardiac-like.

A HIT cell membrane protein of M_r 140,000, which we believe to be the high-affinity sulfonylurea receptor, can be covalently linked to 5^{125}-iodo-2-hydroxyglyburide by ultraviolet irradiation. The receptor has been solubilized and retains binding activity and the same rank order of displacement of the 5^{125}-iodo-2-hydroxyglyburide as observed with the native receptor. The M_r 140,000 protein has been partially purified and the amino acid sequences of three proteolytic fragments have been used to design oligonucleotides to screen HIT cell cDNA libraries. Since the binding constant of glyburide or iodoglyburide is closely correlated with the ability of these compounds to inhibit the ATP-sensitive K^+ channel, increase $[Ca^{2+}]_i$, and elicit insulin secretion, we have identified the M_r 140,000 protein as the sulfonylurea receptor. Expression of the cloned cDNA should allow us to test this hypothesis directly.

ACKNOWLEDGMENTS

This work was supported by United States Public Health Service Grants DK34447 (to A.E.B.), DK41898 (to D.A.N.), and DK27685 (Diabetes and Endocrinology Research Center), funds from The Methodist Research Foundation (to A.E.B.), The John J. Redfern, III Research Foundation (H.R.), a grant from the American Diabetes Association (D.A.N.), a Mentor-Based Postdoctoral Fellowship Grant (to G.C.Y.), from the American Diabetes Association (A.E.B.), a Fellowship Grant from the Juvenile Diabetes Foundation (L.A.B.) and an ATP (Advanced Technology Program) Grant from the State of Texas (to J.B. and L.A.B.)

REFERENCES

Aguilar-Bryan, L., Nelson, D. A., Vu, Q. A., Humphrey, M. B., and Boyd, A. E., III (1990). *J. Biol. Chem.* **265**, 8218–8224.

Ashcroft, S. J. H., and Ashcroft, F. M. (1990). *Cell. Signalling* **2**, 199–214.

Belles, B., Hescheler, J., and Trube, G. (1987). *Pfluegers Arch.* **409**, 582–588.

Bennett, P. H., LeCompte, P. M., Miller, M., and Rushforth, N. B. (1976). *Recent Prog. Horm. Res.* **32**, 333–376.

Bogardus, C., Lillioja, S., Howard, B. V., Reaven, G., and Mott, D. (1984). *J. Clin. Invest.* **74**, 1238–1246.

Boyd, A. E., III (1988). *Diabetes* **37**, 847–850.

Boyd, A. E., III, and Huynh, T. Q. (1990). *Contemp. Intern. Med.* **2**, 13–33.

Brunzell, J. D., Robertson, R. P., Lerner, R. L., Hazzard, W. R., Ensinck, J. W., Bierman, E. L., and Porte, D., Jr. (1976). *J. Clin. Endocrinol. Metab.* **42**, 222–229.

Deckert, T., Lauridsen, U. B., Madsen, S. N., and Mogensen, P. (1972). *Dan. Med. Bull.* **19**, 222–226.

Gaines, K. L., Kunze, D., Hamilton, S., Keahey, H., and Boyd, A. E., III (1987). *Diabetes* **36**, Suppl. 1, 45A (abstr.).

Gaines, K. L., Hamilton, S., and Boyd, A. E., III (1988). *J. Biol. Chem.* **263**, 2589–2592.

Gerich, J. E. (1989). *N. Engl. J. Med.* **321**, 1231–1245.

Halter, J. B., and Porte, D., Jr. (1979). *J. Clin. Endocrinol. Metab.* **48**, 946–954.

Harris, M. I. (1990). *Diabetes/Metab. Rev.* **6**(2), 71–90.

Hsu, W. H., Randolph, U., Sanford, J., Bertrand, P., Olate, J., Nelson, C., Moss, L. G., Boyd, A. E., III, Codina, J., and Birnbaumer, L. (1990). *J. Biol. Chem.* **265**, 11220–11226.

Hsu, W. H., Xiang, H., Rajan, A. S., and Boyd, A. E., III. (1991). *Endocrinology (Baltimore)* **128**, 958–964.

Janbon, M., Chaptal, J., Vedel, A., and Schaap, J. (1942). *Montpellier Med.* **441**, 21–22.

Keahey, H. H., Rajan, A. S., Boyd, A. E., III, and Kunze, D. L. (1989). *Diabetes* **38**, 188–193.

Kramer, W., Oekonomopulos, R., Punter, J., and Summ, H.-D. (1988). *FEBS Lett.* **229**, 355–359.

Loubatieres, A. (1944). *C. R. Seances Soc. Biol. Ses Fil.* **138**, 766–767.

Loubatieres, A. (1957a). *Ann. N. Y. Acad. Sci.* **71**, 4–11.

Loubatieres, A. (1957b). *Ann. N. Y. Acad. Sci.* **71**, 192–206.

Mikami, A., Imoto, K., Tanabe, T., Niidome, T., Mori, Y., Takeshima, H., Narumiya, S., and Numa, S. (1989). *Nature (London)* **340**, 230–233.

Nelson, T. Y., Gaines, K. L., Rajan, A. S., Berg, M., and Boyd, A. E., III (1987). *J. Biol. Chem.* **262**, 2608–2612.

Palmer, J. P., Benson, J. W., Walter, R. M., and Ensinck, J. W. (1976). *J. Clin. Invest.* **58**, 565–570.

Panten, U., Burgfeld, J., Goerke, F., Rennicke, M., Schwanstecher, M., Wallasch, A., Zunkler, B. J., and Lenzen, S. (1989). *Biochem. Pharmacol.* **38**, 1217–1229.

Perez-Reyes, E., Wei, X., Castellano, A., and Birnbaumer, L. (1990). *J. Biol. Chem.* **265**, 20430–20436.

Perley, M. J., and Kipnis, D. M. (1966). *J. Clin. Invest.* **46**, 1954–1962.

Rajan, A. S., Aguilar-Bryan, L., Nelson, D. A., Yaney, G. C., Hsu, W. H., Kunze, D. L., and Boyd, A. E., III (1990). *Diabetes Care* **13**, 340–363.

Reaven, G. M., and Miller, R. (1968). *Diabetes* **17**, 560–569.

Santerre, R. F., Cook, R. A., Crisel, R. M. D., Sharp, J. D., Schmidt, R. J., William, D. C., and Wilson, C. P (1981). *Proc. Natl. Acad. Sci. U.S.A.* **78**, 4339–4343.

Scatchard, D. (1949). *Ann. N. Y. Acad. Sci.* **51**, 660–672.

Schmid-Antomarchi, H., De Weille, J., Fosset, M., and Lazdunski, M. (1987). *J. Biol. Chem.* **262**, 15840–15844.

Seltzer, H. S., Allen, E. W., Herron, A. L., Jr., and Brennan, M. T. (1967). *J. Clin. Invest.* **46**(3), 323–335.

Snutch, T. P., Leonard, J. P., Gilbert, M. M., Lester, H. A., and Davidson, N. (1990). *Proc. Natl. Acad. Sci. U.S.A.* **87**, 3391–3395.

Sturgess, N. C., Kozlowski, R. Z., Carrington, C. A., Hales, C. N., and Ashford, M. J. L. (1988). *Br. J. Pharmacol.* **95**, 83–94.

Ward, W. K., Bolgiano, D. C., McKnight, B., Halter, J. B., and Porte, D., Jr. (1984a). *J. Clin. Invest.* **74**, 1318–1328.

Ward, W. K., Beard, J. C., Halter, J. B., Pfeifer, M. A., and Porte, D., Jr. (1984b). *Diabetes Care* **7**, 491–502.

Zimmet, P. (1983). *In* "Diabetes Mellitus: Theory and Clinical Practice" (Ellenberg and Rifkin, eds.), 3rd ed., pp. 451–468. Medical Examination Publishers, New York.

Zimmett, P., Dowse, G., and Finch, C. (1990). *Diabetes/Metab. Rev.* **6**(2), 91–124.

Zunkler, B. J., Lenzen, S., Manner, K., Panten, U., and Trube, G. (1988). *Naunyn-Schmiedeberg's Arch. Pharmacol.* **337**, 225–230.

DISCUSSION

J. Baxter. Is there any evidence in type II diabetes for down regulation of the calcium channel itself?

A. E. Boyd III. No, there is no evidence in type II diabetes for down regulation of the calcium channel. Graeme Bell has cloned two calcium channels from a human insulinoma library and is now using these as candidate genes to look for RFLP patterns that are unique for patients with type II diabetes, but so far as we know calcium channel function is probably going to be all right in type II diabetes.

J. Baxter. In terms of desensitization, do you know anything about the proteins that down regulate the sulfonylurea receptor?

A. E. Boyd III. Yes, Dr. Hussein Raef is studying desensitization of the β cells to sulfonylureas. We can say that the desensitization is not at the receptor level. If we treat our HIT cells with sulfonylureas, tolbutamide is a good drug to use because of its rapid reversibility. The cells do desensitize. The receptor number and affinity are the same in the desensitized cells. The changes in cytosolic calcium with sulfonylurea are fewer and the cells secrete less insulin. So we think the desensitization is at the level of the calcium channel in the secretory pathway, beyond the receptor. There may also be a distal desensitization site.

J. Baxter. Is it true that there can be independent desensitization to glucose or sulfonylureas? If so, how can the desensitization to sulfonylureas be downstream from the receptor?

A. E. Boyd III. I do not have a good explanation. I was really disappointed when we got this result because of the fact that we do not have a logical explanation for all the desensitization. I thought it would all be proximal.

J. Baxter. Is it possible that the desensitization to sulfonylureas that you observe is different from that which occurs in type II diabetes? For example, you were studying this in HIT cells that may not have the lesion of type II diabetes mellitus.

A. E. Boyd III. Yes. We are studying this in the insulinoma cell line—the HIT cells.

J. Baxter. Do you know if the patient from which the insulinoma was derived had the type II diabetes defect?

A. E. Boyd III. No, the patients were hamsters. This is really going to be the answer because, as I have discussed, these cells, although they secrete insulin to glucose, have some type of abnormality in the calcium signal with glucose in that they do not show an increase in calcium unless cAMP is also increased in concert with glucose. So it may be that we are not even examining the right thing.

A. Powers. At what step do you think that hyperglycemia interferes with insulin secretion or that glucotoxicity occurs?

A. E. Boyd III. According to the "glucose toxicity" theory, if the blood sugar level is high for a period of time, insulin secretion is not as great as would be predicted. Then, if the blood sugar level is lowered for awhile, insulin secretion returns and can be lowered in patients with diet and exercise or sulfonylureas or with insulin itself. So there has been a great deal of interest in "glucose toxicity." We are just starting to examine "glucose toxicity" by growing our cells in high glucose versus lower glucose levels. We do not have any answers as yet. I think it is going to be at a very proximal event in the trigger mechanism, perhaps with this ATP-sensitive potassium channel, but I do not have any direct proof of this. I think it is going to be something that is reversible within a short period of time, say a week or so. Some researchers suspect that it might be something like glycosylation of a channel and that the channels would then turn over and be replaced and then the calcium channel or the ATP-sensitive potassium channel would start working better. These are possibilities. We have no direct data.

A. Powers.   At what level of glucose are your HIT cells grown?

A. E. Boyd III.   The concentration was varied, but normally they are grown in pretty high concentration, around 11 mM.

P. Epstein.   Since an elevation of calcium is central to your insulin secretion hypotheses, how do you deal with the fact that you do get an increase in insulin secretion but no increase in calcium in the HIT cells?

A. E. Boyd III.   There are probably at least three different signal pathways important in triggering the acute release of insulin. Calcium is one. A second is probably activation of protein kinase C; whether it is an independent trigger or whether it potentiates the calcium signal is not clear. I prefer the hypothesis that in general calcium is a primary trigger and that protein kinase C is also activated by glucose, and that in the HIT cells we were seeing the immediate release of insulin without a change in calcium due to the protein kinase C signaling pathway. Finally, there is the cAMP pathway. All interact.

P. Epstein.   Do the HIT cells require extracellular calcium for the glucose response?

A. E. Boyd III.   Yes, if extracellular calcium is removed in cell lines, insulin secretion can be totally shut off by any method, so calcium must be present to get secretion. In islets where it is not possible to probably deplete islets completely of calcium, it is a little more problematic. With calcium-free media, different results are obtained using islets, but insulin secretion cannot be triggered with depolarization or sulfonylureas or any type of secretogogue if there is no calcium outside the HIT cell, particularly if, for example, no calcium is used in the media and EGTA.

P. Epstein.   So perhaps calcium is coming in and channels are opening but you are not seeing it with your measurements?

A. E. Boyd III.   Yes, there could be local changes in calcium that we are still missing with fura2. We have not done single-cell studies to study just the membrane. All studies we have done on calcium are on suspended cells (groups of cells) where we are looking at average levels in large numbers of cells.

Complex Hormone Response Unit Regulating Transcription of the Phosphoenolpyruvate Carboxykinase Gene: From Metabolic Pathways to Molecular Biology

D. GRANNER, R. O'BRIEN, E. IMAI, C. FOREST,
J. MITCHELL, AND P. LUCAS

*Department of Molecular Physiology and Biophysics, Vanderbilt University,
Nashville, Tennessee 37232*

I. Introduction

Mechanisms that provide for the differentiation, replication, and growth of cells, the repair and replacement of their components, and the maintenance of a constant supply of energy in the face of environmental alterations are essential to the survival of multicellular organisms. These processes are regulated by a variety of different classes of molecules, including several hormones. The provision of a constant source of energy is perhaps the most clearly adaptive. Studies of the hormonal regulation of the many processes involved in energy metabolism are based on a series of seminal observations that have culminated in the application of techniques of molecular biology to this topic. In this article we briefly review the history of metabolic regulation and the relevant concepts of molecular biology before showing how these historical antecedents allow for an analysis of a gene critically involved in metabolic regulation. The gene discussed directs the synthesis of phosphoenolpyruvate carboxykinase, the rate-limiting enzyme in gluconeogenesis.

A. DEVELOPMENT OF THE BASIC PRINCIPLES INVOLVED IN HORMONAL REGULATION OF GENE EXPRESSION

Several major discoveries provide the foundation for current efforts directed at discerning how hormones regulate gene expression and, in so doing, control metabolic processes. We have reviewed this subject in some detail earlier (1), but mention here a few key observations (Fig. 1). The concept that the conversion of foodstuffs into cellular constituents or into energy involves an orderly progression of biochemical reactions, each catalyzed by an enzyme, was developed during the early part of this century. By the 1950s the basic metabolic pathways had been defined, although these continue to be refined (2). Subsequent experi-

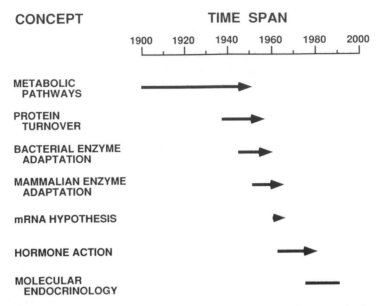

FIG. 1.   Important concepts in metabolic regulation. Approximate time spans for the development of the concepts required for the analysis of how hormones affect critical metabolic processes are illustrated by the arrows.

ments focused on the coordination of these complex pathways. An important event was the formulation, by a number of investigators, of the hypothesis that hormones might provide the means of metabolic coordination (see Ref. 3 for review).

Another important concept emerged in the early 1940s. Prior to that time a cell was thought to consist of a membrane that separated a fixed set of enzymatically catalyzed reactions from the environment. Since the cell was capable of self-replication, this implied a constancy of these reactions from generation to generation. Hence, it was inferred that in a cell the endowment of enzymes did not change. Although energy saving, this arrangement would lack the key feature now recognized as one of the fundamental features of biological systems—flexibility. Schoenheimer, in some of the first experiments to use isotopes, showed in 1942 that cellular constituents were actually in a dynamic state (4). Subsequent studies, in which basic principles of isotopic labeling were developed (5,6), established that various cellular constituents, including proteins, turn over with different rates. Turnover provides the mechanism for flexibility. More rapid and precise changes in response to the environment can be accomplished by making subtle adjustments to an ongoing synthesis/degradation process than by employing an "all or none" mechanism.

The observation that cellular components turn over led to the emergence of the concept that complex organisms could show adaptive responses to their environment. It was soon shown that remarkable changes occur in the amount of enzymes in microorganisms, generally in response to alterations of substrate concentration (7–9). A similar phenomenon was demonstrated in mammalian cells in 1951 when Knox and Mehler showed that administered tryptophan resulted in a six- to eightfold increase in the activity of the hepatic enzyme now known as tryptophan oxygenase (10). This substrate-induced adaptive increase was superficially similar to that noted in microorganisms. Within a few years many examples of changes of the activity of numerous enzymes in response to adrenalectomy, thyroidectomy, hypophysectomy, diabetes, and the replacement with appropriate hormones could be cited (1,11).

The concept of turnover implied that an increased amount of protein could result from an increased rate of synthesis, from a decreased rate of degradation, or from some combination of these processes. The practical basis for such experiments was defined by Schimke and co-workers, who showed that tryptophan slowed hepatic tryptophan oxygenase degradation while hydrocortisone enhanced synthesis of the enzyme (12,13).

How information was transmitted from DNA (the gene) to protein remained an enigma until the seminal work of Jacob and Monod, who first proposed the idea that a gene had structural and regulatory regions, and that genes somehow control protein synthesis through a specific RNA intermediate. Their analysis provided the first suggestion that an unstable intermediate, synthesized rapidly and degraded equally rapidly with respect to total RNA, was involved. Since this unstable intermediate was thought to carry the information from DNA to protein it was called messenger RNA (14). The concept that hormones could regulate the synthesis of a protein by altering the amount of a specific mRNA quickly became a dominant experimental paradigm in molecular endocrinology.

In subsequent decades the general mechanisms of action of various hormones were elucidated. The central role of receptors was a watershed discovery, as was the concept of signal transduction. Hormones of the steroid-thyroid class bind to intracellular receptors, which serve as the intracellular signals. Peptide hormones bind to plasma membrane receptors and generate a number of intracellular signals that mediate the actions of the respective hormones (see Ref. 15 for review).

B.   THE CIS/TRANS MODEL OF TRANSCRIPTION CONTROL

Many hormones regulate the transcription of genes that encode the proteins involved in catalyzing specific metabolic reactions. An understanding of the basic principles of transcription was central to deciphering how hormones affect this process. The fidelity and frequency of initiation of transcription of eukaryotic genes is mediated by the interaction of cis-acting DNA elements with various trans-acting protein factors. In eukaryotic cells a cis-acting element regulates contiguous DNA and does not code for a protein. A trans-acting factor is expressed by a gene not associated with the DNA sequence being regulated. In this cis/trans model, two general types of cis-acting elements are involved (16,17). Some cis-acting elements (TATA, GC, and CCAAT boxes) are located near the transcription initiation site and generally act in an orientation- and position-dependent manner to assure the accuracy of initiation. These basal promoter elements each bind a specific trans-acting transcription factor (TFIID, Sp1, and CBP, respectively, for the examples cited; see Ref. 17). Other cis-acting elements and their associated trans-acting factors are often located farther from the initiation site and act in an orientation- and position-independent manner to promote (enhance) or inhibit (silence) transcription. These elements can function in the context of their cognate promoter or when attached to a heterologous promoter. Most attention has been given to the analysis of enhancers and enhancer-binding proteins, but transcription repression is equally important. Various combinations of positive and negative cis-acting DNA elements are involved in tissue-specific, growth- or differentiation-related, and hormone-mediated control of transcription. Some of the trans-acting proteins that bind to the basal promoter elements and to various enhancer elements have been purified to homogeneity, and several of the corresponding cDNAs have been cloned (18).

C.   HORMONE RESPONSE ELEMENTS

Most HREs fall into the enhancer/silencer class of cis-acting elements since these DNA sequences, and their associated trans-acting factors, can function through heterologous promoters and in a relatively position- and orientation-independent manner (19). However, in some cases HREs can also function as basal promoter elements. Much more is known about HREs acting as enhancers, although in certain instances HREs can act as silencers.

Biochemical (20) and genetic (21) evidence that glucocorticoid receptors bind to DNA was followed by data that showed that certain regions of DNA, containing specific glucocorticoid receptor-binding sites, confer regulation to a gene ordinarily unresponsive to glucocorticoids (19,22). Numerous studies of this type, employing the promoter regions of several hormone-responsive genes ligated to a variety of reporter genes (forming so-called fusion genes) led to the identification of cis-acting elements that mediate the response to several hormones. The strategy used to define such elements has been reviewed in detail elsewhere (23).

Hormone response elements (HREs) for steroid, thyroid, sterol, and retinoid hormones have been identified, as has a cAMP response element (see Fig. 2). The hormone receptor is the trans-acting factor in the case of the steroid-thyroid family of HREs. The observation that a GRE, MRE, PRE, and ARE all have the

Hormone	Hormone Response Element	Consensus Sequence
Glucocorticoids	GRE	GGTACAnnnTGTTCT
Progesterone	PRE	"
Androgens	ARE	"
Mineralocorticoids	MRE	"
Estrogen	ERE	AGGTCAnnnTGACCT
Thyroid Hormones	TRE	TCAGGTCA---TGACCTGA
Retinoic Acid	RRE	"
cAMP	CRE	CTGACGTCAG

FIG. 2.   Hormone response elements. Consensus sequences have been developed for several hormone response elements. n, Any nucleotide will suffice.

same consensus sequence appears to defy the rules of physiology since it implies that the same sequence mediates the effects of all these hormones. This consensus was established through the observation that the sequence GGTACAnnnTGTTCT will mediate the effect of the respective hormone–receptor complexes on the mammary tumor virus gene, but it is not likely to mediate the diverse effects these hormones have on mammalian genes. Various explanations may resolve this enigma. Subtle sequence differences may occur in specific HREs, receptor concentrations may be limiting in different cell types, ligand availability may be limiting, or HREs might not exist as isolated elements but in fact interact with other cis/trans elements to mediate their effect. Such complex hormone response units have been the subject of several recent publications (24–27).

Recently, a protein that interacts with the cAMP-response element (CRE), the CRE-binding protein (CREB), has been purified, and its cDNA has been cloned (28–30). The identification of the CRE was a prerequisite in the purification and characterization of CREB, just as the identification of an insulin-response sequence (IRS) will be the first step in the identification and isolation of the trans-acting factor(s) involved in the regulation of an insulin-responsive gene.

D.   THE GENES OF HEPATIC GLUCOSE METABOLISM

All of the principles discussed above can be brought to bear on the hormonal regulation of hepatic glucose metabolism. This important process is controlled by regulation of the movement of substrates through three major cycles (Fig. 3). The net flux through the glucose/glucose 6-phosphate (G6P), fructose 6-phosphate (Fru 6-P)/fructose 1,6-bisphosphate (Fru 1,6-P$_2$), and phosphoenolpyruvate (PEP)/pyruvate cycles depends on the relative activity of seven enzymes (excluding those involved in glycogen metabolism). Conditions favoring gluconeogenesis (the combination of high plasma glucagon, catecholamine, and glucocorticoid levels with a low plasma insulin level, as occurs when animals are starved or fed a low carbohydrate diet) result in increased activity of phosphoenolpyruvate carboxykinase (PEPCK), fructose-1,6-bisphosphatase (Fru-1,6-P$_2$ase), and glucose-6-phosphatase (G6Pase) and a coordinate decrease in the

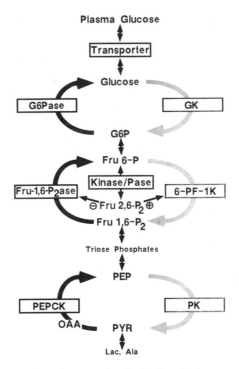

FIG. 3.   The substrate cycles and enzymes involved in hepatic glucose metabolism. Glucokinase (GK), phosphofructokinase (6-PF-1K), pyruvate kinase (PK), phosphoenolpyruvate carboxykinase (PEPCK), fructose-1,6-bisphosphatase (Fru-1,6-P$_2$ase), glucose-6-phosphatase (G6Pase), glucose 6-phosphate (G6P), fructose 6-phospate (Fru 6-P), fructose 2,6-bisphosphate (Fru 2,6-P$_2$), phosphoenolpyruvate (PEP), pyruvate (PYR), oxaloacetate (OAA), 6-PF-2K/Fru-2,6-P$_2$ase (Kinase/Pase), lactate (Lac), alanine (Ala). Reprinted with permission from the *Journal of Biological Chemistry* (31).

activity of pyruvate kinase (PK), 6-phosphofructo-1-kinase (6-PF-1K), and glucokinase (GK) (31). Reciprocal changes in the activities of these enzymes occur when animals are fed a high-carbohydrate diet, particularly after a prolonged fast. In this situation, the plasma insulin concentration increases, levels of the counterregulatory hormones decrease, and glycolysis predominates. The Fru 6-P/Fru 1,6-P_2 substrate cycle is also regulated by a subcycle in which the amount of the regulatory molecule fructose 2,6-bisphosphate (Fru 2,6-P_2) is controlled by the bifunctional enzyme 6-phosphofructo-2-kinase/fructose-2,6-bisphosphatase (6-PF-2K/Fru-2,6-P_2ase) (2). The relative activity of these seven enzymes, established by the interaction of several hormones, determines whether the hepatocyte is a net consumer or producer of glucose. Many other enzymes are also involved in these processes. However, they catalyze equilibrium reactions that are not rate controlling, thus their regulation by hormones, if it occurs, has no metabolic impact.

The direction and magnitude of substrate cycling is modulated by acute and chronic regulatory mechanisms. The acute regulation of hepatic glucose metabolism occurs through hormone-mediated changes in enzymatic activity, principally the phosphorylation or dephosphorylation of two key enzymes, PK and 6-PF-2K/Fru-2,6-P_2ase, and the allosteric regulation of 6-PF-1K and Fru-1,6-P_2ase by Fru 2,6-P_2 (2). Hormones also exert chronic effects on glucose metabolism by changing the rate of enzyme synthesis. These chronic effects are mediated through alterations of the rate of mRNA synthesis, and in some cases by changes in the rate of degradation of specific mRNAs.

Complete cDNA molecules have been obtained for five of the six key enzymes involved in hepatic glucose metabolism (the exception being G6Pase). This has allowed for measurements of the specific mRNA amount and gene transcription in response to various hormone treatments, and it has permitted the isolation and characterization of the cognate genes. In some cases, the promoter regions have also been isolated, and studies of the interaction of specific cis-acting DNA elements with the trans-acting factors that mediate hormone action have begun.

E. HORMONAL REGULATION OF THE PEPCK GENE

Our studies have focused on the hormonal regulation of PEPCK, the rate-limiting enzyme in hepatic gluconeogenesis. Cyclic AMP and glucocorticoids increase gluconeogenesis by increasing the rate of transcription of the PEPCK gene and this results in an enhanced rate of synthesis of the protein (32,33). Insulin exerts an opposite, dominant negative effect (33–36). The studies that led to these conclusions were reviewed in a previous volume in this series (37). The rest of this article deals with our recent effort to define how the HREs in the PEPCK promoter interact to provide coordinate regulation of this gene.

II. Results

A. STRATEGY USED TO DEFINE HORMONE RESPONSE ELEMENTS

Our initial studies involved an analysis of chimeric constructs in which the PEPCK promoter region was ligated to a chloramphenicol acetyltransferase (CAT) reporter gene. The resulting fusion gene was transfected into H4IIE rat hepatoma cells. Expression of the reporter gene was analyzed using transient transfection, during which the transfected DNA remains free in the cell nucleus, and also by stable transfection, in which the transfected DNA becomes integrated into the host cell genome. The details of these techniques and their relative merits have been discussed elsewhere (23).

The transfection approach is a valid means of locating HREs providing the hormonal regulation of the transfected gene mimics that of the endogenous gene. In one set of experiments the modulation of CAT activity by dex-amethasone, cAMP, and insulin was assayed using transient and stable trans-fection of the plasmid pPL1–CAT (which contains the PEPCK promoter se-quence from -2100 to $+69$ relative to the transcription start site) into H4IIE cells. Dexamethasone and cAMP both induced CAT activity and the combination gave a large, synergistic induction (Fig. 4). Insulin, when added simultaneously with dexamethasone, cAMP, or the combination of both, almost totally pre-vented this increase. This regulation is qualitatively similar to that of the endoge-nous gene (33).

The effect of various concentrations of hormones on expression of the trans-gene and the endogenous gene should be comparable if both are regulated by similar mechanisms. For example, insulin, at concentrations ranging from 10^{-11} to 10^{-7} M, prevented dexamethasone- and cAMP-stimulated CAT expression in a concentration-dependent manner in transfected cells (Fig. 5). Total inhibition was seen at 10^{-7} M insulin and the half-maximal effect (EC_{50}) occurred at about 10^{-10} M insulin, effects similar to those observed for the action of insulin on transcription of the endogenous PEPCK gene (33,38). Similar results have been obtained with transiently transfected H4IIE cells (39).

Having demonstrated that the fusion gene shows the responses to hormones

FIG. 4.   Multihormonal regulation of PEPCK–CAT fusion genes. H4IIE cells were transiently or stably transfected with the plasmid pPL1–CAT, which contains the region of the PEPCK promoter from -2100 to $+69$ ligated to the chloramphenicol acetyltransferase (CAT) reporter gene. Cells were exposed to dexamethasone (DEX, 500 nM), 8-(4-chlorophenylthio)-cAMP (cAMP, 0.1 mM) in the absence or presence of insulin (10 nM). CAT activity, corrected for protein concentration in the extract, is expressed as the percentage of control (hormone treated/untreated). Results represent the mean $\pm$ SEM of three to six separate experiments. Reprinted with permission from *Molecular Endocrinology* (43).

that are characteristic of the endogenous gene, the PEPCK HREs were then delineated using 5', 3', or internal deletions of the promoter ligated to the reporter gene. The PEPCK HREs were then further defined by demonstrating an interaction of trans-acting factors with the DNA element (measured by DNase I protection or gel mobility shift assays).

B.   THE PEPCK GENE HAS A UNIQUE GLUCOCORTICOID RESPONSE UNIT

1.   5' Deletion Analysis of the PEPCK Gene Promoter

A set of 5' deletion mutations of the PEPCK promoter (Fig. 6) was used to locate the 5' boundary of the glucocorticoid response unit (GRU). Cotransfection

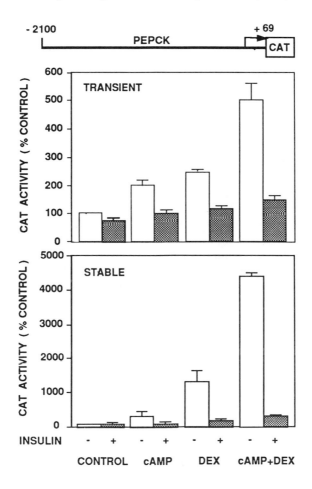

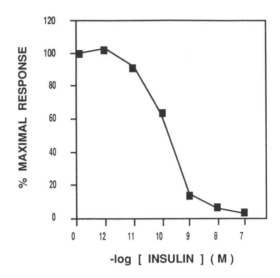

FIG. 5. Insulin concentration response in stably transfected H4IIE cells. H4IIE cells, stably transfected with pPL1–CAT, were treated with dexamethasone and cAMP (see Fig. 4) and insulin was added at the concentration indicated. Results are expressed as a percentage of the fold induction in CAT activity over basal activity in dexamethasone- and cAMP-treated cells.

of the various PEPCK-CAT plasmids with pSVGR1, a GR expression vector, facilitated this analysis, since intracellular levels of glucocorticoid receptor (GR) in H4IIE cells limit the response to glucocorticoids in simple transient transfection analysis of the PEPCK–CAT gene, as has been reported in the analysis of other genes (40). This procedure resulted in a doubling of the response of plasmids containing the full glucocorticoid response unit without affecting basal activity. Plasmids with promoter endpoints at −2100, −1264, −600, or −467 were equally responsive to glucocorticoids (about an 11-fold induction). A prominent decrease in the glucocorticoid response occurred between −467 and −402 (from 11- to 2-fold induction). A second, small diminution in glucocorticoid induction occurred between −306 and −271; no significant glucocorticoid response was obtained when fragments shorter than −271 were fused to the CAT gene. There was no significant difference in basal expression between these various 5′ deletion mutations. We conclude from these studies that the predominant glucocorticoid effect in H4IIE cells requires a DNA sequence whose 5′ end point lies between −467 and −402.

2. Interaction of GR with the PEPCK Promoter

We performed a DNase I footprint analysis to identify precisely the DNA sequence(s) that interacts with the glucocorticoid receptor (GR), expecting to

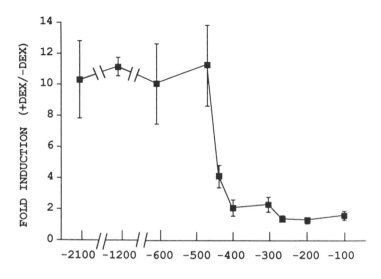

FIG. 6.   An analysis of glucocorticoid induction of CAT expression in various 5' deletion mutations of the PEPCK gene promotor. A series of 5' deletion mutations of the PEPCK promoter were
fused to the CAT reporter gene and were cotransfected with a glucocorticoid receptor expression
vector (pSVGR1) (40) into H4IIE cells. Transfected cells were incubated with or without 500 n*M*
dexamethasone for 24 hours in serum-free medium. Cell lysates were prepared and assayed for CAT
activity. The results are expressed as increase (-fold) in CAT activity over control (+DEX/−DEX).
Each data point represents the mean ± SD of a minimum of three independent experiments. Reprinted with permission from *Molecular Cellular Biology* (24).

find at least some portion of the footprint between −467 and −402. As can be
seen in Fig. 7, there was no evidence of interaction of GR with this region, but
highly purified GR did protect the region between −395 and −349 of the coding
strand against DNase I digestion. This protected region contains two putative
core sequences CACACAnnnTGTGCA (GR1) and AGCATAnnnAGTCCA
(GR2) required for GR binding. GR1 matches the consensus GRE-binding sequence (GGTACAnnnTGTTCT) at only 7/12 positions whereas GR2 matches at
6/12. The location of the GR-binding sites, well 3' or proximal to the region
demonstrated to contain the 5' boundary of the glucocorticoid response unit
(−467 to −402), was surprising.

3.   Physical Evidence of Accessory DNA Elements/Factors

The discrepancy between the boundary of the functional GRU and the
glucocorticoid receptor binding sites raised the possibility that accessory elements and associated trans-acting factors were an integral component of the

A.

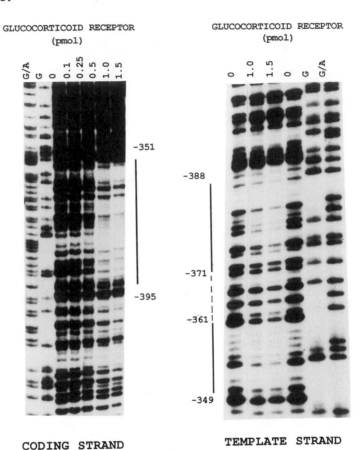

GLUCOCORTICOID RECEPTOR
(pmol)

GLUCOCORTICOID RECEPTOR
(pmol)

CODING STRAND

TEMPLATE STRAND

B.

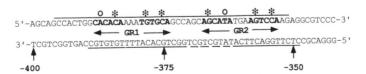

GRU. To ascertain whether accessory factors bind to the region required for glucocorticoid induction, we sought evidence of protein–DNA interactions in the region between −467 and −402, using crude rat liver nuclear extract in the DNase I footprint assay. As shown in Fig. 8, two DNA sequences were protected. The first region, from −455 to −431 was denoted accessory factor site 1 (AF1) and the second, from −420 to −403, was denoted accessory factor site 2 (AF2). There is no substantial DNA sequence homology between AF1 and AF2.

4. Functional Evidence for Accessory Elements/Factors

We characterized the relative functional contributions of the accessory factor-binding sites and the GR-binding sites by analyzing the regulation of CAT expression in plasmids in which one or more element was deleted. The promoter containing the entire GRU (pPL32) gave an 11-fold induction in response to dexamethasone, whereas the promoter lacking either both accessory elements (pPL12), or the entire GRU (pPL33), gave a 2-fold induction (Fig. 9, top). The deletion of either AF1 or AF2 caused 50–60% reduction of CAT expression in response to dexamethasone (Fig. 9, middle). AF1 and AF2 do not function as glucocorticoid responsive enhancers themselves. Plasmids containing AF1 or AF2 alone, or AF1 and AF2 together, were no more responsive to dexamethasone than was the construct that lacked the entire GRU (pPL33) (Fig. 9, bottom).

Deletion of either GR1 or GR2 also resulted in a 50% reduction of the response to dexamethasone (4.3-fold for ΔGR1 and 5.2-fold for ΔGR2) (Fig. 10). The deletion of both GR1 and GR2 in a construct containing both accessory elements reduced glucocorticoid induction to the low level noted when the entire GRU was absent (compare plasmid 468–390/pPL33 to pPL33). These results suggest that each GR-binding site functions independently, and that each accounts for half of the full response, provided both accessory factor sites are present. The GR-binding sites themselves are inert (compare pPL12 with pPL33).

FIG. 7. DNase I protection analysis of specific binding of purified rat liver glucocorticoid receptor to the PEPCK gene promoter. An end-labeled fragment of the PEPCK gene (from −490 to −212 bp) was incubated with 0.1 to 1.5 pmol of purified rat liver glucocorticoid receptor as described (24). After DNase I digestion, the samples were analyzed on a 6% acrylamide/7 M urea gel. The Maxam–Gilbert reactions (G/A and G) are shown in parallel lanes in (A). Homologies to the GRE consensus sequence are labeled GR1 and GR2 in (B). Asterisks (*) indicate nucleotides present in the PEPCK promoter known to be most critical in receptor binding to the consensus GRE. The open circle (O) indicates a nucleotide demonstrated to be important for binding in the tyrosine aminotransferase GRE that is different in the PEPCK promoter. Reprinted with permission from Molecular Cellular Biology (24).

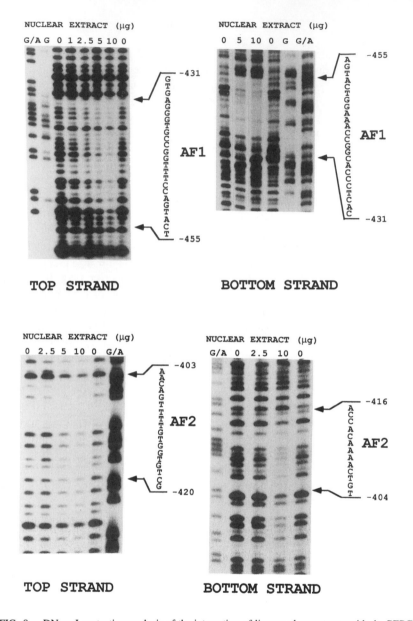

NUCLEAR EXTRACT (µg)

G/A G 0 1 2.5 5 10 0

TOP STRAND

NUCLEAR EXTRACT (µg)

0 5 10 0 G G/A

BOTTOM STRAND

NUCLEAR EXTRACT (µg)

0 2.5 5 10 0 G/A

TOP STRAND

NUCLEAR EXTRACT (µg)

G/A 0 2.5 10 0

BOTTOM STRAND

FIG. 8.   DNase I protection analysis of the interaction of liver nuclear extracts with the PEPCK gene promoter. Rat liver nuclear extracts were prepared and used in a DNase I protection assay as described (24). Five to 10 fmol of either the labeled coding or template strand DNA was incubated with 0–10 µg of the rat liver nuclear extract and then the digestions with DNase I were performed. Samples were analyzed on a 6% acrylamide/7 *M* urea gel. Two protected regions were observed in the region located between −467 and −402 bp relative to the transcription start site. The protected region at the 5′ site is labeled accessory factor 1 (AF1) and that at the 3′ protection site is labeled accessory factor 2 (AF2). Reprinted with permission from *Molecular Cellular Biology* (24).

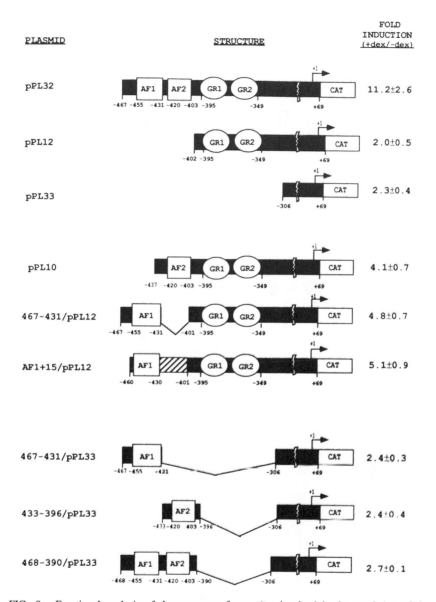

FIG. 9. Functional analysis of the accessory factor sites involved in the regulation of the PEPCK–CAT fusion genes by dexamethasone. H4IIE cells were cotransfected with glucocorticoid receptor expression plasmid pSVGR1 and various other plasmids, and were then incubated in the presence of absence of 500 nM dexamethasone for 24 hours. Cell lysates were prepared and assayed for CAT activity. AF1 and AF2 represent the DNA sequences between −455 and −431 and −420 and −403, respectively (see Fig. 8). GR1 and GR2 represent DNA sequences between −389 and −375 and −367 and −353, respectively (see Fig. 7). The results are expressed as induction (-fold; mean ± SD) of a minimum of four independent transfections with each construct. Reprinted with permission from *Molecular Cellular Biology* (24).

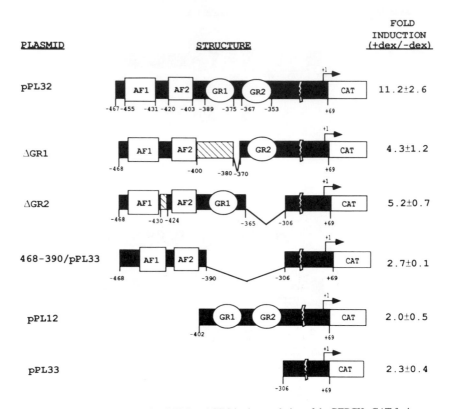

FIG. 10.   Functional analysis of GR1 and GR2 in the regulation of the PEPCK–CAT fusion genes by dexamethasone. H4IIE cells cotransfected with the glucocorticoid receptor expression plasmid pSVGR1 and other plasmids were incubated in the presence or absence of 500 n*M* dexamethasone for 24 hours. GR1, GR2, AF1, and AF2 were described in the legend to Fig. 9. The results are expressed as induction (-fold) over control (mean ± SD) of a minimum of four independent transfections with each construct. Reprinted with permission from *Molecular Cellular Biology* (24).

5.   Unique Proteins Bind to the AF1 and AF2 Elements

When labeled AF1 was incubated with crude rat liver nuclear extract in the gel mobility shift assay, several bands indicative of protein–DNA interactions were detected. Unlabeled oligonucleotide AF1 competed for this binding, but AF2 did not. Conversely, binding to labeled AF2 was displaced by unlabeled AF2 but not by AF1 (Fig. 11). Thus different proteins bind the two elements. The presence of multiple bands suggested that more than one protein binds to the AF1 region (−460 to −430). A binding experiment in which cross-competition was tested with two oligonucleotides, AF1-a (−455 to −438) and AF1-b (−441 to −426), suggests this is so, as AF1-a was unable to compete with AF1-b and vice versa (Fig. 11).

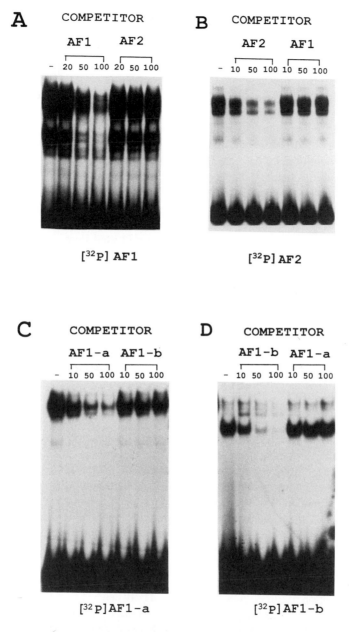

FIG. 11. Unique proteins bind to AF1 and AF2. A gel retardation assay was performed as described (24). The DNA fragments used in the gel retardation assay were as follows: (A) 8 fmol of DNA fragment −460 to −430 (AF1); (B) 5 fmol of DNA fragment −420 to −402 (AF2); (C) 18 fmol of DNA fragment −454 to −438 (AF1-a); and (D) 20 fmol of DNA fragment −441 to −426 (AF1-b). Each was end labeled and incubated with 4 µg of rat liver nuclear extract with or without competitor DNA. The molar excess of competitor DNA is indicated at the top of each panel. Samples were subjected to electrophoresis through 6% native acrylamide gels. Reprinted with permission from *Molecular Cellular Biology* (24).

6. Brief Description of the PEPCK Gene GRU

The DNA sequence required for stimulation of PEPCK gene transcription by glucocorticoids is uniquely complex (24). Rather than being the simple 15-mer consensus sequence required for GR binding, and designated a GRE (Fig. 2), the region extends over at least 110 base pairs (bp) of the PEPCK promoter and it consists of binding sites for several different proteins. Thus, we define this region as a glucocorticoid response unit (GRU) to distinguish it from the canonical GRE. Two contiguous sites, each of which presumably binds a glucocorticoid receptor dimer, compose the 3'-most portion of the region. These have little similarity to the canonical GRE. In H4IIE hepatoma cells these receptor-binding sites are, by themselves, functionally inert in the context of the PEPCK promoter, which also differentiates them from the typical GRE. Two rather extended DNA elements, designated AF1 and AF2, are located immediately adjacent to the 5' boundary of the receptor-binding sites. AF1 is composed of at least two distinct elements (AF1-a and AF1-b) and also serves as a retinoic acid response element (RARE) (41). Mutations that destroy the ability of AF1 to promote the glucocorticoid effect also disable it as an RARE (41a). AF2 also has multiple functions (see below).

C.   LOCATION OF AN INSULIN RESPONSE SEQUENCE (IRS)

1. 5' Deletion Mutations of the PEPCK Gene Promoter

Having demonstrated the validity of the transfection approach to locate an IRS (Figs. 4 and 5) we next attempted to further delineate the region of the PEPCK promoter responsible for the inhibitory effect of insulin. CAT activity was induced in a series of transfectants containing 5' deletion mutations by stimulation with a combination of cAMP and dexamethasone to amplify the inhibitory effect of insulin. All the stable transfectants responded to insulin; however, the effect varied in magnitude (Fig. 12). Inhibition was almost complete (a mean value of 90%) for deletions with 5' end points at −2100 (HL1), −600 (HL9), −467 (HL32), and −437 (HL10). When shorter fragments were used, e.g., −402 (HL12), −306 (HL33), or −271 (HL45), this decreased to a mean value of 57%. Thus the region of the PEPCK gene situated between −437 and +69, relative to the transcription start site, provides the full insulin effect and the reduction in the magnitude of the insulin effect between deletion end points −437 and −402 represents the loss of at least one insulin responsive sequence (IRS). One or more IRSs presumably reside within −402 and +69, and, since there is no apparent reduction in the insulin effect between −402 and −271, the other element(s) must be 3' from the latter. The location of other IRSs is being pursued.

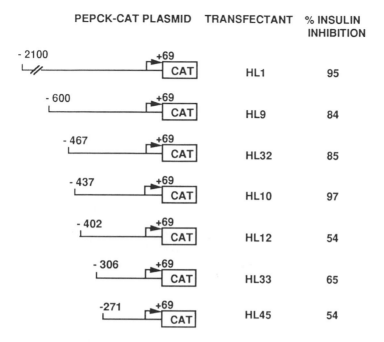

PEPCK-CAT PLASMID	TRANSFECTANT	% INSULIN INHIBITION
−2100 ... +69 CAT	HL1	95
−600 ... +69 CAT	HL9	84
−467 ... +69 CAT	HL32	85
−437 ... +69 CAT	HL10	97
−402 ... +69 CAT	HL12	54
−306 ... +69 CAT	HL33	65
−271 ... +69 CAT	HL45	54

FIG. 12. Analysis of insulin inhibition of CAT expression in stable transfectants containing various 5′ deletion mutations of the PEPCK gene promoter. Cells were treated for 18 hours with 8-CPT cAMP (0.1 mM) and dexamethasone (500 nM) in the presence or absence of insulin (10 nM). The data represent the percentage inhibition by insulin of the induction in CAT activity by dexamethasone and cAMP. Results represent the mean of three to seven separate experiments in which one to three stable transfectants for each construct were tested.

2. Further Definition of the Distal IRS

We made use of a vector (TKC-VI) containing the herpes simplex virus thymidine kinase (TK) promoter for further delineation of the distal IRS. A series of overlapping, double-stranded oligomers spanning the region between −453 and −271 of the PEPCK gene promoter were synthesized and each was ligated into the *Bam*HI site of TKC-VI (Fig. 13A), a vector similar to that previously used to identify the inhibitory sterol responsive element in the low-density lipoprotein (LDL) receptor gene promoter (42). Each oligomer contained 38 bp of PEPCK promoter sequence flanked by *Bam*HI-compatible sticky ends (GATC). The effect of insulin on CAT expression by these constructs was initially analyzed using transient expression of the chimeric DNA in H4IIE cells. The vector lacking an insert, and most of the constructs spanning the region from position −403 to −271, showed a slight stimulation of CAT expression in response to insulin.

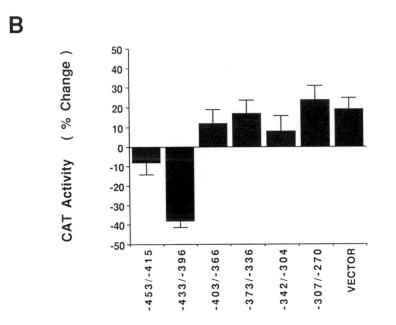

PEPCK Insert

FIG. 13. Use of the herpes simplex virus thymidine kinase (TK) promoter to identify a PEPCK gene IRS. A schematic representation of the TKC-VI vector, which contains the TK promoter sequence from -480 to $+51$ [similar to that previously described (42)], is shown in (A). A series of 42-mers containing 38 bp of PEPCK sequence with GATC ends was inserted into the *Bam*HI site of the TKCAT vector. The analysis of PEPCK/TKCAT constructs by transient transfection is illustrated in (B). Results are the ratio of CAT activity in insulin-treated vs. control cells (expressed as percentage change) and represent the mean ± SEM of nine separate transfections. Reprinted with permission from *Science* (44).

Constructs containing the PEPCK sequence between -433 and -396 showed an insulin-responsive inhibition of CAT expression (Fig. 13B). This construct worked equally well in both orientations (data not shown).

H4IIE cells stably transfected with either the insulin-responsive $-433/-396$ construct or, as a negative control, the $-403/-366$ construct were used to

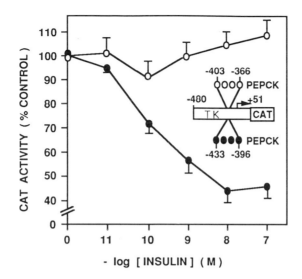

FIG. 14. Analysis of PEPCK/TKCAT constructs by stable transfection. Stable cell lines were established using the −433/−396 and −403/−366 PEPCK/TKCAT constructs (see Fig. 13). Cultures of individual transfectants were incubated in serum-free medium for 18 hours with the concentrations of insulin indicated on the abscissa. Cells were harvested and CAT activity was assayed. Results are expressed as a percentage of the CAT activity in control cells and represent the mean ± SEM of six separate experiments. Reprinted with permission from John Libbey & Co. (42a).

confirm that this effect was not simply a function of the transient transfection system. A −433/−396-containing transfectant showed a concentration-dependent inhibition of CAT expression in response to insulin; the EC_{50} was about 10^{-10} M and maximal inhibition was obtained with 10^{-8} M insulin (Fig. 14), concentrations that provide similar degrees of inhibition of the endogenous gene (33,38). In contrast, a −403/−366-containing transfectant showed a slight increase of CAT expression in response to insulin, a characteristic shared by the native vector.

3. Mutational Analysis of the IRS

The transient and stable transfection analyses revealed that the PEPCK sequence −433/−396 contains an IRS, so we sought to further delineate the boundaries of this element. The sequences with the end points −425/−421, −411/−407, and −401/−397 were changed to produce mutations designated M1, M2, and M3, respectively. These were inserted into the TKC-VI vector and the transient expression of CAT activity was analyzed in H4IIE cells. The M2 mutation abolished the insulin effect but plasmids containing both orientations of the M1 and M3 mutations still responded to insulin (Fig. 15). Another mutation,

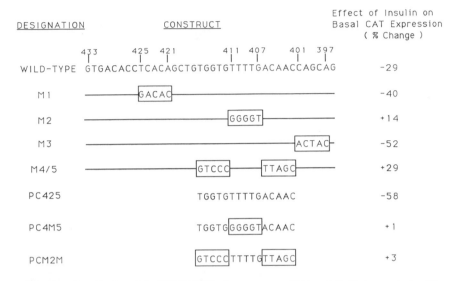

FIG. 15.   Localization of the PEPCK IRS in the sequence −433 to −396. Various mutants of the PEPCK sequence between −433 and −396 were synthesized with *Bam*HI (GATC) ends and cloned into the TKC-VI vector. The effect of insulin on CAT expression was analyzed by transient transfection. Results represent the ratio of CAT activity in insulin-treated versus control cells (expressed as percentage change) and are the mean ± SEM of 6–14 separate transfections for each construct. The boxed areas contain the mutant sequences. Reprinted with permission from *Science* (44).

designed M4/5, in which the 5-bp sequences between −416/−412 and −406/−402 were changed on either side of the wild-type −411/−407 sequence, failed to give an insulin-dependent inhibition of CAT expression in either orientation (Fig. 15).

A 15-bp core sequence that spans −416 and −402 (PC425) showed insulin-dependent, orientation-independent, inhibition of CAT expression. Mutations equivalent to M2 and M4/5 within PC425, designated PC4M5 and PCM2M, respectively, abolished the inhibitory effect of insulin, as did identical mutations in the full-length oligomer (Fig. 16). The 15-bp sequence, −416 to −402, is therefore a functional IRS.

4.   Protein Binding to the IRS

Protein–DNA complexes indicative of specific interactions were detected when the wild-type −433/−396 sequence was used as the labeled probe in the mobility shift assay (Fig. 16). Nuclear extracts prepared from rat liver or H4IIE cells gave qualitatively identical patterns. There was an excellent correlation between protein binding and the insulin response. Sequences that conferred a

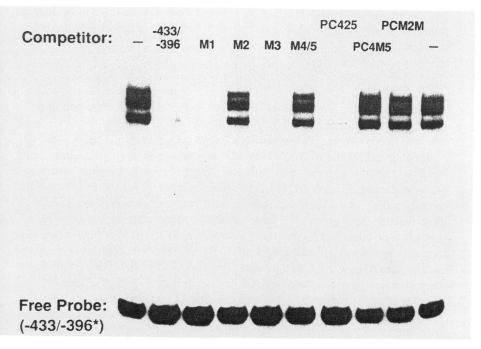

FIG. 16.   Correlation of protein binding with the insulin response. The −433 to −396 PEPCK sequence, radiolabeled with [α-^{32}P]dATP using the Klenow fragment of *E. coli* DNA polymerase I, was incubated with 10 μg of a rat liver nuclear extract as described (44). Unlabeled competitor DNA fragments (100-fold molar excess) were added for competition analysis. Reprinted with permission from *Science* (44).

response to insulin in transient transfection, including −433/−396, M1, M3, and the 15-bp −416/−402 sequence (PC425), effectively competed for binding with the labeled probe. Sequences that did not confer the insulin response, i.e., the M2, M4/5, PC4M5, and PCM2M mutations, failed to compete (Fig. 16). Thus, the DNA–protein interactions correlate with the insulin response.

5.   *Brief Description of the IRS*

Chimeric constructs containing the entire PEPCK gene promoter ligated to the CAT reporter gene responded to insulin in a manner analogous to that of the endogenous PEPCK gene. Using a series of 5′ deletions we located one insulin-responsive sequence (IRS) in the −437 to −402 region of PEPCK gene promoter (43). Using a heterologous promoter and a series of mutations, we have narrowed this to a 15-bp segment located between −416 and −402 (44). This IRS functions as an insulin-dependent silencing element when attached to a heterologous promoter, works in an orientation-independent fashion, and selectively binds one or more proteins.

III.　Discussion

Hormones affect gluconeogenesis by increasing or decreasing the rate of transcription of the PEPCK gene. We have analyzed the PEPCK gene promoter in an attempt to define the hormonal regulation of the rate-limiting step of gluconeogenesis at the molecular level. Hormone response elements have been identified for the agonists that stimulate the gene such as glucocorticoids, glucagon (acting through cAMP), and retinoic acid, and for the inhibitory hormone, insulin. The cAMP and retinoic acid response elements are similar to those found in other genes (45); however, the glucocorticoid response "element" is much more complex. It consists of a tandem array (5' to 3') of two accessory factor-binding sites designated AF1 and AF2, and two glucocorticoid receptor-binding sites (Fig. 17). This entire complex spans about 110 bp and is obviously more complex than the canonical GRE described in Fig. 2. Because of this complex array, we designated this a glucocorticoid response unit, or GRU (24).

The recognition of the essential role of accessory elements/factors in glucocorticoid-mediated transcriptional control raises interesting questions. Are accessory factors unique to the action of glucocorticoids? Do accessory elements provide cross-signaling between different steroid hormone–receptor complexes and their

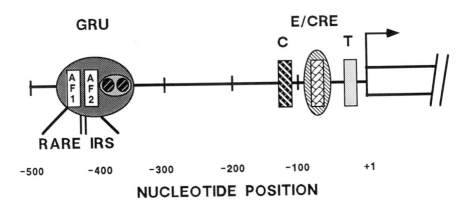

FIG. 17.　A schematic representation of the PEPCK gene promoter. The arrow marks the transcription initiation site. Glucocorticoid response unit (GRU), accessory factors 1 and 2 (AF1 and AF2), retinoic acid response element (RARE), insulin response sequence (IRS), CAAT box (C), basal enhancer-cAMP response element (E/CRE), and TATA box (T).

respective HREs? How do accessory factors influence the interaction between a steroid hormone receptor and its cognate HRE?

The first evidence that accessory factors could augment the function of a GRE did not involve a natural promoter. Schule *et al.* showed that a variety of transcription elements, including the CACC box and the SP1, NF1, and octamer-binding sites, increased the function of a consensus GRE (46). In addition to our studies with the PEPCK gene GRU, there is evidence that several other natural HREs require the presence of accessory factors and that many of these systems are significantly more complex than originally imagined (25–27). Whether the proliferin gene GRE is inert, positive, or negative depends on the intracellular concentrations of c-*fos* and c-*jun*, the AP-1 site-binding proteins (25). The GRE is inactive in the absence of c-*jun*, it acts as a positive element when c-*jun* is present, and it acts as a negative element in the presence of c-*jun* and high levels of c-*fos* (25). The AP-1 site is also part of a complex hormone response unit in the osteocalcin gene promoter (26). In this promoter the vitamin D response element also confers responsiveness to retinoic acid. The element contains an AP-1 site, and cotransfection of cells with c-*fos* and c-*jun* expression vectors supresses basal expression of the osteocalcin gene, and the stimulation of transcription of this gene by vitamin D or retinoic acid (26). The distal enhancer in the prolactin promoter requires an interaction of both the estrogen receptor and the Pit-1 POU protein in order to be fully active (27). There is indirect evidence that other hormone response elements are also complex. For example, the binding of both the thyroid hormone receptor and the progesterone receptor to their respective HREs is enhanced by the addition of nuclear extracts to the reaction mixture (47,48).

Accessory elements could provide cross-signaling between different hormones and their respective HREs, particularly if they are located in the context of one another. This appears to be the case with the GRU. For example, AF1 coincides with a retinoic acid response element (RARE) (41). Whether the RARE is AF1, or this represents another example of a common DNA element capable of binding two distinct proteins, remains to be established. Two observations are interesting in this regard: (1) Retinoic acid acts synergistically with dexamethasone to stimulate the transcription of the endogenous PEPCK gene, and of transfected genes bearing an intact GRU. (2) Mutations of AF1 that destroy its ability to function as an RARE also disable it with respect to promoting the glucocorticoid effect (J. A. Mitchell, unpublished observations).

The functions of AF2 and the IRS also involve coincident DNA sequences, and thus provide an opportunity for cross-signaling. The former was defined between positions -420 and -403 and the latter between -416 and -402. An analysis by site-directed mutagenesis indicates that a core sequence, -TTTTG-, is required for the function of both the AF2 and IRS elements (44; J. A. Mitchell, unpublished observations). This coincident location of regulatory elements is the probable explanation of the dominant negative effect insulin has with respect to the action of glucocorticoids on the PEPCK gene (33–36). A schematic representation of these elements is illustrated in Fig. 17.

The nature of the interactions between the proteins that bind to these elements in the PEPCK gene GRU must still be established. Accessory factors may promote hormone receptor–HRE interactions when the affinity of the receptor for the DNA element is low. In the case of the PEPCK gene, the two GR-binding sites in the GRU are not homologous with the consensus GRE and we have evidence that these sites bind GR 5–10 times less avidly than does the MMTV GRE (E. Imai, unpublished observations). Efforts to demonstrate cooperative binding between the two GR sites, or between the accessory factor sites and the GR sites, have, to date, been unsuccessful. The recent observation that AF1 binds the retinoic acid receptor (RAR) raises the interesting possibility that the function of the GRU involves the interaction of two heterologous receptors, i.e., the RAR and GR.

The protein(s) that binds to the AF2/IRS element is of obvious importance to the function of the GRU and PEPCK promoter. In the absence of insulin the protein(s) bound to the AF2/IRS element could have a positive effect on the immediately adjacent GRE complexes. Insulin could promote the binding of a unique protein to the AF2/IRS element, and thus preclude the binding of the AF2 protein. Or, it could cause a conformational change of a common protein(s), possibly by a phosphorylation–dephosphorylation mechanism, which would prevent its interaction with the GR. Alternatively, such a conformational change could enable the binding of a different protein, to the promoter or through protein–protein interaction, which either exerts a negative effect on transcription or occludes the GR-binding site. These possibilities can be tested once the protein(s) that binds to this region is purified or the cognate mRNA is made available by cDNA cloning.

IV. Conclusion

The regulation of the PEPCK gene is a paradigm for past, current, and future studies of metabolic regulation. The enzyme plays a central role in gluconeogenesis. Given the importance of gluconeogenesis to the survival and well-being of an organism, it is not surprising that several hormones act in concert to provide the positive and negative regulation of this process. In recent years it has become obvious that these hormones primarily regulate the amount of PEPCK enzyme (hence gluconeogenesis) by either increasing or decreasing the rate of synthesis of PEPCK mRNA (transcription). It now appears as though much of the regulation of PEPCK gene transcription occurs through an ~110-bp segment of the PEPCK gene promoter. The next challenge is to determine how the various proteins that bind to components of this metabolic control domain interact.

ACKNOWLEDGMENTS

The authors are grateful to Deborah Caplenor for her help in preparing this manuscript. The work was supported by NIH Grant DK35107, the Vanderbilt Diabetes and Endocrinology Research Center (DK20593), Juvenile Diabetes Foundation fellowships (R.O'B. and C.F.), and an American Diabetes Association Mentor-Based Training Grant (E.I.).

REFERENCES

1. Granner, D. K., and Beale, E. G. (1985). In "Biochemical Actions of Hormones" (G. Litwack, ed.), Vol. 12, pp. 89–138. Academic Press, New York.
2. Pilkis, S. J., El-Maghrabi, M. R., and Claus, T. H. (1988). Annu. Rev. Biochem. 57, 755–783.
3. Levine, R. (1981). Diabetes Care 4, 38–44.
4. Schoenheimer, R. (1942). "The Dynamic State of Body Constituents." Harvard Univ. Press, Cambridge, Massachusetts.
5. Shemin, D., and Rittengerg, D. (1946). J. Biol. Chem. 166, 627–636.
6. Thompson, R. C., and Ballou, J. E. (1956). J. Biol. Chem. 223, 795–809.
7. Dubos, R. J. (1940). Bacteriol. Rev. 4, 1–16.
8. Monod, J. (1947). Growth 11, 223–289.
9. Spiegelman, S. (1946). Cold Spring Harbor Symp. Quant. Biol. 11, 256–277.
10. Knox, W. E., and Mehler, A. H. (1951). Science 113, 237–238.
11. Knox, W. E., Auerbach, V. H., and Lin, E. C. C. (1956). Physiol. Rev. 36, 164–254.
12. Schimke, R. T. (1973). Adv. Enzymol. 37, 135–187.
13. Berlin, C. M., and Schimke, R. T. (1965). Mol. Pharmacol. 1, 149–156.
14. Jacob, F., and Monod, J. (1961). J. Mol. Biol. 3, 318–356.
15. Granner, D. K. (1990). In "Harper's Biochemistry" (R. Murray, D. Granner, P. Mayes, and V. Rodwell, eds.), 22nd ed., pp. 467–477. Appleton & Lange, East Norwalk, Connecticut.
16. Mitchell, P. J., and Tjian, R. (1989). Science 245, 371–378.
17. Jones, N. C., Rigby, P. W. J., and Ziff, E. B. (1988). Genes Dev. 2, 267–281.
18. Singh, H., LeBowitz, J. H., Baldwin, A. S., and Sharp, P. A. (1988). Cell (Cambridge, Mass.) 52, 415–423.
19. Chandler, V. L., Maler, B. A., and Yamamoto, K. R. (1983). Cell (Cambridge, Mass.) 33, 489–499.
20. Rousseau, G. G., and Baxter, J. D. (1979). In "Glucocorticoid Hormone Action" (J. D. Baxter and G. G. Rousseau, eds.), pp. 49–78. Springer-Verlag, Berlin and New York.
21. Yamamoto, K. R., Gehring, U., Stampfer, M. R., and Sibley, C. J. (1976). Recent Prog. Horm. Res. 32, 3–32.
22. Lee, F., Mulligan, R., Berg, P., and Ringold, G. (1981). Nature (London) 294, 228–232.
23. O'Brien, R. M., and Granner, D. K. (1990). Diabetes Care 13, 327–339.
24. Imai, E., Stromstedt, P.-E., Quinn, P. G., Carlstedt-Duke, J., Gustafsson, J.-Å., and Granner, D. K. (1990). Mol. Cell. Biol. 10, 4712–4719.
25. Diamond, M. I., Miner, J. M., Yoshinaga, S. K., and Yamamoto, K. R. (1990). Science 249, 1266–1272.
26. Schule, R., Umesono, K., Mangelsdorf, D. J., Bolado, J., Pike, J. W., and Evans, R. M. (1990). Cell (Cambridge, Mass.) 61, 497–504.
27. Day, R. N., Koike, S., Sakai, M., Muramatsu, M., and Maurer, R. A. (1990). Mol. Endocrinol. 4, 1964–1976.

28. Gonzalez, G. A., Yamamoto, K. K., Fischer, W. H., Karr, D., Menzel, P., Biggs, W., Vale, W. W., and Montminy, M. R. (1989). *Nature (London)* **337,** 749–752.

29. Hoeffler, J. P., Meyer, T. E., Yun, Y., Jameson, J. L., and Habener, J. F. (1988). *Science* **242,** 1430–1433.

30. Quinn, P. G., and Granner, D. K. (1990). *Mol. Cell. Biol.* **10,** 3357–3364.

31. Granner, D. K., and Pilkis, S. J. (1990). *J. Biol. Chem.* **265,** 10173–10176.

32. Lamers, W. H., Hanson, R. W., and Meisner, H. M. (1982). *Proc. Natl. Acad. Sci. U.S.A.* **79,** 5137–5141.

33. Sasaki, K., Cripe, T. P., Koch, S. R., Andreone, T. L., Petersen, D. D., Beale, E. G., and Granner, D. K. (1984). *J. Biol. Chem.* **259,** 15242–15251.

34. Granner, D. K., Andreone, T. L., Sasaki, K., and Beale, E. (1983). *Nature (London)* **305,** 549–551.

35. Granner, D. K., and Andreone, T. L. (1985). *Diabetes Metab. Rev.* **1,** 139–170.

36. Granner, D. K. (1987). *Kidney Int.* **32,** Suppl. 23, 82–93.

37. Granner, D. K., Sasaki, K., Andreone, T. L., and Beale, E. (1986). *Recent Prog. Horm. Res.* **42,** 111–141.

38. Chu, D. T., Davis, C. M., Chrapkiewicz, N. B., and Granner, D. K. (1988). *J. Biol. Chem.* **263,** 13007–13011.

39. Magnuson, M. A., Quinn, P. G., and Granner, D. K. (1987). *J. Biol. Chem.* **262,** 14917–14920.

40. Vanderbilt, J. N., Miesfeld, R., Maler, B. A., and Yamamoto, K. R. (1987). *Mol. Endocrinol.* **1,** 68–74.

41. Lucas, P. C., O'Brien, R. M., Mitchell, J. A., Davis, C. M., Imai, E., Forman, B. M., Samuels, H. H., and Granner, D. K. (1991). *Proc. Natl. Acad. Sci. U.S.A.* **88,** 2184–2188.

41a. Mitchell, J. A. (1991). In preparation.

42. Sudhof, T. C., Russell, D. W., Brown, M. S., and Goldstein, J. L. (1987). *Cell (Cambridge, Mass.)* **48,** 1061–1069.

42a. Forest, C. D., O'Brien, R. M., Lucas, P. C., and Granner, D. K. (1990). *In* "Genetics and Human Nutrition" (P. Randle, J. Bell, J. Scott, eds.), pp. 111–120. John Libbey & Co., Ltd., London.

43. Forest, C., O'Brien, R. M., Lucas, P. C., Magnuson, M. A., and Granner, D. K. (1990). *Mol. Endocrinol.* **4,** 1302–1310.

44. O'Brien, R., Lucas, P. C., Forest, C. D., Magnuson, M. A., and Granner, D. K. (1990). *Science* **249,** 533–537.

45. Beato, M. (1989). *Cell (Cambridge, Mass.)* **56,** 335–344.

46. Schule, R., Muller, M., Otsuka-Murakami, H., and Renkawitz, R. (1988). *Nature (London)* **332,** 87–90.

47. Murray, M. B., and Towle, H. C. (1989). *Mol. Endocrinol.* **3,** 1434–1442.

48. Edwards, D. P., Kuhnel, B., Estes, P. A., and Nordeen, S. K. (1989). *Mol. Endocrinol.* **3,** 381–391.

DISCUSSION

J. Baxter. Have you tested to see whether the glucocorticoid response unit works through a heterologous promoter?

D. Granner. We are in the process of testing that now. Most GREs are not very promoter specific. They will work on a lot of different promoters and in different orientations and positions. We have not finished this work, but it appears that if we attach this GRU to a simple promoter, such as that found in the thymidine kinase promoter, it will not work as well as it does in the context of the PEPCK promoter.

J. Baxter. Why do you think the PEPCK gene GRU is so different from other GREs?

D. Granner. Why is this GRU so unusual? Maybe it is not so unusual. If you consider that the function of this enzyme (PEPCK) is to regulate gluconeogenesis and that glucocorticoids are named for their ability to promote gluconeogenesis, maybe this is a standard GRE and the others are up-mutants.

A. Means.   Since you have now identified that one of your accessory factors is for retinoic acid, have you checked whether retinoic acid has any effect on the transcription of the endogenous gene? If so, is it developmentally controlled? And, finally, does retinoic acid influence the activity of any of the factors that up or down regulate the normal gene?

D. Granner.   Retinoic acid, by itself, causes three- or fourfold stimulation of the transcription of the PEPCK gene in H4IIE cells. It acts with glucocorticoids to synergistically enhance transcription of the endogenous gene. We have not checked its role in the developmental expression of the PEPCK gene.

A. Means.   What about cAMP? Can you substitute retinoic acid for the glucocorticoid and see the cAMP response?

D. Granner.   We have not done that experiment yet.

A. Powers.   Glucocorticoids induce the PEPCK gene and insulin suppresses it, so what is the role of insulin? Does it cause accessory factor 2 (AF2) to come off the gene, or does it modulate another protein that competes for AF2?

D. Granner.   Both are possible mechanisms. The parsimonious explanation is that there is a single protein that binds to the AF2 site. Let us hypothesize that this is a phosphoprotein. In its phosphorylated state it promotes glucocorticoid receptor binding, or stabilizes this binding, and thereby promotes the glucocorticoid response. One might imagine that insulin activates a phosphatase that dephosphorylates that protein. In its dephosphorylated form it is unable to promote the glucocorticoid response but it also inhibits the response. Of course the phosphorylation/dephosphorylation could be reversed vis-à-vis these activities. Alternatively, one could imagine that insulin changes the covalent state of another protein which then is capable of binding to a protein on the AF2 site and that this interaction prevents the glucocorticoid response and inhibits the gene. We do not have any way of discriminating between these mechanisms as yet, and there are other possibilities. Our model obviously is based on other known actions of insulin which involve its ability to change the phosphorylation state of proteins.

A. Powers.   Can you show that insulin treatment of H4IIE cells induces activity of a binding protein?

D. Granner.   No. I did not mention this, but nuclear extracts isolated from livers of diabetic rats, or such rats treated with insulin, or from untreated or insulin-treated H4IIE cells, were all the same in mobility shift assay. This is not a surprising observation. We showed some years ago that the insulin effect occurred in the absence of protein synthesis and that this effect occurred within 5 minutes. The rapidity of the effect, coupled with the fact that protein synthesis is not required, excludes an induction effect. We believe that covalent modification, probably phosphorylation/dephosphorylation, is involved. Such a modification would probably not survive the rather prolonged and harsh procedures involved in isolating the nuclear extracts.

W. Bardin.   Is glucose involved in the regulation of the PEPCK gene?

D. Granner.   Glucose apparently is not involved in regulating the PEPCK gene. The gene behaves normally in glucose-free medium. Ron Kahn (Joslin Clinic, Boston) and I grew H4IIE cells for a year in the absence of glucose, and the gene functioned just as it does in cells grown in normal medium (*Mol. Endocrinol.* **3,** 840–845, 1989). I think we have excluded glucose as a factor. It is about the only thing we have excluded.

F. Murdoch.   Have you been able to dissect which pathway of second messenger is necessary for this insulin effect?

D. Granner.   No. This is obviously a big question. Insulin action presumably starts with activation of the receptor tyrosine kinase, but what happens after that is anybody's guess. It is reasonable to presume that a phosphorylation cascade is involved, but the kinase(s) and substrate(s) have not been identified. The problem involved in working from the receptor forward is that one can never be certain that a certain kinase or phosphoprotein is really involved in some biologic action of insulin. Several years ago, when we first designed this project, we said "Let's identify an absolutely well-defined action of insulin at the gene level and then work backward toward the receptor." At each step of the way we would be working with something that was actually involved. Now we know how to find the ultimate protein involved in regulating the effect of insulin on the PEPCK gene. The trans-acting factor that binds to the IRS is this protein. After isolating this protein we can begin searching for the kinase that modifies it, etc.

J. Baxter.   Why do you call this an IRS rather than an IRE?

D. Granner.   It is true that convention would have required that it be called an IRE. However, there already are two IREs. There is an iron response element and an interferon response element. A third, the insulin response element, would have added to the confusion. The real reason for calling it the IRS, though, is that it transmits a negative, dominant effect. It is arcane and finding it has been a tremendous money sink.

Mammalian Glucose Transporters: Structure and Molecular Regulation

CHARLES F. BURANT,* WILLIAM I. SIVITZ,† HIROFUMI FUKUMOTO,‡
TOSHIAKI KAYANO,§ SHINYA NAGAMATSU,‖ SUSUMO SEINO,*,**
JEFFREY E. PESSIN,¶ AND GRAEME I. BELL**,††

*Section of Endocrinology, Department of Medicine, The University of Chicago, Chicago, Illinois
60637, Departments of †Internal Medicine and ¶Physiology and Biophysics, The University of
Iowa, Iowa City, Iowa 52242, ‡Shiga Prefectural Sogo-Iryo-Center Hospital, Moriyama, Shiga,
Japan, §National Institute of Agrobiological Research, Tsukuba, Japan, ‖Department of
Biochemistry and Molecular Biology, Kyorin University School of Medicine, Tokyo, Japan, and
**Howard Hughes Medical Institute and ††Department of Biochemistry and Molecular Biology,
The University of Chicago, Chicago, Illinois 60637*

I. Introduction

Glucose represents a major source of energy for mammalian cells and its metabolism provides a steady supply of ATP under both anaerobic and aerobic conditions. Moreover, its storage as glycogen in the liver during periods of excess provides a mechanism by which plasma glucose levels can be rapidly buffered by the catabolism of glycogen and release of glucose into the circulation. Glycogen stores also represent the primary source of energy for exercising muscle. In adipocytes, glucose can be converted to triglycerides, a more energetically compact fuel store that can be mobilized in the form of fatty acids and glycerol during periods of stress. As well as being a major source of metabolic energy, glucose is part of or provides a source of structural moieties for other biological macromolecules, including glycoproteins, proteoglycans, glycolipids, and nucleic acids. Because of the diverse metabolic roles played by glucose, defects in its uptake or metabolism, such as occur in diabetes mellitus, can lead to dramatic alterations in cellular functions resulting in severe morbidity and mortality.

In the past few years, an explosion of information regarding the molecular structures of the proteins involved in the transport of glucose across the plasma membrane and their regulation in normal and altered metabolic states has provided new insight into the physiological basis of glucose homeostasis. These studies have revealed the presence of two distinct families of glucose transport proteins that have evolved to transport this important metabolite across the plasma membrane (Table I). One family of glucose carriers, the sodium-dependent

349

TABLE I

Mammalian Glucose Transporters: Major Sites of Expression and Physiological Functions[a]

Designation	Major sites of expression	Function
Sodium-dependent glucose transporters		
SGLT1	Small intestine and kidney	Active uptake of dietary glucose from the lumen of the small intestine, and reabsorption of filtered glucose in the proximal tubule of the kidney
Facilitative glucose transporters		
GLUT1	Placenta, brain, kidney, and colon	Basal uptake of glucose by cells, transport of glucose across blood–tissue barriers
GLUT2	Liver, pancreatic β cell, small intestine, and kidney	Uptake and release of glucose by hepatocyte β cell glucose sensor Release of absorbed glucose across the basolateral surface of absorptive epithelial cells of small intestine and kidney
GLUT3	Many tissues in humans, including brain, placenta, and kidney; in other species may only be expressed at high levels in the brain	Basal uptake of glucose by all cells in humans, including those of the brain; uptake of glucose by cells of the brain in other species
GLUT4	Skeletal and cardiac muscle, and brown and white adipose tissue	Insulin-stimulated glucose uptake
GLUT5	Small intestine (jejunum)	Absorption of glucose from the lumen of the small intestine (?)

[a]The cellular distribution and intracellular localization of several isoforms have been determined by immunocytochemistry or *in situ* hybridization. GLUT1 is expressed at high levels in microvessels of the brain (Gerhart *et al.*, 1989; Pardridge *et al.*, 1990a), and at the blood–nerve (Froehner *et al.*, 1988; Gerhart and Drewes, 1990) and blood–eye (Harik *et al.*, 1990a) barriers. In many tissues in which it is expressed, GLUT1 is concentrated in cells of blood–tissue barriers (Takata *et al.*, 1990; Harik *et al.*, 1990b). GLUT1 is also expressed at variable levels in different segments of the nephron (Thorens *et al.*, 1990c), being highest in connecting segments and collecting ducts. It is also expressed in perivenous hepatocytes (Tal *et al.*, 1990). GLUT2 has been localized to the plasma membrane of the β cell (Orci *et al.*, 1989), the sinusoidal membrane of hepatocytes (Thorens *et al.*, 1990b), the basolateral membrane of fully differentiated absorptive epithelial cells of the small intestine (Thorens *et al.*, 1990b), and the basolateral membrane of proximal tubule cells of the kidney (Thorens *et al.*, 1990b,c). GLUT4 has been localized in the trans Golgi reticulum of myocytes and adipocytes in the basal state, and in adipocytes, to the plasma membrane after insulin stimulation (Slot *et al.*, 1990). In another study of GLUT4 distribution in skeletal muscle, it was observed within the triad (on terminal cisternae and transverse tubules) and an intracellular compartment, possibly sarcoplasmic tubules (Friedman *et al.*, 1991).

glucose transporters or Na^+/glucose cotransporters, mediates the active uptake of glucose against its concentration gradient. In this transport system, the uptake of glucose is directly coupled to Na^+ transport and will not occur in the absence of an Na^+ gradient. In contrast to the Na^+/glucose cotransporters, the facilitative glucose transporter family mediates the saturatable, bidirectional passive transport of glucose across the cell membrane down its concentration gradient. Together these two families of structurally and functionally distinct glucose carrier proteins ensure efficient uptake of dietary glucose and its distribution to other tissues. This article will focus on recent advances in the molecular biology of glucose transport and on the different mechanisms by which it may be regulated.

II. Sodium-Dependent Glucose Transporters

The sodium-dependent glucose transporter or Na^+/glucose cotransporter is a secondary active transport system that utilizes the energy from an extracellular to intracellular Na^+ electrochemical gradient, generated by Na^+,K^+-ATPases, to drive the intracellular accumulation of glucose against its concentration gradient. The Na^+/glucose cotransporters are located in the brush-border membranes of intestinal and kidney epithelial cells and are responsible for the absorption of glucose from the lumen of the small intestine and proximal tubule of the kidney.

Utilizing a novel expression cloning strategy, Wright and co-workers isolated cDNA clones encoding the Na^+/glucose cotransporter expressed in rabbit small intestinal mucosa (Hediger et al., 1987). The expression assay measured phorizin-sensitive α-methyl-D-glucopyranoside uptake in Xenopus oocytes stimulated by the injection of poly(A)$^+$ RNA from rabbit small intestinal mucosa and in vitro-transcribed RNA synthesized from cDNA clones. The hexose α-methyl-D-glucopyranoside is a specific substrate for the Na^+/glucose transporter and is not transported by facilitative glucose transport proteins. The sequence of the cDNA clone that was isolated using this approach indicated that the rabbit small intestinal Na^+/glucose cotransporter, designated SGLT1, was a 662-amino acid protein. SGLT1 is a member of a superfamily that also includes the bacterial Na^+/proline cotransporter (Hediger et al., 1989b). SGLT1 mRNA is expressed throughout the rabbit small intestine (Coady et al., 1990), and its relative abundance determined by hybridization is jejunum (1):ileum (0.75):duodenum (0.5), with very low levels of mRNA in colon and undetectable levels in gastric mucosa.

The rabbit cDNA clone was used as a probe to isolate cDNAs encoding the corresponding human Na^+/glucose cotransporter from a human ileal cDNA library (Hediger et al., 1989b). The human protein was 664 amino acids in length and possessed 85% identity with rabbit SGLT1 (Fig. 1). The sequence of a cDNA clone encoding part of the pig SGLT1 has been recently reported, and there is 86% identity between this sequence and those of human or rabbit SGLT1

```
                         <------- M1 -------->
Human    1 MDSSTWSPKTTAVTRPVETHELIRNAADISIIVIYFVVVMAVGLWAMFSTNRGTVGGFFLAGRSMVWWPIGA  72
Rabbit   1 -----L--L--STAA-L-SY-R--------V-----L-------------------------------------  72
Pig

                         <------- M2 -------->                      <-
Human      SLFASNIGSGHFVGLAGTGAASGIAIGGFEWNALVLVVVLGWLFVPIYIKAGVVTMPEYLRKRFGGQRIQVY 144
Rabbit     ------------------------T--------IM------V------R----------Q-----K---I- 144
Pig        -----Y--------------A---T--------IW-------------------------------K-----

           ------- M3 ------->                 <------- M4 -------->          <----------
Human      LSLLSLLLYIFTKISADIFSGAIFINLALGLNLYLAIFLLLAITALYTITGGLAAVIYTDTLQTVIMLVGSL 216
Rabbit     --I---------------------Q-T---DI-V--II---V--G------------------A--M---V 216
Pig        --I---M-----------------T-----D-----------G--------------------A------F

           -- M5 --->                                              <---------
Human      ILTGFAFHEVGGYDAFMEKYMKAIPTIVSDGNTTFQEKCYTPRADSFHIFRDPLTGDLPWPGFIFGMSILTL 288
Rabbit     -------------E--T----R---SQI-Y---SIPQ------E-A------AI---I----LV-------- 288
Pig        ----------------I----N----VI-D--I-IKKE--A-------------K-------LT--L---A-

           --- M6 --->                 <------- M7 -------->
Human      WYWCTDQVIVQRCLSAKNMSHVKGGCILCGYLKLPMFIMVMPGMISRILYTEKIACVVPSECEKYCGTKVG 360
Rabbit     ------------------L----A---------V----LI--M--V------D-V---------R----R-- 360
Pig        --------------------A--VM---F--L---VI-------V--------T--------------

                         <------- M8 -------->                      <-------
Human      CTNIAYPTLVVELMPNGLRGLMLSVMLASLMSSLTSIFNSASTLFTMDIYAKVRKRASEKELMIAGRLFILV 432
Rabbit     -----F----------------------M--------------------T-I--K-------------M-F 432
Pig        -S-----------------------------I------------------T------V---I---------

           ---- M9 ---->                 <------- M10 ------->      <------ M11 -------
Human      LIGISIAWVPIVQSAQSGQLFDYIQSITSYLGPPIAAVFLLAIFWKRVNEPGAFWGLILGLLIGISRMITEF 504
Rabbit     ---------------------------------------------------------V--F----------- 504
Pig        ------------------------V----------------C-----E------VI-CM--LA------

           ->                 <------- M12 ------->
Human      AYGTGSCMEPSNCPTIICGVHYLYFAIILFAISFITIVVISLLTKPIPDVHLYRLCWSLRNSKEERIDLDAE 576
Rabbit     --------------------------------V--I--V--V--F-------------------------G 576
Pig        -------V------------------V--I-IVL-V--F--------------------------

Human      EENIQEGPKETIEIETQVPEKKKGIFRRAYDLFCGLEQHGAPKMTEEEEKAMKMKMTDTSEKPLWRTVLNVN 648
Rabbit     --D---A-E-ATDT-::--K----F-----------D-DKG----K---A----L-L-----H------V-I- 646
Pig        --D---A-E------::--E---C---T------D-QKG----K---A---L--------------V-I-

Human      GIILVTVAVFCHAYFA - 664
Rabbit     -V--LA-----Y---- - 662
Pig        ----L-----------
```

FIG. 1. Comparison of amino acid sequences of the Na⁺-dependent glucose cotransporter, SGLT1. References for the sequences of human and rabbit SGLT1 and the partial sequence of pig SGLT1 are indicated in the text. Amino acids are indicated by their single-letter abbreviations and residues that are identical to those in human SGLT1 are noted by dashes. Gaps introduced to generate this alignment are represented by colons. The 12 putative transmembrane segments (M1–M12) predicted from our analysis of the sequence of human SGLT1 are indicated. The potential N-glycosylation site in the extracellular segment connecting M5 and M6 is noted in bold type. The number of the amino acid residue at the end of each line is indicated. Single-letter abbreviations for amino acids are as follows: A, alanine; C, cysteine; D, aspartic acid; E, glutamic acid; F, phenylalanine; G, glycine; H, histidine; I, isoleucine; K, lysine; L, leucine; M, methionine; N, asparagine; P, proline; Q, glutamine; R, arginine; S, serine; T, threonine; V, valine; W, tryptophan; and Y, tyrosine.

(Fig. 1) (Ohta *et al.*, 1990). Computer analysis of the predicted amino acid sequence of the human intestinal Na⁺/glucose cotransporter revealed 13 hydrophobic segments that could represent membrane-spanning regions; similar analyses of the sequence of the rabbit protein suggested that it had 11 transmembrane domains (Hediger *et al.*, 1987). At the present time the topology of the Na⁺

/glucose cotransporter in the plasma membrane is uncertain. A model for the orientation of SGLT1 in the plasma membrane is shown in Fig. 2. This model is based on our interpretation of hydropathy plots of human SGLT1 and is different from those proposed for human or rabbit SGLT1 by Hediger et al. (1987, 1989b). We prefer this model because the presence of 12 membrane-spanning segments with intracellularly located NH_2- and COOH-terminal domains is a feature of many integral membrane transport proteins, including the lac permease (an H^+/lactose cotransporter) (Kaback, 1987), the facilitative glucose transporter superfamily (this family also includes bacterial and plant H^+/sugar symporters) (Mueckler et al., 1985; Maiden et al., 1987; Bell et al., 1990), the γ-aminobutyric acid (GABA) transporter (Guastella et al., 1990), and the P-glycoprotein/multidrug transporter superfamily (Juranka et al., 1989). The features of this model remain to be confirmed. However, rabbit SGLT1 is glycosylated at Asn-248, a residue predicted to be in the extracellular loop connecting membrane-spanning segments M5 and M6 (Figs. 1 and 2) (Hediger et al., 1989b).

Determination of the functional molecular size of the Na^+/glucose cotransporter by radiation inactivation studies suggests that it may function in situ as a tetramer of four identical monomer subunits, each of which has both Na^+ and glucose binding sites (Stevens et al., 1990). The region(s) of individual monomers that participate in oligomerization are unknown. However, the long intracellular COOH-terminal domain might be one such region, perhaps stabilizing monomer interactions through ionic and hydrophobic interactions since this domain contains stretches of charged and hydrophobic amino acids (Figs. 1 and 2).

The molecular mechanism by which the Na^+/glucose cotransporter translocates glucose across the plasma membrane is unknown. Sodium is believed to increase the affinity of the protein for glucose, and in the presence of bound Na^+ and glucose there is a change in conformation that exposes these binding sites to the interior of the cell, resulting in the release of Na^+ and glucose (Peerce, 1990). In contrast to the facilitative glucose transporters described below, the Na^+/glucose cotransporter cannot mediate the bidirectional transport of glucose, suggesting that the outward and inward facing conformations of this protein are not functionally equivalent.

The intestinal Na^+/glucose cotransporter has been expressed in Xenopus oocytes (Hediger et al., 1987, 1989b; Ikeda et al., 1989) and in cultured monkey kidney, COS-7, cells (Birnir et al., 1990). The properties of the protein expressed in these heterologous systems are similar to those of the protein present in native brush-border membranes, including affinity for different sugars, sugar analogs, and sensitivity to phlorizin.

In addition to mediating the active uptake of dietary glucose from the lumen of the small intestine, Na^+-dependent glucose transporters are also responsible for the reabsorption of filtered glucose in the proximal tubule of the kidney. Kinetic

A

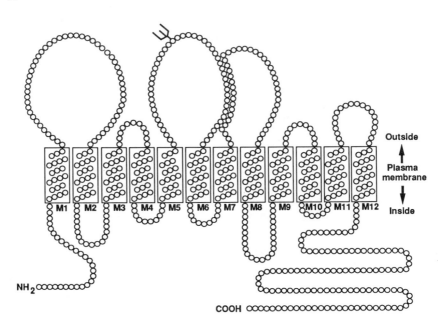

B

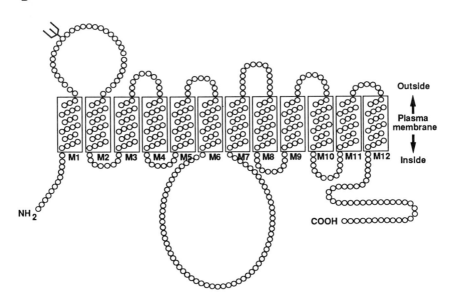

and physical studies suggest that there may be two proteins responsible for the active transport of glucose in the kidney (reviewed in Coady *et al.*, 1990). These proteins have different affinities for glucose with apparent K_m values (at 40 mM NaCl) of 0.35 and 6 mM, and are preferentially localized in different regions of the kidney. The outer portion of the renal cortex contains the low-affinity transporter and the outer medulla has the high-affinity protein. A partial cDNA clone that likely represents the cortical (i.e., low-affinity) renal Na$^+$/glucose cotransporter isoform has been isolated from a rabbit renal cortex cDNA library (Coady *et al.*, 1990). There was >99% identity between the cDNA sequences, and 100% identity between the amino acid sequences, of the rabbit renal and small intestinal cDNA clones, indicating that SGLT1 is expressed in the renal cortex as well as the small intestine. Thus, SGLT1 is responsible, at least in part, for reabsorption of filtered glucose in the kidney.

SGLT1 is a member of a superfamily of proteins that includes Na$^+$-dependent amino acid transporters, and may also include other sugar cotransporters as well as the iodide symporter (Hediger *et al.*, 1989b; Coady *et al.*, 1990). The presence of a family of Na$^+$/glucose cotransporters in mammalian tissues is suggested from biochemical studies and from studies of patients with glucose–galactose malabsorption syndrome and familial renal glycosuria "Mendelian Inheritance in Man," #231600 and 233100, respectively; McKusick, 1990). These disorders are characterized by defects in the uptake of glucose by the intestinal mucosa and renal tubule (Desjeux, 1989; McKusick, 1990). The differences in phenotypes of patients with glucose–galactose malabsorption syndrome and familial renal glycosuria suggest that they result from mutations in two different Na$^+$/glucose cotransporters. Glucose–galactose malabsorption syndrome is a rare (<50 cases described), life-threatening autosomal recessive disorder in which the intestinal mucosa is unable to take up glucose. It is characterized by neonatal onset of severe, watery, acidic diarrhea that leads to dehydration. Mutations in the *SGLT1* gene, which resides on human chromosome 22 (Hediger *et al.*, 1989a), may result in the expression of an abnormal protein and, as a consequence, the absorptive epithelial cells cannot actively absorb glucose and galactose from the lumen of the small intestine. Studies are in

←————————————————————————————

FIG. 2. Models for the orientation of (A) the sodium-dependent glucose transporter (SGLT1) and (B) members of the facilitative glucose transporter family (GLUT1 to 5) in the plasma membrane. The 12 potential membrane-spanning α helices are shown as boxes and are numbered M1 to M12. The location of the potential N-linked glycosylation site in these two classes of glucose transporters is shown. The hydrophobic membrane-spanning segments are shown as α-helical segments. At this time, it is not possible to exclude that some of the hydrophobic segments might not exist as membrane-spanning β sheet rather than α helix. We represent the Na$^+$-dependent glucose transporter with 12 transmembrane segments; however, Hediger *et al.* (1987, 1989b) have proposed that it spans the plasma membrane 11 times.

progress to determine if mutations in *SGLT1* are indeed responsible for this disorder. Familial renal glycosuria is an autosomal dominant disorder (an autosomal recessive mode of inheritance has not been excluded in all instances) affecting 0.2–0.6% of the general population and is characterizated by the excretion of large amounts of glucose (10–100 g/24 hours) in the urine in the presence of normal amounts of glucose in the blood. It is a benign condition, usually without symptoms or physical consequences. Subjects with benign renal glycosuria are able to store and utilize carbohydrates normally. The molecular basis for benign renal glycosuria has not been determined. However, perhaps mutations in a second Na^+/glucose cotransporter, such as the protein expressed in the renal medulla, may be responsible for the defect in renal reabsorption of filtered glucose. Turk *et al.* (1991) have shown that glucose/galactose malabsorption is caused by a mutation in the Na^+/glucose cotransporter, SGLT1, gene. Sequence analysis of the SGLT1 gene of two sisters with glucose/galactose malabsorption syndrome revealed a mutation, Asp 28 → Asn 28, which resulted in complete loss of Na^+-dependent glucose transport.

Physiological measurements have indicated that the Na^+-dependent glucose transporters appear prenatally and, as a consequence, the intestine is ready to absorb the first ingested glucose (Buddington and Diamond, 1989). The cloned cDNAs and specific antibodies for the different Na^+/glucose transporter isoforms will be valuable tools for identifying the specific cells in the intestine and kidney that express these proteins, and for studying the regulation of their expression during development and in altered metabolic states such as diabetes mellitus or pregnancy. Such studies may provide new insight into molecular mechanisms regulating the uptake of dietary nutrients.

III. Facilitative Glucose Transporters

Except for the active uptake of glucose from the lumen of the small intestine and proximal tubule of the kidney, the transport of glucose across the cell membrane occurs by facilitated diffusion. This energy-independent process is mediated by a family of structurally related proteins (Table I and Fig. 3) encoded by distinct genes that are expressed in a tissue-specific manner (Gould and Bell, 1990; Mueckler, 1990; Bell *et al.*, 1990). Transport of glucose by these proteins is saturable, stereoselective, and bidirectional [the kinetics of glucose transport

---

FIG. 3. Comparison of the amino acid sequences of human facilitative glucose transporters. Amino acids are indicated by their single-letter abbreviations. Gaps introduced to generate this alignment are represented by colons. Amino acids that are identical in all five transporters as well as those that are identical in four of the proteins are indicated in bold type. The various domains of this family of structurally related proteins are noted. The potential N-linked glycosylation site in the extracellular loop connecting M1 and M2 in each of these proteins is underlined. Adapted from Kayano *et al.* (1990).

```
                      <-------- M1 --------><--- Extracellular Loop --
GLUT1         1 MEPSSKKLTGRLMLAVGGAVLGS:LQFGYNTGVINAPQKVIEEFYNQTWVHRYGESILPT  59
GLUT2         1 MTEDKVTGTLVFTVITAVLGS:FQFGYDIGVINAPQQVIISHYRHVLGVPLDDRKAIN   57
GLUT3         1 MGTQKVTPALIFAITVATIGS:FQFGYNTGVINAPEKIIKIFINKTLTDKGNAPPSEV   57
GLUT4 1 MPSGFQQIGSEDGEPPQQRVTGTLVLAVFSAVLGS:LQFGYNIGVINAPQKVIEQSYNETWLGRQGPEGPSS   71
GLUT5         1 MEQQDQSMKEGRLTLVLALATLIAAFGSSFQYGYNVAAVNSPALLMQQFYNETYYGRTQEFMEDF   65


        --------------------------------------><------- M2 --------->         <-
GLUT1   T::::::::::::::::::::::::::::::::::::::LTTLWSLSVAIFSVGGMIGSFSVGLFVNRFGRRNSML   97
GLUT2   NYVINSTDELPTISYSMNPKPTPWAEEETVAAAQLITMLWSLSVSSFAVGGMTASFFGGWLGDTLGRIKAML  129
GLUT3   L::::::::::::::::::::::::::::::::::::::LTSLWSLSVAIFSVGGMIGSFSVGLFVNRFGRRNSML   95
GLUT4   IPPGT:::::::::::::::::::::::::::::::::::LTTLWALSVAIFSVGGMISSFLIGIISQWLGRKRAML  113
GLUT5   P:::::::::::::::::::::::::::::::::::::LTLLWSVTVSMFPFGGFIGSLLVGPLVNKFGRKGALL  103


        ------- M3 ------->          <------- M4 --------->          <------ M5 ---
GLUT1   MMNLLAFVSAVLMGFSKLGKSFEMLILGRFIIGVYCGLTTGFVPMYVGEVSPTALRGALGTLHQLGIVVGII  169
GLUT2   VANILSLVGALLMGFSKLGPSHILIIAGRSISGLYCGLISGLVPMYIGEIAPTALRGALGTFHQLAIVTGIL  201
GLUT3   IVNLLAVTGGCFMGLCKVAKSVEMLILGRLVIGLFCGLCTGFVPMYIGEISPTALRGAFGTLNQLGIVVGIL  167
GLUT4   VNNVLAVLGGSLMGLANAAASYEMLILGRFLIGAYSGLTSGLVPMYVGEIAPTHLRGALGTLNQLAIVIGIL  185
GLUT5   FNNIFSIVPAILMGCSRVATSFELIIISRLLVGICAGVSSNVVPMYLGELAPKNLRGALGVVPQLFITVGIL  175


        ------>          <------- M6 --------><---------- Intracellular Loop ---
GLUT1   IAQVFGLDSIMGNKDLWPLLLSIIFIPALLQCIVLPFCPESPRFLLINRNEENRAKSVLKKLRGTADVTHDL  241
GLUT2   ISQIIGLEFILGNYDLWHILLGLSGVRAILQSLLLFFCPESPRYLYIKLDEEVKAKQSLKRLRGYDDVTKDI  273
GLUT3   VAQIFGLEFILGSEELWPLLLGFTILPAILQSAALPFCPESPRFLLINRKEENAKQILQRLWGTQDVSQDI  239
GLUT4   IAQVLGLESLLGTASLWPLLLGLTVLPALLQLVLLPFCPESPRYLYIIQNLEGPARKSLKRLTGWADVSGVL  257
GLUT5   VAQIFGLRNLLANVDGWPILLGLTGVPAALQLLLLPFFPESPRYLLIQKKDEAAAKKALQTLRGWDSVDREV  247


        --------------------------><------- M7 --------->          <------
GLUT1   QEMKEESRQMMREKKVTILELFRSPAYRQPILIAVVLQLSQQLSGINAVFYYSTSIFEKAGVQQP::VYATI  311
GLUT2   NEMRKEREEASSEQKVSIIQLFTNSSYRQPILVALMLIIVAQQFSGINGIFYYSTSIFQTAGISKP::VYATI  343
GLUT3   QEMKDESARMSQEKQVTVLELFRVSSYRQPIIISIVLQLSQQLSGINAVFYYSTGIFKDAGVQEP::IYATI  309
GLUT4   AELKDEKRKLERERPLSLLQLLGSRTHRQPLIIAVVLQLSQQLSGINAVFYYSTSIFETAGVGQP::AYATI  327
GLUT5   AEIRQEDEAEKAAGFISVLKLFRMRSLRWQLLSIIVLMGGQQLSGVNAIYYYADQIYLSAGVPEEHVQYVTA  319


        --- M8 -------->          <------- M9 --------->          <---- M10 -----
GLUT1   GSGIVNTAFTVVSLFVVERAGRRTLHLIGLAGMAGCAILMTIALALLEQLPWMSYLSIVAIFGFVAFFEVGP  383
GLUT2   GVGAVNMVFTAVSVFLVKEAGRRSLFLIGMSGMFVCAIFMSVGLVLLNKFSWMSYVSMIAIFLFVSFFEIGP  415
GLUT3   GAGVVNTIFTVVSLFLVERAGRRTLHMIGLGGMAFCSTLMTVSLLLKDNYNGMSFVCIGAILVFVAFFEIGP  381
GLUT4   GAGVVNTVFTLVSVLLVERAGRRTLHLLGLAGMCGCAILMTVALLLLERVPAMSYVSIVAIFGFVAFFEIGP  399
GLUT5   GTGAVNVVMTFCAVFVVELLGRRLLLLLGFSICLIACCVLTAALALQDTVSWMPYISIVCVISYVIGHALGP  391


        ----->          <------- M11 ------->          <------- M12 -------><----
GLUT1   GPIPWFIVAELFSQGPRPAAIAVAGFSNWTSNFIVGMCFQYVEQLCGPYVFIIFTVLLVLFFIFTYFKVPET  455
GLUT2   GPIPWFMVAEFFSQGPRPAALAIAAFSNWTCNFIVALCFQYIADFCGPYVFFLFAGVLLAFTLFTFFKVPET  487
GLUT3   GPIPWFIVAELFSQGPRPAAMAVAGCSNWTSNFLVGLLFPSAAHYLGAYVFIIFTGFLITFLAFTFFKVPET  453
GLUT4   GPIPWFIVAELFSQGPRPAAMAVAGFSNWTSNFIIGMGFQYVAEAMGPYVFLLFAVLLLGFFIFTFLRVPET  471
GLUT5   SPIPALLITEIFLQSSRPSAFMVGGSVHWLSNFTVGLIFPFIQEGLGPYSFIVFAVICLLTTIYIFLIVPET  463


        -- Intracellular COOH-Terminal Domain ---->
GLUT1   KGRTFDEIASGFRQGGASQSDKTPEELFHPLGADSQV - 492
GLUT2   KGKSFEEIAAEFQKKSGSAHRPKAAVEMKFLGATETV - 524
GLUT3   RGRTFEDITRAFEGQAHGADRSGKDGVMEMNSIEPAKETTTNV - 496
GLUT4   RGRTFDQISAAFHRTPSLLEQEVKPSTELEYLGPDEND - 509
GLUT5   KAKTFIEINQIFTKMNKVSEVYPEKEELKELPPVTSEQ - 501
```

inward and outward need not be identical and, in fact, are not for the erythrocyte, GLUT1, glucose transporter (Carruthers, 1990)]. The mechanism of protein-mediated facilitated diffusion of sugars across the plasma membrane has not been established and several models have been proposed to account for the transport of glucose, including the alternating conformer and fixed site models (the ability of these models and their variants to describe the characteristics of glucose transport has been recently reviewed by Carruthers, 1990). In protein-mediated facilitated diffusion, the direction of net flux of glucose is determined by its relative concentration on either side of the plasma membrane. In most cells, cytoplasmic glucose is rapidly phosphorylated by hexokinase or glucokinase, resulting in the absence of an appreciable intracellular pool of free glucose. Thus there is a net inward flux of glucose. However, in some instances, glucose must be transported out of the cell under normal physiological conditions. For example, the absorptive epithelial cells of the small intestine and kidney must release absorbed glucose into the circulation. In these cells, the Na^+/glucose cotransporter is responsible for the active uptake of glucose and a facilitative glucose transporter mediates its efflux from the cell into the interstitium. In hepatocytes, facilitative glucose transporters participate in the uptake of glucose from the portal circulation following a meal, and in the release of glucose generated by glycogenolysis or gluconeogenesis in the postabsorptive state and during fasting.

cDNAs encoding five functional facilitative glucose transporter isoforms have been isolated and characterized from human tissues (Table I) (Mueckler et al., 1985; Fukumoto et al., 1988b, 1989; Kayano et al., 1988, 1990). They have been designated GLUT (the gene symbol for facilitative glucose transporter) followed by a number based on the chronological order of publication of the cDNA sequence: GLUT1, GLUT2, GLUT3, GLUT4, and GLUT5. Another designation in which each isoform is described based on its major site of expression and which conveys a minimum of physiological information has been suggested by Pilch (1990): erythrocyte, liver, brain, muscle–fat, and small intestine, respectively. In addition to five functional facilitative glucose transport isoforms, an expressed facilitative glucose transporter pseudogene (*GLUT6*) has been identified in human tissues (Kayano et al., 1990). cDNAs encoding four of five facilitative glucose transporter isoforms, GLUT1, GLUT2, GLUT3, and GLUT4, have also been isolated from other species: GLUT1, rat (Birnbaum et al., 1986), mouse (Kaestner et al., 1989; Reed et al., 1990), rabbit (Asano et al., 1988), and pig (Weiler-Guttler et al., 1989); GLUT2, rat (Thorens et al., 1988) and mouse (Suzue et al., 1989; Asano et al., 1989b); GLUT3, mouse (Nagamatsu et al., 1991); and GLUT4, rat (James et al., 1989a; Birnbaum, 1989; Charron et al., 1989) and mouse (Kaestner et al., 1989). The human GLUT5 cDNA readily cross-hybridizes with mouse DNA, suggesting that this isoform will be found in other species. However, the human GLUT6 probe (*GLUT6* is an expressed pseudogene) does not cross-hybridize with mouse DNA under normal conditions, suggesting that this pseudogene may not be present in all species.

TABLE II
K_m *Values for 2-Deoxy-*D*-Glucose of Mammalian Facilitative Glucose Transporters*[a]

Isoform	K_m (mM)
Rat GLUT1	6.9 ± 1.5
Human GLUT2	13.2 ± 2.4
Human GLUT3	1.8 ± 0.6
Human GLUT4	4.6 ± 0.3

[a]Synthetic RNA encoding each isoform was prepared by *in vitro* transcription of cloned cDNAs. The RNA was injected into *Xenopus* oocytes and uptake of 2-deoxy-D-glucose (zero-trans entry) measured 48 hours later as described in Kayano *et al.* (1990). Each value is the mean of three to five determinations. The K_m of human GLUT5 has not been determined. These data are from Burant and Bell (1991).

Southern blotting studies of other species, especially primates, could indicate the evolutionary history of this pseudogene.

The GLUT1, GLUT2, GLUT3, GLUT4, and GLUT5 proteins have each been expressed in heterologous systems, including the bacterium *Escherichia coli* [GLUT1 (Sarkar *et al.*, 1988) and GLUT2 (Thorens *et al.*, 1988)], *Xenopus* oocytes [GLUT1 to 5 (Birnbaum, 1989; Gould and Lienhard, 1989; Keller *et al.*, 1989; Vera and Rosen, 1989; Permutt *et al.*, 1989; Kayano *et al.*, 1990; Gould *et al.*, 1991)], and cultured mammalian cells [GLUT1 (Gould *et al.*, 1989; Asano *et al.*, 1989a; Harrison *et al.*, 1990a,b) and GLUT3 and GLUT4 (Takeda *et al.*, 1991)]. The proteins expressed in these systems have the properties of facilitative glucose transporters and can be distinguished based on their affinity for hexoses such as 2-deoxy-D-glucose, a nonmetabolizable analog of glucose (Table II). Transport is specific for D-glucose but not L-glucose, indicating that each is a stereoselective glucose carrier. Furthermore, glucose transport can be inhibited by the fungal toxin cytochalasin B, a specific high-affinity competitive inhibitor of facilitative glucose transport that does not inhibit glucose transport by the Na$^+$/glucose cotransporter. The selective inhibition of facilitative glucose transport in mammalian cells by cytochalasin B provides a means of distinguishing facilitative from Na$^+$-dependent glucose transporters. In addition to functioning as sugar carriers, the facilitative glucose transporters may also served as membrane water channels (Fischbarg *et al.*, 1990); however, for an alternative view of these studies see Zhang *et al.* (1990).

A. GLUT1/ERYTHROCYTE ISOFORM

GLUT1, the erythrocyte-type facilitative glucose transporter isoform, was the first glucose transporter to be cloned. Using antibodies raised against the purified erythrocyte glucose transporter to screen antigen-expression cDNA libraries prepared using RNA from the human HepG2 hepatoblastoma cell line or rat brain, two groups independently isolated clones encoding this isoform (Mueckler *et al.*, 1985; Birnbaum *et al.*, 1986). Subsequently, cDNAs encoding rabbit (Asano *et al.*, 1988), mouse (Kaestner *et al.*, 1989; Reed *et al.*, 1990), and pig (Weiler-Guttler *et al.*, 1989) GLUT1 were isolated. Biochemical studies of this glucose transporter isoform have revealed that it is heterogeneously glycosylated and that the pattern of glycosylation can vary in different tissues and cell lines (Sogin and Hinkle, 1978; Gorga *et al.*, 1979; Haspel *et al.*, 1985; Bramwell *et al.*, 1990; Endo *et al.*, 1990). GLUT1 has a K_m for glucose of ~2 mM when assayed by zero-trans uptake and approximately 20 mM under equilibrium exchange conditions with a high degree of stereospecificity for hexose and pentose monosaccharides in the pyranose ring form (Carruthers, 1990). The C-1 chair conformation, in which the hydroxyl groups are in an equatorial position, appears to be the structure that is most efficiently transported; e.g., the K_m for D-glucose is 2 mM, whereas for L-glucose the K_m is greater than 3 M. The erythrocyte glucose transporter *in situ* exhibits asymmetric transport kinetics with efflux having an approximately fourfold higher K_m and V_{max} compared to influx (Carruthers, 1990).

The amino acid sequence of GLUT1 is highly conserved (Fig. 4) and there is 98% identity between the sequences of human and rat GLUT1, and 97% identity between the sequences of human and mouse, rabbit or pig GLUT1. This high degree of sequence conservation implies that all domains of this 492-residue protein are functionally important.

Although SGLT1, the Na$^+$/glucose cotransporter, and GLUT1 both catalyze the transport of glucose across the plasma membrane, their sequences are unrelated. Nor are there any short regions of sequence homology between SGLT1 and GLUT1 that might suggest residues that participate in the transport of glucose and thus provide clues to the location of the channel or pore through which glucose moves.

Computer analyses of the primary sequence of GLUT1, including hydropathy and secondary-structure determinations, revealed that GLUT1 was an extremely hydrophobic protein and suggested that about 50% of the protein might be in the lipid bilayer (Mueckler *et al.*, 1985). Based on their analysis of the sequence of GLUT1, Mueckler *et al.*, (1985) proposed a model for the topology of GLUT1 in the plasma membrane (Fig. 2) in which the protein spanned the plasma membrane 12 times with its NH$_2$ and COOH termini being internally oriented. A large extracellular segment of 33 amino acids connects transmembrane domains 1 and 2, M1 and M2, and a very hydrophilic intracellular segment of 65 amino acids joins M6 and M7. Short regions of 7–14 amino acids connect the remaining membrane-spanning regions. There are two potential N-linked glycosylation sites in GLUT1 at Asn-45 and Asn-411 (Fig. 4). The former site is in the large extracellular loop connecting M1 and M2 and has been shown to be glycosylated (Mueckler and Lodish, 1986). The latter site is predicted to be within M11 and is

```
                <------- M1 --------><----- Extracellular Loop ------><-----
Human   1 MEPSSKKLTGRLMLAVGGAVLGSLQFGYNTGVINAPQKVIEEFYNQTWVHRYGESILPTTLTTLWSLSVAIF 72
Rat     1 -------V----------------------------------------N-------PS------------- 72
Mouse   1 -D-----V----------------------------------------N-----P-PS------------- 72
Rabbit  1 -------V--------------------------------------I-----R----------------- 72
Pig                                              -------L-------S-A------------
```

```
          ---- M2 ------>          <------- M3 --------->              <------ M4 -------
Human   SVGGMIGSFSVGLFVNRFGRRNSMLMMNLLAFVSAVLMGFSKLGKSFEMLILGRFIIGVYCGLTTGFVPMYV 144
Rat     -------------------------------------------------------------------- 144
Mouse   -----------------------------------A-------------------------------- 144
Rabbit  --------------------------------------------A----------------------- 144
Pig     --------------------------------I----------------------------------- 
```

```
          -->         <------- M5 -------->            <------- M6 --------><---------
Human   GEVSPTAFRGALGTLHQLGIVVGILIAQVFGLDSIMGNKDLWPLLLSIIFIPALLQCIVLPFCPESPRFLLI 216
Rat     -------L-----------------------------A--------V----------L------------- 216
Mouse   -------L-----------------------------A--------V----------L------------- 216
Rabbit  -------L---------------------------E--------V--V----------L----------- 216
Pig     -------L---------------------------EE--------V----------VL------------- 
```

```
          ----------------- Intracellular Loop -----------------><------ M7 ------
Human   NRNEENRAKSVLKKLRGTADVTHDLQEMKEESRQMMREKKVTILELFRSPAYRQPILIAVVLQLSQQLSGIN 288
Rat     -------------------R--------G----------------------------------------- 288
Mouse   -------------------R--------G----------------------------------------- 288
Rabbit  ---------------N----R------------------------------S--------------- 288
Pig     -------------------R--------------------------A-------------------
```

```
          --->          <------- M8 --------->          <------- M9 -------->
Human   AVFYYSTSIFEKAGVQQPVYATIGSGIVNTAFTVVSLFVVERAGRRTLHLIGLAGMAGCAILMTIALALLEQ 360
Rat     -------------------------------------------------------V---------- 360
Mouse   -------------------------------------------------------V----------R 360
Rabbit  -----------------------------------------------------A--V---------- 360
Pig     -------------------------------------------------------V----------
```

```
          <------- M10 ------->          <------- M11 ------->          <--
Human   LPWMSYLSIVAIFGFVAFFEVGPGPIPWFIVAELFSQGPRPAAIAVAGFSNWTSNFIVGMCFQYVEQLCGPY 432
Rat     -----------------------------------------V------------------------- 432
Mouse   ----------------------------------------------------------------- 432
Rabbit  -------------------------          --------V----------------------- 432
Pig     -----------------------------------------------------------------
```

```
          ------ M12 ------><-- Intracellular COOH-Terminal Domain -->
Human   VFIIFTVLLVLFFIFTYFKVPETKGRTFDEIASGFRQGGASQSDKTPEELFHPLGADSQV - 492
Rat     -------------------------------------------------- - 492
Mouse   -------------------------------------------------- - 492
Rabbit  -------------------          ------------------------- - 492
Pig     -------------------------------------------------
```

FIG. 4. Comparison of amino acid sequences of GLUT1/erythrocyte facilitative glucose transporters. References for the sequences of human, mouse, and rabbit GLUT1 and the partial sequence of pig GLUT1 are indicated in the text. Residues that are identical to human GLUT1 are noted by dashes. The various domains are indicated and the potential N-linked glycosylation site in the extracellular loop connecting M1 and M2 is shown in bold type.

therefore unlikely to be modified. Several features of the model for the orientation of GLUT1 in the plasma membrane shown in Fig. 2A have been confirmed. The glycosylation of Asn-45 has been shown by expressing GLUT1 *in vitro* in the presence of dog pancreas microsomes (Mueckler and Lodish, 1986). The cytoplasmic orientation of the intracellular loop connecting M6 and M7 and of the COOH terminus have been confirmed by using site-specific peptide antibodies as well as proteolytic digestion to identify regions of the protein not present in the lipid bilayer (Davies *et al.*, 1990; Carruthers, 1990). The accessibility of specific residues to chemical modification (May *et al.*, 1990) is also consistent with this model. Studies are underway in a number of laboratories examining the biochemical properties of variants of GLUT1 generated by *in vitro* mutagenesis with the aim of refining this model as well as identifying amino acid residues that participate in the transport of glucose (Oka *et al.*, 1990b).

Like the Na^+/glucose cotransporter, GLUT1 may also oligomerize and radiation target size analyses and biochemical studies suggest that GLUT1 may exist in the plasma membrane as a dimeric or tetrameric assembly (Carruthers, 1990).

B.   GLUT2/LIVER ISOFORM

Biochemical studies of glucose transport by hepatocytes revealed that the kinetics of glucose transport were significantly different from those of erythrocytes, which we now know express the GLUT1 isoform. The hepatocyte glucose transporter had an ~10-fold higher K_m (lower affinity) for monosaccharides and a 10-fold lower affinity for cytochalasin B (Craik and Elliott, 1979; Axelrod and Pilch, 1983). With the cloning of the GLUT1 isoform, it was quickly recognized that the liver did not express appreciable levels of GLUT1 mRNA (Birnbaum *et al.*, 1986; Flier *et al.*, 1987a), consistent with the presence of another glucose transporter in this tissue. Low-stringency hybridization screening of human liver and kidney (Fukumoto *et al.*, 1988b) and rat liver (Thorens *et al.*, 1988) cDNA libraries using the human GLUT1 cDNA as a probe resulted in the isolation of cDNAs encoding novel human and rat glucose transporters, termed GLUT2, of 524 or 522 amino acids, respectively, having 55% amino acid identity with sequence of GLUT1 (Figs. 3 and 5). Computer analysis of the predicted amino acid sequence of GLUT2 suggests that its overall structure and orientation in the plasma membrane are very similar to that of GLUT1. The difference in size between GLUT1 and GLUT2 is primarily due to the marked difference in size of the extracellular loop connecting M1 and M2, which is 33 and 65 residues in GLUT1 and GLUT2, respectively (Fig. 3). GLUT2 is a glycoprotein and there is a potential site for N-linked glycosylation in the large extracellular loop in this isoform as there is in GLUT1. In contrast to GLUT1, the sequence of which is highly conserved, there is only 81% identity between the sequences of human and rat GLUT2 (Fig. 5). The most divergent regions of GLUT2 are the large extracellular loop and the intracellular COOH-terminal domain. In addition to human and rat GLUT2, cDNA clones encoding mouse GLUT2 have been isolated (Asano *et al.*, 1989b; Suzue *et al.*, 1989); there is 81 and 94% amino acid identity between mouse GLUT2 and human and rat GLUT2, respectively (Fig. 5).

Besides the liver, GLUT2 is expressed in the small intestine, kidney, and the insulin-secreting β cell of the endocrine pancreas (Fukumoto *et al.*, 1988b; Thorens *et al.*, 1988). Western blotting studies have revealed that the size of the protein expressed in these tissues differs (Thorens *et al.*, 1988, 1990b). This

```
              <------- M1 --------><------- Extracellular Loop -------------
Human 1 MTEDKVTGTLVFTVITAVLGSFQFGYDIGVINAPQQVIISHYRHVLGVPLDDRKAINNYVINSTDELLTISY  72
Rat   1 -S---I----A---F---------------------E-----------------R-TI--D--G--TP-IVTP  72
Mouse 1 -S---I----A---F----S---------------E-----------------AI--DV-G--TP--VTP    72


        ------------------------><------- M2 --------->      <------ M3 ------
Human   SMNPKPTPWAEEETVAAAQLITMLWSLSVSSFAVGGMTASFFGGWLGDTLGRIKAMLVANILSLVGALLMGF  144
Rat     :AHTT-DA-:----EGS-HIV----------------V----------K--------A--S----T------C  142
Mouse   :AYTT-A--D----EGS-HIV----------------V----------K--------A--S---T------C   143


         --->       <------- M4 --------->        <------- M5 -------->
Human   SKLGPSHILIIAGRSISGLYCGLISGLVPMYIGEIAPTALRGALGTFHQLAIVTGILISQIIGLEFILGNYD  216
Rat     --F--A-A-------V---------------------T-------L----L---------A--S-----Q-    214
Mouse   --F--A-A-------V---------------------NT-------L----L---------A--S-----Q-   215


          <------- M6 -------><-------------- Intracellular Loop --------------
Human   LWHILLGLSGVRAILQSLLLFFCPESPRYLYIKLDEEVKAKQSLKRLRGYDDVTKDINEMRKEREEASSEQK  288
Rat     Y--------A-P-L--C---L-------------LN-E---R--K-------TE-I----------K----T---  286
Mouse   H--------A-P-L--C---L-------------E---R--K-------TE---------K--K----T---   287


        ---------------><------- M7 ---------><-------- M8 --------->
Human   VSIIQLFTNSSYRQPILVALMLHVAQQFSGINGIFYYSTSIFQTAGISKPVYATIGVGAVNMVFTAVSVFLV  360
Rat     --V-----DPN-----V------L-----------------Q----------I--I------L--          358
Mouse   --V-----DAN-----------M-----------------Q----------I--IL-----L--          357


           <------- M9 --------->        <------- M10 ------->
Human   EKAGRRSLFLIGMSGMFVCAIFMSVGLVLLNKFSWMSYVSMIAIFLFVSFFEIGPGPIPWFMVAEFFSQGPR  432
Rat     ------T---A---I----F--V---L-----D--T-------T-----------------------       430
Mouse   ------T---T--I---F-T-----------D--A-------T----------------------        431


           <------- M11 ------->        <------- M12 -------><- Intracellular COOH-
Human   PAALAIAAFSNWTCNFIVALCFQYIADFCGPYVFFLFAGVLLAFTLFTFFKVPETKGKSFEEIAAEFQKKSG  504
Rat     -T---L------V----I----------L-----------V-V-----------------D-----R----   502
Mouse   ST--L------V---VI----------L-----------V-V-----------------------R----    503


        Terminal Domain --->
Human   SAHRPKAAVEMKFLGATETV - 524
Rat     --PPR--T-Q-E---SS--- - 522
Mouse   --PPR----Q-E--ASS-S- - 523
        |
       (P)
```

FIG. 5. Comparison of amino acid sequences of GLUT2/liver glucose transporters. References for the sequences of human, rat, and mouse GLUT2 are indicated in the text. Residues that are identical to human GLUT2 are noted by dashes. Gaps introduced to generate this alignment are represented by colons. The potential N-linked glycosylation site in the extracellular loop connecting M1 and M2 is shown in bold type. (P), A potential cAMP-dependent protein kinase A phosphorylation site in the intracellular COOH-terminal domain the sequence of which is conserved in human, rat, and mouse GLUT2.

variation in apparent size is most likely due to differences in glycosylation, since the sequences of cDNA clones isolated from liver, kidney, and β cells are identical (Fukumoto et al., 1988b; Permutt et al., 1989; Thorens et al., 1988; Johnson et al., 1990a). The functional consequences of the tissue-specific differences in glycosylation of GLUT2 are unknown.

C. GLUT3/BRAIN ISOFORM

Screening of a human fetal skeletal muscle cDNA library under low-stringency conditions using a human GLUT1 cDNA probe identified a third member of the

```
                <------- M1 --------><----- Extracellular Loop ------><-------
Human 1 MGTQKVTPALIFAITVATIGSFQFGYNTGVINAPEKIIKEFINKTLTDKGNAPPSEVLLTSLWSLSVAIFSV  72
Mouse 1 ---T----S-V--V--------------------T-L-D-L-Y--EERLEDL---G---A----C------  72

        -- M2 ------>          <------- M3 -------->           <------- M4 -------
Human   GGMIGSFSVGLFVNRFGRRNSMLIVNLLAVTGGCFMGLCKVAKSVEMLILGRLVIGLFCGLCTGFVPMYIGE 144
Mouse   ---------------------L-----IIA--L--FA-I-E----------L--I-------------- 144

        >           <------- M5 -------->              <------- M6 --------><-----------
Human   ISPTALRGAFGTLNQLGIVVGILVAQIFGLEFILGSEELWPLLLGFTILPAILQSAALPFCPESPRFLLINR 216
Mouse   V----------------------------D----------G---L--I--------------------K 216

        ---------------- Intracellular Loop ---------------><------- M7 -------
Human   KEEENAKQILQRLWGTQDVSQDIQEMKDESARMSQEKQVTVLELFRVSSYRQPIIISIVLQLSQQLSGINAV 288
Mouse   ---DQ-TE--------S--V-E--------V-----------------SPN-V--LL--------------- 288

        ->           <------- M8 -------->          <------- M9 -------->
Human   FYYSTGIFKDAGVQEPIYATIGAGVVNTIFTVVSLFLVERAGRRTLHMIGLGGMAFCSTLMTVSLLLKDNYN 360
Mouse   ---------------------------------------------------V--VF--I------D-E 360

        <------- M10 ------->           <------- M11 ------->        <----
Human   GMSFVCIGAILVFVAFFEIGPGPIPWFIVAELFSQGPRPAAMAVAGCSNWTSNFLVGLLFPSAAHYLGAYVF 432
Mouse   A------V---IY------------------------I-----C---------M------A------- 432

        ----- M12 -----><----- Intracellular COOH-Terminal Domain ------>
Human   IIFTGFLITFLAFTFFKVPETRGRTFEDITRAFEGQAHGADRSGKD:GVMEMNSIEPAKETTTNV - 496
Mouse   ---AA---F--I---------K-------A--------::::---GPAGV-L--MQ-V---PG-A - 493
```

FIG. 6.   Comparison of amino acid sequences of GLUT3/brain glucose transporters. References for the sequences of human and mouse GLUT3 are indicated in the text. Residues that are identical to human GLUT3 are noted by dashes. Gaps introduced to generate this alignment are represented by colons. The potential N-linked glycosylation site in the extracellular loop connecting M1 and M2 is shown in bold type.

facilitative transporter gene family, GLUT3 (Kayano *et al.*, 1988). Human GLUT3 is 496 amino acids in length and has 64 and 52% identity with human GLUT1 and GLUT2, respectively (Fig. 3). Recently, cDNA clones encoding mouse GLUT3 have been isolated and characterized (S. Nagamatsu *et al.*, in preparation). There is 83% amino acid sequence identity with the sequences of human and mouse GLUT3 (Fig. 6). Thus, as with GLUT2, the sequence of GLUT3 is not as highly conserved as is that of GLUT1. The greatest degree of sequence divergence in GLUT3 occurs in the extracellular loop and intracellular COOH-terminal domains (Fig. 6). These are the same regions that are the most divergent between isoforms (Fig. 3). The orientation of GLUT3 in the lipid bilayer is predicted to be similar to that proposed for GLUT1.

D.   GLUT4/MUSCLE–FAT ISOFORM

Insulin causes a rapid and reversible increase in glucose transport activity in isolated adipocytes and skeletal muscle (reviewed in Simpson and Cushman, 1986). This increase results primarily from translocation of a latent pool of glucose transporters from an intracellular site to the plasma membrane (Cushman and Wardzala, 1980; Suzuki and Kono, 1980). The isolation and characterization of a monoclonal antibody that specifically recognized this "insulin-regulatable" glucose transporter indicated that it was a unique isoform different from the glucose transporters present in erythrocytes, brain, kidney, jejunum, and liver (James *et al.*, 1988). Simultaneously, four groups using low-stringency cross-hybridization strategies and GLUT1 cDNA probes identified human (Fukumoto *et al.*, 1989), rat (James *et al.*, 1989a; Birnbaum, 1989; Charron *et al.*, 1989), and mouse (Kaestner *et al.*, 1989) cDNA clones encoding the "insulin-regulatable" glucose transporter isoform, termed GLUT4. Human GLUT4 is 509 amino acids and has 65, 54, and 58% identity with human GLUT1, GLUT2, and GLUT3, respectively (Fig. 3). The sequence of GLUT4 is highly conserved (Fig. 7) and there is 95 and 94% identity between the sequences of human and rat or mouse GLUT4, respectively. Analysis of the sequence of GLUT4 suggests that

```
                            <------- M1 -------><--- Extracellular Loop ---
Human 1 MPSGFQQIGSEDGEPPQQRVTGTLVLAVFSAVLGSLQFGYNIGVINAPQKVIEQSYNETWLGRQGPEGPSSI  72
Rat   1 -------------------------------------------------------A--------G--D---  72
Mouse 1 -----------E----R--------------------------------------A--------G--D--   72

        --------><------- M2 --------->        <------- M3 -------->        <-
Human   PPGTLTTLWALSVAIFSVGGMISSFLIGIISQWLGRKRAMLVNNVLAVLGGSLMGLANAAASYEMLILGRFL 144
Rat     -Q----------------------------------A---------A--------------I-------    144
Mouse   -Q----------------------------------A---------A-------V----I-------     144

        ------ M4 -------->         <------- M5 -------->        <------- M6 ---
Human   IGAYSGLTSGLVPMYVGEIAPTHLRGALGTLNQLAIVIGILIAQVLGLESLLGTASLWPLLLGLTVLPALLQ 216
Rat     ----------------------------------------V--------M----T------AI-------- 216
Mouse   ----------------------------R--------V--------M----T------A--------    216

        -----><--------------------- Intracellular Loop -------- ----------->< 
Human   LVLLPFCPESPRYLYIIQNLEGPARKSLKRLTGWADVSGVLAELKDEKRKLERERPLSLLQLLGSRTHRQPL 288
Rat     -L------------------R------------------DA---------------------------- 288
Mouse   -I--------------------R----------P--------DA-------------------M-------- 288

        ------- M7 -------->            <------- M8 -------->        <------ 
Human   IIAVVLQLSQQLSGINAVFYYSTSIFETAGVGQPAYATIGAGVVNTVFTLVSVLLVERAGRRTLHLLGLAGM 360
Rat     ----------------------------L---E---------------------------------- 360
Mouse   ----------------------------S--------------------------------------    360

        -- M9 ------->        <------- M10 -------->        <------- M11 -- 
Human   CGCAILMTVALLLLERVPAMSYVSIVAIFGFVAFFEIGPGPIPWFIVAELFSQGPRPAAMAVAGFSNWTSNF 432
Rat     -------------------S---------------------------------------------C-- 432
Mouse   -----------------------------------------:----------------- ----C-- 431

        -- -->        <------- M12 --------><-- Intracellular COOH-Terminal Domain 
Human   IIGMGFQYVAEAMGPYVFLLFAVLLLGFFIFTFLRVPETRGRTFDQISAAFHRTPSLLEQEVKPSTELEYLG 504
Rat     -V-------D------------------------------------T-R----------------- 504
Mouse   -V-------DR----------------------K--------------R----------------- 503
                                                          |
                                                          P
        ---->
Human   PDEND - 509
Rat     ----- - 509
Mouse   ----- - 508
```

FIG. 7. Comparison of amino acid sequences of GLUT4/muscle–fat glucose transporters. References for the sequences of human, rat, and mouse GLUT4 are indicated in the text. Residues that are identical to human GLUT4 are noted by dashes. Gaps introduced to generate this alignment are represented by colons. The potential N-linked glycosylation site in the extracellular loop connecting M1 and M2 is shown in bold type. P, The major site phosphorylated by cAMP-dependent protein kinase A *in vitro* and in response to isoproterenol *in vivo* in rat GLUT4, Ser488 (Lawrence *et al.*, 1990a); the sequence around this site is conserved in the human, rat, and mouse proteins.

its secondary structure and orientation in the plasma membrane will be similar to other members of this family. One of the most striking differences between GLUT4 and the other isoforms is its extended NH_2 terminus, which is 7–14 residues longer than the corresponding region of the other facilitative glucose transporters (Fig. 3). It is tempting to speculate that the intracellular NH_2-terminal domain preceding M1 may be responsible for specific targeting of GLUT4 to the unique vesicle population with which it is associated.

E. GLUT5/SMALL INTESTINE ISOFORM

The most recent member of the facilitative glucose transporter gene family to be identified is GLUT5 (Kayano et al., 1990). cDNA clones encoding GLUT5 were isolated from a human jejunal cDNA library by low-stringency cross-hybridization. GLUT5 is 501 amino acids and has 42, 40, 39, and 42% identity with human GLUT1, GLUT2, GLUT3, and GLUT4, respectively (Fig. 3). Although GLUT5 is the most divergent member of the facilitative glucose transporter gene family identified to date, its orientation in the lipid bilayer is predicted to be similar to that proposed for GLUT1 and other members of this gene family. GLUT5 cDNA clones have not yet been isolated and characterized from other species.

IV. Tissue Distribution of Facilitative Glucose Transporters

Studies of the tissue distribution of mRNAs encoding the five facilitative glucose transporters have revealed distinct but overlapping patterns of expression (Fig. 8) that may reflect the specific role of each isoform in the maintenance of glucose homeostasis. The GLUT1 isoform is the predominant facilitative glucose transporter expressed in cultured cells, and may be the only isoform present in many cell lines (Mueckler et al., 1985; Kahn and Flier, 1990). In vivo, the highest levels of GLUT1 are found in fetal tissues, brain, and placenta (Fig. 8) (Flier et al., 1987a; Fukumoto et al., 1988a; Werner et al., 1989). In many tissues in which it is expressed, GLUT1 is concentrated in cells of blood–tissue barriers (Takata et al., 1990; Harik et al., 1990b), indicating a role for GLUT1 in

→

FIG. 8. Expression of facilitative glucose transporter mRNAs in human tissues. The patterns of expression in placenta, and various adult human tissues and tumors of mRNAs encoding GLUT1 to GLUT5, are shown. The sizes of the hybridizing transcripts are indicated on the right. Twenty micrograms of total cellular RNA was denatured with glyoxal, separated by agarose gel electrophoresis, and blotted onto a nitrocellulose filter. The filter was hybridized with the appropriate human cDNA. This filter was repeatedly hybridized after removal of bound radioactivity. (C) GLUT3 mRNA is readily evident in skeletal muscle with a longer exposure of the autoradiogram shown here. These data are adapted from Fukumoto et al. (1988a,b, 1989) and Kayano et al. (1988, 1990).

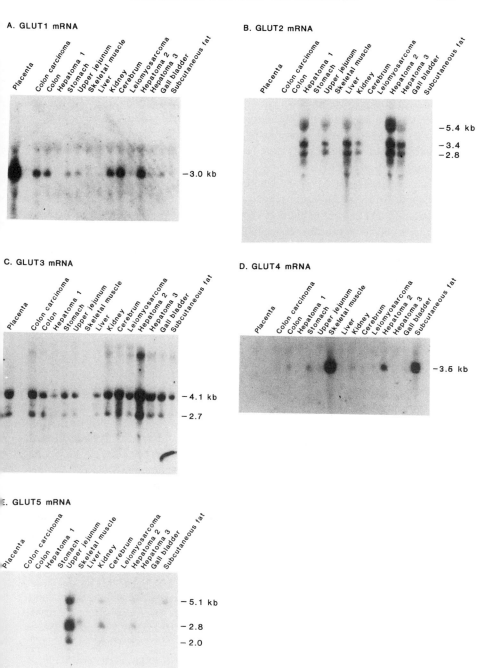

A. GLUT1 mRNA

−3.0 kb

B. GLUT2 mRNA

−5.4 kb
−3.4
−2.8

C. GLUT3 mRNA

−4.1 kb
−2.7

D. GLUT4 mRNA

−3.6 kb

E. GLUT5 mRNA

−5.1 kb
−2.8
−2.0

the movement of glucose across these barriers and delivery to the underlying tissues. In addition, since low levels of GLUT1 protein or mRNA can be detected in most tissues, this glucose transporter isoform may be responsible, at least in part, for constitutive, noninsulin-stimulated glucose uptake.

GLUT2 is expressed at highest levels in the β cells of the endocrine pancreas and in liver and at lower levels in the small intestine and kidney (Fig. 8) (Thorens et al., 1988; Fukumoto et al., 1988b). Histochemical studies have shown that GLUT2 is present in insulin-producing β cells and it is found at highest levels in microvilli acting adjacent endocrine cells; GLUT2 has not been found in other islet endocrine cells (Orci et al., 1989). In the liver, kidney, and small intestine, GLUT2 has been localized to the sinusoidal membrane of hepatocytes and to the basolateral surface of fully differentiated absorptive intestinal epithelial cells and proximal tubule cells of the kidney (Thorens et al., 1990b,c). The tissue distribution of GLUT2 suggests that it mediates the uptake and release of glucose by hepatocytes and that it participates in the transepithelial transport of absorbed and reabsorbed glucose by the small intestine and kidney, respectively. Its presence in the β cell suggests that it may function in the regulation of glucose-stimulated insulin secretion.

GLUT3 mRNA is present at variable levels in all adult human tissues that have been examined (Fig. 8) (Kayano et al., 1988). It is present at highest levels in brain and kidney, as well as placenta. The ubiquitous distribution of GLUT3 in human tissues suggests that it may be responsible together with GLUT1 for noninsulin-stimulated glucose transport. Interestingly, in monkeys, rabbits, rats, and mice the pattern of expression of GLUT3 is much different than observed in humans (Yano et al., 1991; Nagamatsu et al., 1991). In these animals, GLUT3 mRNA is found at high levels in brain and the levels of GLUT3 are very low or undetectable in other tissues. The sites of GLUT3 mRNA expression in adult mouse brain have been identified by in situ hybridization and GLUT3 mRNA has been shown to be present at high levels in the hippocampus (Nagamatsu et al., 1991). Thus, two facilitative glucose transporters may be involved in the uptake and disposal of glucose in the brain. GLUT1 is responsible for the transport of glucose across the blood–brain barrier and GLUT3 effects its uptake into neuronal cells. The presence of GLUT3 mRNA in glial cells (Sadiq et al., 1990) suggests that GLUT3 may also mediate glucose uptake by these cells as well. The biochemical properties of GLUT1 and GLUT3 are also compatible with this dual glucose transporter system for uptake of glucose by the brain. GLUT3 has a higher affinity for sugars than does GLUT1 (Table II). Since glucose concentrations are higher at the blood–brain barrier than in the brain itself, this difference in affinity ensures efficient uptake of glucose by neuronal cells even at low extracellular glucose concentrations.

GLUT4 mRNA is found at highest levels in adipose and muscle tissue (Fig. 8) (James et al., 1989a; Birnbaum, 1989; Charron et al., 1989; Fukumoto et al.,

1989; Kaestner et al., 1989). This includes brown and white adipose tissue as well as cardiac and skeletal muscle. GLUT4 may also be expressed in smooth muscle tissue. The human cDNA was isolated from a jejunal cDNA library (Fukumoto et al., 1989). Since the mRNA used to prepare this library was from a segment of jejunum that included both mucosa and smooth muscle, we believe that the GLUT4 cDNA may have been derived from cells of smooth muscle origin. GLUT4 mRNA also has been detected in human uterus (Kayano et al., 1990). GLUT4 mediates insulin-stimulated glucose uptake. Under basal conditions, GLUT4 is associated with a specific intracellular vesicular population and does not contribute appreciably to glucose transport activity in the absence of insulin stimulation (Zorzano et al., 1989; Holman et al., 1990). Acute insulin stimulation induces the fusion of these vesicles with the plasma membrane and transiently increases the number of plasma membrane-associated glucose transporters. The recruitment or translocation of glucose transporters from an intracellular pool to the cell surface appears to be the primary mechanism responsible for the increase in glucose transport seen in the presence of insulin.

GLUT5 is expressed predominantly in the jejunal region of the small intestine, although low levels can be detected in human kidney, skeletal muscle, and adipose tissue (Fig. 8) (Kayano et al., 1990). The subcellular distribution and function of GLUT5 in these tissues is unknown. It is expressed in the same region of the small intestine as are the Na^+/glucose cotransporter isoform, SGLT1, and the facilitative glucose transporter isoform, GLUT2. One possibility is that GLUT5 may participate in the uptake of dietary glucose from the lumen of the small intestine. It could function as a facilitative glucose carrier in this capacity, or perhaps it may be an H^+/glucose symporter and participate in the active uptake of glucose. This latter hypothesis may not be unreasonable since bacterial and plant H^+/sugar symporters are structurally related to mammalian facilitative glucose transporters (Maiden et al., 1987; Sauer et al., 1990). At the present time, it is not possible to distinguish a facilitative glucose carrier from an active H^+/sugar symporter based on amino acid sequence information alone. The fact that members of this superfamily of sugar transport proteins can function as facilitative or active glucose transporters highlights their functional versatility. Moreover, it suggests that some members of this expanding family of mammalian glucose transporters may be active transporters.

V. Insulin Regulation of Glucose Transport Activity

Insulin rapidly (maximal activation within 10 minutes) and reversibly stimulates glucose transport activity in adipose and muscle tissue primarily by increasing the maximum transport velocity (V_{max}) rather than by altering the apparent affinity (K_m) of the carrier for glucose (Pessin and Czech, 1985; Simpson and Cushman, 1986). This increase in the rate of glucose uptake occurs in the absence of new protein synthesis. The pioneering studies of Cushman, Kono, and colleagues in 1980 described a novel molecular mechanism by which insulin increased glucose transport in adipocytes (Cushman and Wardzala, 1980; Suzuki and Kono, 1980). These two groups independently demonstrated that the major effect of insulin on isolated rat adipocytes was to induce the translocation of an intracellular pool of glucose transporters to the plasma membrane; i.e., insulin increased the number of functional transport proteins at the surface of the cell. The increase in plasma membrane-associated glucose transporters was accompanied by a concomitant decrease in their abundance in the intracellular low-density microsomal fraction. Although the transport of glucose itself does not require ATP, the translocation of glucose transporters in response to insulin and the reversal of this process are ATP dependent (Kono et al., 1981). Insulin-stimulated glucose transport in skeletal and cardiac muscle also occurs by insulin-induced translocation of preformed glucose transporters from an intracellular pool to the plasma membrane (Wardzala and Jeanrenaud, 1981, 1983; Watanabe et al., 1984; Klip et al., 1987; Hirshman et al., 1990). Thus, the mechanism for the stimulatory action of insulin on facilitated glucose transport may be similar in adipose and muscle tissue.

The isolation and characterization of a monoclonal antibody, 1F8, that recognized the insulin-regulated glucose transporter of adipocytes (James et al., 1988) suggested that tissue-specific, insulin-regulated glucose transport was conferred by the expression of a unique facilitative glucose transporter isoform, a suggestion subsequently confirmed with the cloning of GLUT4. Using the monoclonal antibody 1F8 and polyclonal antibodies to the erythrocyte glucose transporter, now designated GLUT1, Zorzano et al., (1989) demonstrated that GLUT4 represented about 90% of the glucose transporters present in rat adipocytes. Under basal conditions, GLUT4 is predominantly localized to an intracellular vesicle population with very little being present in the plasma membrane. GLUT1 is also expressed by adipocytes and represents ~3% of the cytochalasin B-photolabeled glucose transporters in these cells (Oka et al., 1988; Zorzano et al., 1989). However, in contrast to GLUT4, GLUT1 is distributed approximately equally between plasma membrane and intracellular low-density microsomal fractions. GLUT4 and GLUT1 also appear to be associated with different populations of vesicles in primary rat adipocytes. In response to insulin, the amount of GLUT4 present at the cell surface increased 15- to 20-fold whereas GLUT1 increased ~5-fold (Zorzano et al., 1989; Holman et al., 1990). Since insulin stimulates glucose transport activity in primary rate adipocytes 20- to 30-fold, and the levels of GLUT4 in adipocytes are ~20- to 30-fold greater than those of GLUT1, recruitment of GLUT4 can explain the majority of insulin-stimulated increase in glucose transport activity. The small difference between the fold increase in glucose transport activity (20- to 30-fold) compared to the extent of GLUT4 translocation (15- to 20-fold) may be accounted for by differences in the turnover numbers of GLUT1 and GLUT4 (see discussion in Holman *et al.*, 1990).

Several cultured cell lines have been studied as model systems for investigating the effects of insulin on glucose transport. The best characterized of these are 3T3-L1 adipocytes in which insulin stimulates glucose transport ~10- to 20-fold (Frost and Lane, 1985; Calderhead *et al.*, 1990). However, in contrast to isolated primary rat adipocytes, GLUT1 is more abundant than GLUT4 in these cells. Quantitative immunoblotting studies indicate that there are about 950,000 and 280,000 molecules of GLUT1 and GLUT4, respectively, per cell (Calderhead *et al.*, 1990). Moreover, GLUT1 and GLUT4 appear to be colocalized within the same intracellular vesicles in 3T3-L1 adipocytes whereas they are in different vesicle populations in primary rat adipocytes (Zorzano *et al.*, 1989). This difference may be a consequence of the larger amounts of GLUT1 in 3T3-L1 adipocytes compared to primary adipocytes. Insulin treatment of 3T3-L1 adipocytes causes 6- and 17-fold increases in the amounts of GLUT1 and GLUT4, respectively, at the surface of the cell (Calderhead *et al.*, 1990). These results are qualitatively similar to those determined for insulin-stimulated translocation in primary rat adipocytes described above, and suggest that translocation may account for the full effect of insulin on glucose transport in both systems.

Numerous studies utilizing primary and cultured adipocytes demonstrate that both GLUT1 and GLUT4 are translocated in response to insulin. Thus, in terms of translocation, there is not a single insulin-regulatable transporter. In insulin-stimulated primary rat adipocytes, the fold increase in plasma membrane-associated GLUT4 correlates with the observed increase in glucose transport activity and may account for all the effects of insulin on glucose transport (Holman *et al.*, 1990). Although GLUT1 is also translocated to the plasma membrane, the intracellular pool of this transporter is quantitatively insufficient under normal conditions to make a significant contribution to insulin-stimulated glucose transport activity. Interestingly, in 3T3-L1 adipocytes, the ~21-fold increase in insulin-stimulated glucose uptake correlates with the ~17-fold increase in GLUT4 present in the plasma membrane (Calderhead *et al.*, 1990). This is in spite of the fact that GLUT1 levels in these cells are 3.4-fold higher than those of GLUT4 and that there is an ~6.5-fold increase in GLUT1 in the plasma membrane of insulin-treated 3T3-L1 adipocytes. This suggests that the intrinsic activity or turnover number of GLUT4 may be considerably greater than that of GLUT1 or, alternatively, that GLUT1 may require activation after translocation, whereas GLUT4 does not.

Although translocation of GLUT4 may account for all the effects of insulin on glucose transport activity, changes in intrinsic activity may also affect glucose transporter function. Studies examining the effects of counter regulatory hormones on insulin-stimulated glucose transport have indicated that adipocyte glucose transport activity can be modulated independently of recruitment (Simpson and Cushman, 1986). The mechanism(s) responsible for modulating the intrinsic activity of the facilitative glucose transporters are unclear. However, both GLUT1 and GLUT4 can be phosphorylated: GLUT1 by protein kinase C (Witters *et al.*, 1985) and GLUT4 by protein kinase A (James *et al.*, 1989b; Lawrence *et al.*, 1990a). The site of protein kinase C-mediated phosphorylation of GLUT1 has not been determined. The site of phosphorylation of rat GLUT4 by protein kinase A is restricted to the region of the putative intracellular COOH-terminal domain and Ser-488 (Fig. 7) (Lawrence *et al.*, 1990a). This residue is the major site phosphorylated both by cAMP-dependent protein kinase A *in vitro* and in response to isoproterenol *in vivo* and is conserved in the sequences of the rat, human, and mouse GLUT4 (Fig. 7). It will now be possible by site-directed mutagenesis to change this residue and to determine whether its phosphorylation is of regulatory significance, perhaps by decreasing the intrinsic activity of the transporter or promoting its internalization (Lawrence *et al.*, 1990b).

VI.   *In Vitro* Regulation of Glucose Transporter Expression

Most studies of glucose transporter regulation in cultured cells have focused primarily on the expression of GLUT1, in part because it was the first isoform cloned, but also because GLUT1 is the predominant and, in many cultured cells, the only isoform expressed. GLUT1 expression in cultured cells can be altered by growth factors, insulin, transformation, glucose, agents that activate protein kinases A and C, hypoglycemic agents such as sulfonylureas and vanadate, glucocorticoids, and cellular differentiation (reviewed in Kahn and Flier, 1990). In many instances, the effects of these factors on GLUT1 expression are cell type specific and do not alter the expression of other glucose transporter isoforms that may be coexpressed in the same cell line. These agents may regulate GLUT1 expression at multiple levels and include transcriptional, posttranscriptional, and posttranslational mechanisms.

VII.   *In Vivo* Regulation of Glucose Transporter Expression

It is now clear that the expression of the different mammalian glucose transporters is both tissue and cell type specific. That this is the case should not be surprising in view of the differing requirements of tissues for glucose and of the different contributions of tissues to the maintenance of glucose homeostasis. The effects of fasting, diabetes mellitus, obesity, and chronic infusion of glucose or insulin on glucose transporter expression has been examined in human subjects and in animal models with the goal of understanding how glucose uptake is regulated in normal and altered metabolic states (Kahn and Flier, 1990). These studies have shown that the nature of the response of a specific glucose transporter isoform to hormonal or metabolic changes depends on the tissue in which it is expressed.

A. ADIPOSE TISSUE

Adipocytes require glucose for the synthesis of triglycerides, which represent the major storage form of metabolic energy in humans and other mammals. Since adipose tissue functions as an energy reservoir, catabolic states would be expected to be associated with decreased glucose uptake. In fact, there is a chronic decrease in glucose transport activity and glucose transporter protein levels in adipocytes isolated from rats that have been fasted or made diabetic by destruction of their insulin-producing β cells with the drug streptozotocin (STZ) (Karnielli et al., 1981; Kahn and Cushman, 1987; Kahn et al., 1988). Glucose transport activity and glucose transporter mRNA and protein levels are also reduced in adipocytes isolated from obese and diabetic human subjects (Garvey et al., 1988; Foley, 1988). Studies in rats have shown that these changes are reversible and that refeeding or insulin treatment of diabetic rats resulted in a concomitant restoration of glucose transport activity and glucose transporter protein levels. The use of isoform-specific antibodies and cDNA probes has established that fasting and diabetes mellitus are associated with an ~90% decrease in the levels of GLUT4 mRNA (Fig. 9) and protein (Berger et al., 1989; Sivitz et al., 1989a; Garvey et al., 1989; Kahn et al., 1989; Charron and Kahn, 1990). Refeeding or insulin treatment increased mRNA and protein levels and resulted in a transient, approximately twofold overexpression of GLUT4 mRNA and protein above the levels present in nontreated control animals. GLUT4 mRNA and protein levels and glucose transport activity declined gradually to normal control values over a period of about 7 days (Sivitz et al., 1990). Since GLUT4 is the predominant isoform found in adipocytes, the specific depletion of the intracellular pool of GLUT4 in fasted and diabetic animals as a result of decreased expression can directly account for the dramatic reduction in insulin-stimulated glucose uptake. In rat adipocytes, GLUT1 represents only a minor fraction of the glucose transporter mRNA and protein, and its levels are not significantly altered in these states of insulin deficiency.

Recently, Garvey et al., (1991) have examined glucose transport activity and levels of GLUT4 mRNA and protein in adipocytes from lean and obese nondiabetic human subjects and in obese patients with noninsulin-dependent diabetes mellitus (NIDDM). They have shown that the decreased insulin-stimulated glucose transport activity seen in adipocytes from obese, nondiabetic and NIDDM subjects is associated with decreased levels of GLUT4 mRNA and protein and that the magnitude of the decrease is much greater in NIDDM subjects than in their obese, nondiabetic controls. They concluded that in obesity, the insulin resistance in adipocytes is due to a chronic depletion of GLUT4, and in NIDDM there is a further reduction in GLUT4 that is not attributable to obesity. Their studies suggest that suppression of GLUT4 expression may be an important mechanism for producing and maintaining the diabetic state in human subjects.

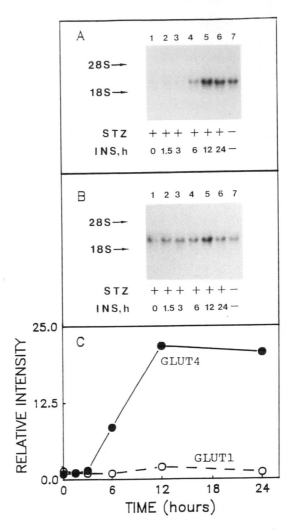

FIG. 9.   Effects of insulin on levels of GLUT4 and GLUT1 mRNA in adipose tissue of strep-tozotocin-induced diabetic male rats. Male Sprague-Dawley rats were made acutely diabetic by injection of streptozotocin as described in Sivitz *et al.* (1989a). After 3 days, diabetic animals were treated with insulin and then killed at the times indicated. Epididymal fat pads were excised and RNA was extracted. Total cellular RNA (30 μg/lane) was size fractionated on a 1% agarose/formaldehyde gel and then transferred to a nitrocellulose filter. The filters were hybridized with GLUT4 (A) or GLUT1 cDNA (B) probes. The relative abundance of GLUT4 and GLUT1 mRNA determined by laser scanning densitometry is shown in (C). Diabetes caused a specific depletion of GLUT4 mRNA in rat adipose tissue and the decrease could be reversed by insulin treatment. The $t_{1/2}$ for this recovery was about 7 hours. Diabetes and insulin treatment had no significant effect on GLUT1 mRNA levels. Adapted from Sivitz *et al.* (1989a).

Messenger RNAs encoding GLUT3 and GLUT5 are also present in human adipose tissue (Fig. 8). However, GLUT3 mRNA cannot be detected by Northern blotting in rat adipose tissue (W. I. Sivitz, unpublished; Yano et al., 1991). The tissue distribution of GLUT5 mRNA has not been examined in species other than humans. Currently, there is no information regarding the effects of diabetes or fasting on expression of GLUT3 and GLUT5 in either humans or rodents.

Diabetes mellitus and fasting are both insulinopenic states, yet are associated with opposite changes in glycemia. However, they both result in a specific decrease in GLUT4 expression in adipose tissue, suggesting that the expression of this gene may be more responsive to the levels of circulating insulin than glucose. This is supported by studies using phlorizin, a compound that inhibits renal reabsorption of filtered glucose and thus reduces blood glucose levels without affecting circulating insulin levels. Normalization of blood glucose levels in diabetic rats with phlorizin does not increase the levels of GLUT4 mRNA or protein in adipose tissue (Sivitz et al., 1990; Kahn et al., 1991). However, treatment with phlorizin does restore insulin-stimulated glucose transport. These results are consistent with a role for insulin in regulating transcription of the GLUT4 gene and, in addition, suggest that glucose levels may affect the functional activity of GLUT4. The recent demonstration that chronic hyperinsulinemia increases the abundance of GLUT4 mRNA in white adipose tissue of rats is consistent with a role for insulin in regulating its expression (Cusin et al., 1990). The relative contribution of transcriptional and posttranscriptional events (i.e., mRNA stabilization) in regulating GLUT4 mRNA levels is currently an active area of investigation. Preliminary experiments suggest that insulin increases GLUT4 mRNA levels in adipose tissue of diabetic rats, at least in part, by increasing the rate of transcription of the GLUT4 gene (A. L. Olson and J. E. Pessin, unpublished).

In contrast to insulin-resistant states in both rats and human described above, the levels of GLUT4 mRNA and protein in adipose tissue are not significantly different between obese, hyperglycemic, insulin-resistant db/db mice compared to their lean (db/+) littermates (Koranyi et al., 1990). The reasons for the differences in the regulation of GLUT4 expression in genetically obese, insulin-resistant and diabetic mice, compared to STZ-diabetic rats or obese, insulin-resistant and diabetic human subjects are unknown.

B. MUSCLE

Most studies of the regulation of glucose transport activity by insulin and other agents have utilized adipocytes as a model system due to their relative ease of manipulation and extreme sensitivity and responsiveness to insulin. However, skeletal muscle is the major site of insulin-mediated peripheral glucose disposal (DeFronzo, 1988) and glucose transport appears to be the rate-limiting step in glucose utilization by this tissue (Ziel *et al.*, 1988). GLUT4 is the major facilitative glucose transporter isoform expressed in skeletal muscle with substantially lower levels of GLUT1 mRNA (Fig. 8). In addition, low levels of GLUT3 and GLUT5 mRNA are present in human skeletal muscle (Fig. 8).

Recently several laboratories have begun to examine the regulation of glucose transporter expression in skeletal muscle because of the central role of this tissue in the maintenance of glucose homeostasis. However, such studies are complicated by the presence of multiple muscle fiber types with differing insulin sensitivity in any given muscle. For example, white, fast-twitch glycolytic fibers are significantly less insulin responsive than red slow-twitch oxidative muscle fibers (James *et al.*, 1986). Recent studies have shown that there is a strong correlation between insulin-stimulated glucose transport, contractile activity, and GLUT4 protein content (Henriksen *et al.*, 1990; Kern *et al.*, 1990). It has also been suggested that differences in insulin-stimulated glucose transport observed in muscles of different fiber type may be dependent on the amount of GLUT4 protein; i.e., red, slow-twitch oxidative muscles have higher levels of GLUT4 protein than white, fast-twitch glycolytic muscles. The differences in levels in GLUT4 protein, together with possible differences in the regulation of expression of this gene in muscles of different fiber types, may account in part for the discordant results that are described below regarding the regulation of GLUT4 expression in skeletal muscle.

In contrast to the profound >90% decrease in GLUT4 mRNA and protein levels seen in adipose tissue of STZ-diabetic rats, there is only a modest ~30% decrease in skeletal muscle (Garvey *et al.*, 1989; Bourey *et al.*, 1990; Richardson *et al.*, 1991). However, similar to adipose tissue, the decreased expression of GLUT4 mRNA in the muscle of STZ-diabetic rats can be reversed with insulin therapy. In fasted rats, there is a two- to threefold increase in GLUT4 protein in a mixed soleus and gastrocnemius muscle preparation (Charron and Kahn, 1990) and soleus muscle (Bourey *et al.*, 1990) relative to fed rats. This increase is associated with a threefold increase (Charron and Kahn, 1990) or with no change (Bourey *et al.*, 1990) in GLUT4 mRNA levels. Charron and Kahn (1990) also examined the effects of fasting on GLUT1 expression in the same mixed soleus and gastrocnemius muscle preparation and noted a coordinate twofold increase in both mRNA and protein. Thus, GLUT1 and GLUT4 appear to be coordinately regulated and both increased in skeletal muscle of fasted animals, whereas in adipose tissue GLUT4 expression is decreased with no change in GLUT1. Chronic insulin-induced hypoglycemia caused a 50 and 240% increase in GLUT4 mRNA and protein levels, respectively (Bourey *et al.*, 1990), in one study, and was associated with a significant decrease in GLUT4 mRNA and protein levels in another (Cusin *et al.*, 1990). It is unclear if the differences between these two studies are methodological or related to the strains of rats used. In studies of genetically obese and insulin-resistant diabetic animals,

GLUT4 protein levels were similar in gastrocnemius muscle of Zucker (fa/fa) rats and their lean (Fa/−) littermates (Friedman et al., 1990). GLUT4 protein levels were also similar in heart, diaphragm, and red and white quadriceps muscle of obese (db/db) mice and their lean (db/+) littermates; however, GLUT4 mRNA levels in quadriceps muscle of obese (db/db) mice were reduced ~30% and protein levels were not significantly different (Koranyi et al., 1990).

Glucose transporter expression has also been examined in muscle tissue of human subjects (Pedersen et al., 1990; Handberg et al., 1990). There were no significant differences in the levels of GLUT4 and GLUT1 mRNA and protein in the vastus lateralis of lean and obese subjects compared to lean or obese non-diabetic subjects.

In addition to the effects of insulin on glucose transport, exercise also increases uptake of glucose into muscle and induces the translocation of GLUT4 (Douen et al., 1990). Exercise training has also been shown to increase GLUT4 protein levels in rat skeletal muscle; however, the effects of training appear to be specific to certain muscle groups with a 60% increase in GLUT4 levels in plantaris muscle but no change in soleus muscle (Rodnick et al., 1990). In another study, GLUT4 levels in gastrocnemius muscle of obese Zucker (fa/fa) rats increased approximately twofold with exercise training (Friedman et al., 1990).

Studies of the regulation of glucose transporter expression in skeletal muscle have revealed only modest changes in expression of GLUT4 in altered metabolic states or with exercise even through several of these states are characterized by a profound impairment in insulin-stimulated glucose uptake. This is in marked contrast to the situation in adipose tissue where fasting and diabetes cause a specific depletion of GLUT4 mRNA and protein, which can account for the decreased insulin-stimulated glucose uptake. At this time, the molecular basis for decreased insulin-stimulated glucose transport in skeletal muscle of humans or rodents with genetic or acquired forms of insulin resistance is unclear. However, the data suggest that the defect lies not with the glucose transporter but elsewhere in the insulin-signaling pathway.

C. LIVER

The liver plays a central role in the regulation of glucose homeostasis having the capacity to store glucose in the form of glycogen in times of excess and to break down these stores and release glucose into the circulation in times of need. The liver is also the major site of gluconeogenesis, using lactate and alanine from muscle, glycerol from adipose tissue, and glucogenic amino acids from the diet to form glucose. It is a metabolically unique tissue in that the intracellular concentration of glucose in the hepatocyte may exceed that in the circulation. GLUT2 is the predominant glucose transporter isoform expressed in liver and mediates both the uptake and release of glucose. GLUT2 has a high K_m and high V_{max} for glucose (Craik and Elliott, 1979; Elliott and Craik, 1982). These kinetic parameters mean that the rate of flux through this transporter will be directly proportional to glucose concentration and that transport should not limit metabolism at physiological glucose concentrations.

GLUT2 expression in liver has been examined under conditions of fasting, diabetes, and chronic hypoglycemia and hyperinsulinemia. Thorens *et al.* (1990a) have reported that GLUT2 mRNA levels decreased 45% without a change in protein levels in rats fasted for 48 hours, and refeeding caused a 75 and 500% increase in GLUT2 protein and mRNA levels, respectively. Fasting was also associated with an approximate three- and fourfold increase in GLUT1 mRNA and protein levels, respectively. GLUT2 mRNA and protein levels were unchanged in STZ-diabetic rats, whereas GLUT1 mRNA levels increased 2.4-fold and protein levels were unchanged. In contrast to these latter results, Oka *et al.* (1990a) observed a twofold increase in both GLUT2 mRNA and protein in STZ-diabetic rats, which returned to normal after 5 days of insulin treatment, and they suggested that increased synthesis of GLUT2 may contribute to the increased hepatic glucose output that occurs in diabetes mellitus. The reason for the discrepancy between these two studies is unknown. Chen *et al.* (1990) examined hepatic GLUT2 mRNA levels in rats following 4 or 12 days of chronic insulin-induced hypoglycemia and noted no consistent changes in liver tissue although the levels in β cells of the islets of Langerhans increased by 46%. Together, these studies suggest that hepatic GLUT2 expression is not under insulin-dependent regulatory control and that the levels of GLUT2 mRNA and protein do not change dramatically with altered metabolic conditions.

D. β CELLS OF THE ISLETS OF LANGERHANS

The GLUT2 isoform is expressed at high levels in the insulin-secreting β cells of the islets of Langerhans (Thorens *et al.*, 1988; Orci *et al.*, 1989; Johnson *et al.*, 1990a). The presence of the high-K_m (low-affinity) glucose transporter allows the β cell to precisely regulate insulin secretion in response to glucose, since the rate of glucose uptake changes in direct proportion to the extracellular glucose concentration over the range of normal glucose concentrations (5–10 m*M*) that stimulates insulin secretion. Recent studies have demonstrated reduced levels of GLUT2 mRNA and protein in islets from rats with acquired and genetic forms of NIDDM (Thorens *et al.*, 1990d; Johnson *et al.*, 1990b; Orci *et al.*, 1990a,b). GLUT2 protein levels were also decreased in the islets of BB/Wor rats, a model of insulin-dependent diabetes mellitus (IDDM), on the first day of overt diabetes, which is characterized by a loss of glucose-stimulated insulin secretion but which precedes the autoimmune depletion of β cells (Orci *et al.*, 1990a). Unger and colleagues have also reported that insulin-induced hypoglycemia for

12 days resulted in the loss of GLUT2 mRNA and transport activity in islets, whereas 5 days of glucose-induced hyperglycemia increased GLUT2 mRNA levels by 46% (Chen *et al.*, 1990). The temporal correlation between reduction of GLUT2 protein and mRNA and loss of glucose-stimulated insulin secretion suggests that GLUT2 is an integral component of the β cell glucose sensor and that decreased expression of this transporter accounts at least in part for β cell failure in diabetes mellitus. As discussed in Johnson *et al.* (1990a), the reduction in GLUT2 is not secondary to hyperglycemia, but rather represents the most proximal abnormality identified in NIDDM. Thus, identification of the factors regulating GLUT2 expression could lead to a better understanding of the causes of diabetes.

E. BRAIN

Glucose is the primary source of energy for brain, and since it lacks fuel stores, it requires a continuous flow of glucose, except during periods of prolonged starvation when ketone bodies are utilized. At least two facilitative glucose transporters are involved in glucose transport in brain. GLUT1 may be primarily responsible for the transport of glucose across the blood–brain barrier (Flier *et al.*, 1987a; Gerhart *et al.*, 1989; Boada and Pardridge, 1990; Takata *et al.*, 1990; Pardridge *et al.*, 1990a; Harik *et al.*, 1990b) and GLUT3 may be the isoform that mediates glucose uptake by neuronal and glial cells (Yano *et al.*, 1991; Sadiq *et al.*, 1990; Nagamatsu *et al.*, 1991).

Studies of glucose transport activity in the brain in diabetes and other altered metabolic states have shown that hyperglycemia decreases glucose transport (Gjedde and Crone, 1981) and hypoglycemia increases glucose transport activity (McCall *et al.*, 1986). In the case of hyperglycemia, it has been demonstrated that there is a decrease in GLUT1 protein at the blood–brain barrier (Pardridge *et al.*, 1990b). The effects of diabetes, fasting, or other changes in metabolic status on GLUT3 expression in the brain have not been described. It will be important to examine GLUT3 expression in neurons to determine if its altered expression or activity may contribute to the neuropathy associated with diabetes.

F. DEVELOPMENTAL REGULATION

Not surprisingly, facilitative glucose transporter expression is also developmentally regulated. Generally, the highest levels of GLUT1 mRNA appear to be in fetal tissues (Werner *et al.*, 1989; Sivitz *et al.*, 1989b), including placenta (Fig. 8A). In heart, lung, liver, kidney, and stomach, the levels of GLUT1 mRNA are highest in fetal tissues and decline postnatally. There is a moderate postnatal decrease in GLUT1 mRNA levels in kidney and stomach, with adult levels being about 50% of those in the corresponding fetal tissues. In heart and lung, the adult levels are about 10–20% of those in the fetal tissues, and in liver, they are <5% of those in fetal liver. Brain shows a slightly different pattern of expression. GLUT1 mRNA and proteins levels are high prior to birth, transiently decrease neonatally, then increase as the animal matures, with the highest levels in adult brain. Testes show a third pattern of development, with lower levels of GLUT1 mRNA in neonatal and adult animals, and highest levels at 22 days after birth.

In contrast to GLUT1, there is little information regarding changing patterns of expression of the other facilitative glucose transporters during development. In regenerating rat liver, GLUT2 mRNA levels increase transiently 2.7-fold at 8 hours following partial hepatectomy and return to normal by 48 hours (Yamada *et al.*, 1990), and GLUT3 mRNA levels are higher in adult than fetal brain tissue (Sadiq *et al.*, 1990).

Although descriptive, these studies suggest that glucose transporters may serve specific roles during development just as they do in adult tissues.

G. NEOPLASTIC TISSUES

Increased rates of glucose uptake are a characteristic marker of the transformed phenotype (Hatanaka and Hanafusa, 1970). Transformation of cultured cells with activated *ras*, *src*, *fps*, and *neu* oncogenes results in a marked 3- to 5-fold stimulation of glucose transport activity and a 5- to 10-fold increase in GLUT1 mRNA and protein levels (Salter *et al.*, 1982; Flier *et al.*, 1987b; Birnbaum *et al.*, 1987; Sistonen *et al.*, 1989). However, glucose uptake and GLUT1 expression are not increased in cells transfected with the *myc* oncogene, suggesting that only oncogenes that are involved in the transduction of extracellular signals such as *ras*, *src*, *fps*, and *neu* increase glucose uptake on transformation, whereas those oncogenes, such as *myc*, that are weakly transforming and specify transcription factors do not.

Tumor cells also have increased rates of glucose uptake and metabolism (Warburg, 1956). Yamamoto *et al.* (1990) have examined the levels of glucose transporter mRNAs in a series of gastrointestinal tumors. They observed that GLUT1 and GLUT3 mRNA levels were elevated in most human tumor samples compared to normal tissues and that GLUT2 mRNA levels were often increased in hepatomas. They found no evidence for increased expression of GLUT4 or GLUT5 in any of the human tumors they examined. Su *et al.* (1990) have reported that GLUT1 mRNA levels were consistently elevated in human hepatomas. The results of these two studies suggest that increased glucose transporter expression may account in part for the increased glucose uptake. Since GLUT1 and GLUT3 have low K_m values (i.e., high affinity) for glucose and glucose analogs (Table II) (Gould *et al.*,1991; Burant and Bell, 1991), their increased expression may provide a growth advantage by assuring a supply of glucose under conditions where it could be limiting, as might be the case when the growth of the tumor cells exceeds that of neovascularization. The increased expression of GLUT1 in both fetal tissues and tumors suggests that GLUT1 may be the transporter of rapidly proliferating cells. The expression of GLUT1 by apparently all cultured cells regardless of origin is consistent with this notion.

VIII. Summary and Overview

The regulation of glucose transport *in vivo* is complex and involves two classes of membrane proteins. One class, the Na+-dependent glucose transporters (the SLGT family), is involved in the active transport of dietary glucose from the lumen of the small intestine and reabsorption of filtered glucose in the proximal tubule of the kidney. cDNA clones encoding one member of this family have been isolated and characterized. However, it seems likely there are other members of this family that have yet to be cloned. The regulation of Na+/glucose cotransporter expression in the small intestine and kidney has not yet been systematically investigated. With the availability of the SGLT1 cDNA clone and antibodies to this protein, these studies should be forthcoming.

The second class of glucose transporters found in mammalian cells, the facilitative glucose transporters, represents ubiquitous proteins that appear to be present on all cells. They have distinct but overlapping patterns of tissue distribution and are responsible for mediating the passive transport of glucose across the cell membrane down its concentration gradient. The facilitative glucose carriers are a family of structurally related membrane proteins and currently cDNA clones encoding five members of this family (the GLUT family) have been isolated and characterized and the search for additional members of this family is continuing.

The facilitative glucose transporters have acquired distinct physiological and biochemical properties that allow them to serve specific functions in the tissues in which they are expressed. The GLUT1 and GLUT3 isoforms have a very high affinity for glucose that allows them to transport glucose efficiently even under conditions of low glucose levels. Moreover, the rate of glucose uptake by GLUT1 and GLUT3 will be determined by their concentration in the plasma membrane rather than by glucose concentration. The GLUT2 isoform has a low affinity for glucose, but a high capacity, which means that the rate of flux through this transporter will be directly proportional to glucose concentration. This property allows GLUT2 to participate in glucose sensing by the β cell, the transepithelial movement of glucose in the small intestine and kidney, and the regulation of circulating glucose levels by the liver. The GLUT4 isoform has a distinct subcellular localization and resides not on the cell surface but in association with an intracellular vesicle population. This distribution provides the basis for a novel but very specific mechanism by which insulin can rapidly and reversibly stimulate glucose uptake by increasing the concentration of GLUT4 at the cell surface. Studies are underway in a number of laboratories to precisely define the kinetic properties (K_m, V_{max}, and turnover number), substrate specificities, and the concentration of each isoform in specific tissues or cells. These studies will lead to a better understanding of how this family of proteins participates in the maintenance of glucose homeostasis.

Further studies of the regulation of expression of the facilitative glucose transporters in altered metabolic states such as fasting and diabetes mellitus, as well as uremia, sepsis, and hypertension, are required in order to determine the metabolic basis for the alterations in glycemic control seen in these conditions. The possible contribution of posttranslational modifications of the facilitative glucose transporters, which may regulate their intrinsic activity, needs to be unequivocally demonstrated, and the underlying molecular mechanisms defined. The mechanisms by which glucose alone can alter glucose transporter levels, and the role of glucose "toxicity" in altering glucose uptake, likewise need to be determined. Finally, an understanding of the mechanism(s) by which obesity and diabetes mellitus decrease glucose transport is critical for our future ability to effectively treat the most two common disorders of carbohydrate metabolism. Zeller *et al.* (1991) have shown altered glucose transporter mRNA levels in tissues of a rat model of septic shock.

ACKNOWLEDGMENTS

The studies from our laboratories presented in this review were supported by: the Howard Hughes Medical Institute; Juvenile Diabetes Foundation, International; Northern Illinois Affiliate of the American Diabetes Association; and the National Institutes of Health (Research Grants: DK-20595, -25295, and HL-14388).

REFERENCES

Asano, T., Shibasaki, Y., Kasuga, M., Kanazawa, Y., Takaku, F., Akanuma, Y., and Oka, Y. (1988). *Biochem. Biophys. Res. Commun.* **154,** 1204–1211.
Asano, T., Shibasaki, Y., Ohno, S., Taira, H., Lin, J.-L., Kasuga, M., Kanazawa, Y., Akanuma, Y., Takaku, F., and Oka, Y. (1989a). *J. Biol. Chem.* **264,** 3416–3420.
Asano, T., Shibasaki, Y., Lin, J.-L., Akanuma, Y., Takaku, F., and Oka, Y. (1989b). *Nucleic Acids Res.* **17,** 6386.
Axelrod, J. D., and Pilch, P. F. (1983). *Biochemistry* **22,** 2222–2227.
Bell, G. I., Kayano, T., Buse, J. B., Burant, C. F., Takeda, J., Lin, D., Fukumoto, H., and Seino, S. (1990). *Diabetes Care* **13,** 198–208.
Berger, J., Biswas, C., Vicario, P. P., Strout, H. V., Saperstein, R., and Pilch, P. F. (1989). *Nature (London)* **340,** 70–72.
Birnbaum, M. J. (1989). *Cell (Cambridge, Mass.)* **57,** 305–315.
Birnbaum, M. J., Haspel, H. C., and Rosen, O. M. (1986). *Proc. Natl. Acad. Sci. U.S.A.* **83,** 5784–5788.
Birnbaum, M. J., Haspel, H. C., and Rosen, O. M. (1987). *Science* **235,** 1495–1498.
Birnir, B., Lee, H.-S., Hediger, M. A., and Wright, E. M. (1990). *Biochim. Biophys. Acta* **1048,** 100–104.

Boado, R. J., and Pardridge, W. M. (1990). *Biochem. Biophys. Res. Commun.* **166,** 174–179.

Bourey, R. E., Koranyi, L., James, D. E., Mueckler, M., and Permutt, M. A. (1990). *J. Clin. Invest.* **86,** 542–547.

Bramwell, M. E., Davies, A., and Baldwin, S. A. (1990). *Exp. Cell Res.* **188,** 97–104.

Buddington, R. K., and Diamond, J. M. (1989). *Annu. Rev. Physiol.* **51,** 601–619.

Burant, C. F., and Bell, G. I. (1991). In preparation.

Calderhead, D. M., Kitagawa, K., Tanner, L. I., Holman, G. D., and Lienhard, G. E. (1990). *J. Biol. Chem.* **265,** 13800–13808.

Carruthers, A. (1990). *Physiol. Rev.* **70,** 1135–1176.

Charron, M. J., and Kahn, B. B. (1990). *J. Biol. Chem.* **265,** 7994–8000.

Charron, M. J., Brosius, F. C., Alper, S. L., and Lodish, H. F. (1989). *Proc. Natl. Acad. Sci. U.S.A.* **86,** 2535–2539.

Chen, L., Alam, T., Johnson, J. H., Hughes, S., Newgard, C. B., and Unger, R. H. (1990). *Proc. Natl. Acad. Sci. U.S.A.* **87,** 4088–4092.

Coady, M. J., Pajor, A. M., and Wright, E. M. (1990). *Am. J. Physiol.* **259,** C605–C610.

Craik, J. D., and Elliott, K. R. F. (1979). *Biochem. J.* **182,** 503–508.

Cushman, S. W., and Wardzala, L. J. (1980). *J. Biol. Chem.* **255,** 4758–4762.

Cusin, I., Terrettaz, J., Fohner-Jeanrenaud, F., Zarjevski, N., Assimacopoulos-Jeannet, F., and Jeanrenaud, B. (1990). *Endocrinology (Baltimore)* **127,** 3246–3248.

Davies, A., Ciardelli, T. L., Lienhard, G. E., Boyle, J. M., Whetton, A. D., and Baldwin, S. A. (1990). *Biochem. J.* **266,** 799–808.

DeFronzo, R. F. (1988). *Diabetes* **37,** 667–687.

Desjeux, J.-F. (1989). *In* "The Metabolic Basis of Inherited Disease" (C. R. Scriver, A. L. Beaudet, W. S. Sly, and D. Valle, eds.), 6th ed., Vol. II, pp. 2463–2478. McGraw-Hill, New York.

Douen, A. G., Ramlal, T., Rastogi, S., Bilan, P. J., Cartee, G. D., Vranic, M., Holloszy, J. O., and Klip, A. (1990). *J. Biol. Chem.* **265,** 13247–13430.

Elliott, K. R. F., and Craik, J. D. (1982). *Biochem. Soc. Trans.* **10,** 12–13.

Endo, T., Kasahara, M., and Kobata, A. (1990). *Biochemistry* **29,** 9126–9134.

Fischbarg, J., Kuang, K., Vera, J. C., Arant, S., Silverstein, S. C., Loike, J., and Rosen, O. M. (1990). *Proc. Natl. Acad. Sci. U.S.A.* **87,** 3244–3247.

Flier, J. S., Mueckler, M., McCall, A. L., and Lodish, H. F. (1987a). *J. Clin. Invest.* **79,** 657–661.

Flier, J. S., Mueckler, M. M., Usher, P., and Lodish, H. F. (1987b). *Science* **235,** 1492–1495.

Foley, J. E. (1988). *Diabetes/Metab. Rev.* **4,** 487–505.

Friedman, J. E., Sherman, W. M., Reed, M. J., Elton, C. W., and Dohm, G. L. (1990). *FEBS Lett.* **268,** 13–16.

Friedman, J. E., Dudek, R. W., Whitehead, D. S., Downes, D. L., Frisell, W. R., Caro, J. F., and Dohm, G. L. (1991). *Diabetes* **40,** 150–154.

Froehner, S. C., Davies, A., Baldwin, S. A., and Lienhard, G. E. (1988). *J. Neurocytol.* **17,** 173–178.

Frost, S. C., and Lane, M. D. (1985). *J. Biol. Chem.* **260,** 2646–2652.

Fukumoto, H., Seino, S., Imura, H., Seino, Y., and Bell, G. I. (1988a). *Diabetes* **37,** 657–661.

Fukumoto, H., Seino, S., Imura, H., Seino, Y., Eddy, R. L., Fukushima, Y., Byers, M. G., Shows, T. B., and Bell, G. I. (1988b). *Proc. Natl. Acad. Sci. U.S.A.* **85,** 5434–5438.

Fukumoto, H., Kayano, T., Buse, J. B., Edwards, Y., Pilch, P. F., Bell, G. I., and Seino, S. (1989). *J. Biol. Chem.* **264,** 7776–7779.

Garvey, W. T., Huecksteadt, T. P., Matthaei, S., and Olefsky, J. M. (1988). *J. Clin. Invest.* **81,** 1528–1536.

Garvey, W. T., Huecksteadt, T. P., and Birnbaum, M. J. (1989). *Science* **245,** 60–63.

Garvey, W. T., Maianu, L., Huecksteadt, T. P., Birnbaum, M. J., Molina, J. M., and Ciaraldi, T. P. (1991). *J. Clin. Invest.* **87,** 1072–1081.

Gerhart, D. Z., and Drewes, L. R. (1990). *Brain Res.* **508,** 46–50.

Gerhart, D. Z., LeVasseur, R. J., Broderius, M. A., and Drewes, L. R. (1989). *J. Neurosci. Res.* **22,** 464–472.

Gjeddi, A., and Crone, C. (1981). *Science* **214,** 456–457.

Gorga, F. R., Baldwin, S. A., and Lienhard, G. E. (1979). *Biochem. Biophys. Res. Commun.* **91,** 955–961.

Gould, G. W., and Bell, G. I. (1990). *Trends Biochem. Sci.* **15,** 18–23.

Gould, G. W., and Lienhard, G. E. (1989). *Biochemistry* **28,** 9447–9452.

Gould, G. W., Derechin, V., James, D. E., Tordjman, K., Ahern, S., Gibbs, E. M., Lienhard, G. E., and Mueckler, M. (1989). *J. Biol. Chem.* **264,** 2180–2184.

Gould, G. W., Thomas, H. M., Jess, T. J., and Bell, G. I. (1991). *Biochemistry* **30,** 5139–5145.

Guastella, J., Nelson, N., Nelson, H., Czyzyk, L., Keynan, S., Miedel, M. C., Davidson, N., Lester, H. A., and Kanner, B. I. (1990). *Science* **249,** 1303–1306.

Handberg, A., Vaag, A., Damsbo, P., Beck-Nielsen, H., and Vinten, J. (1990). *Diabetologia* **33,** 625–627.

Harik, S. I., Kalaria, R. N., Whitney, P. M., Andersson, L., Lundahl, P., Ledbetter, S. R., and Perry, G. (1990a). *Proc. Natl. Acad. Sci. U.S.A.* **87,** 4621–4264.

Harik, S. I., Kalaria, R. N., Andersson, L., Lundahl, P., and Perry, G. (1990b). *J. Neurosci.* **10,** 3862–3872.

Harrison, S. A., Buxton, J. M., Helgerson, A. L., MacDonald, R. G., Chlapowski, F. J., Carruthers, A., and Czech, M. P. (1990a). *J. Biol. Chem.* **265,** 5793–5801.

Harrison, S. A., Buxton, J. M., Clancy, B. M., and Czech, M. P. (1990b). *J. Biol. Chem.* **265,** 20106–20116.

Haspel, H., C., Birnbaum, M. J., Wilk, E. W., and Rosen, O. M. (1985). *J. Biol. Chem.* **260,** 7219–7225.

Hatanaka, M., and Hanafusa, H. (1970). *Virology* **41,** 647–652.

Hediger, M. A., Coady, M. J., Ikeda, T. S., and Wright, E. M. (1987). *Nature (London)* **330,** 379–381.

Hediger, M. A., Budarf, M. L., Emanuel, B. S., Mohandas, T. K., and Wright, E. M. (1989a). *Genomics* **4,** 297–300.

Hediger, M. A., Turk, E., and Wright, E. M. (1989b). *Proc. Natl. Acad. Sci. U.S.A.* **86,** 5748–5752.

Henriksen, E. J., Bourey, R. E., Rodnick, K. J., Koranyi, L., Permutt, M. A., and Holloszy, J. O. (1990). *Am. J. Physiol.* **259,** E593–E598.

Hirshman, M. F., Goodyear, L. J., Wardzala, L. J., Horton, E. D., and Horton, E. S. (1990). *J. Biol. Chem.* **265,** 987–991.

Holman, G. D., Kozka, I. J., Clark, A. E., Flower, C. J., Saltis, J., Habberfield, A. D., Simpson, I. A., and Cushman, S. W. (1990). *J. Biol. Chem.* **265,** 18172–18179.

Ikeda, T. S., Hwang, E-S., Coady, M. J., Hirayama, B. A., Hediger, M. A., and Wright, E. M. (1989). *J. Membr. Biol.* **110,** 87–95.

James, D. E., Zorzano, A., Boni-Schnetzler, M., Nemenoff, R. A., Powers, A., Pilch, P. F., and Ruderman, N. B. (1986). *J. Biol. Chem.* **261,** 14939–14944.

James, D. E., Brown, R., Navarro, J., and Pilch, P. F. (1988). *Nature (London)* **333,** 183–185.

James, D. E., Strube, M., and Mueckler, M. M. (1989a). *Nature (London)* **338,** 83–87.

James, D. E., Hiken, J., and Lawrence, J. C. (1989b). *Proc. Natl. Acad. Sci. U.S.A.* **86,** 8368–8372.

Johnson, J. H., Newgard, C. B., Milburn, J. L., Lodish, H. F., and Thorens, B. (1990a). *J. Biol. Chem.* **265,** 6548–6551.

Johnson, J. H., Ogawa, A., Chen, L., Orci, L., Newgard, C. B., Alam, T., and Unger, R. H. (1990b). *Science* **250,** 546–549.

Juranka, P. F., Zastawny, R. L., and Ling, V. (1989). *FASEB J.* **3**, 2583–2592.

Kaback, H. R. (1987). *Biochemistry* **26**, 2071–2076.

Kaestner, K. H., Christy, R. J., McLenithan, J. C., Braiterman, L. T., Cornelius, P., Pekala, P. H., and Lane, M. D. (1989). *Proc. Natl. Acad. Sci. U.S.A.* **86**, 3150–3154 and 4937.

Kahn, B. B., and Cushman, S. W. (1987). *J. Biol. Chem.* **262**, 5118–5124.

Kahn, B. B., and Flier, J. S. (1990). *Diabetes Care* **13**, 548–564.

Kahn, B. B., Simpson, I. A., and Cushman, S. W. (1988). *J. Clin. Invest.* **82**, 691–699.

Kahn, B. B., Charron, M. J., Lodish, H. F., Cushman, S. W., and Flier, J. S. (1989). *J. Clin. Invest.* **84**, 404–411.

Kahn, B. B., Shulman, G. I., DeFronzo, R. A., Cushman, S. W., and Rossetti, L. (1991). *J. Clin. Invest.* **87**, 561–570.

Karnielli, E., Hissin, P. J., Simpson, I. A., Salans, L. B., and Cushman, S. W. (1981). *J. Clin. Invest.* **68**, 811–814.

Kayano, T., Fukumoto, H., Eddy, R. L., Fan, Y.-S., Byers, M. G., Shows, T. B., and Bell, G. I. (1988). *J. Biol. Chem.* **263**, 15245–15248.

Kayano, T., Burant, C. F., Fukumoto, H., Gould, G. W., Fan, Y.-S., Eddy, R. L., Byers, M. G., Shows, T. B., Seino, S., and Bell, G. I. (1990). *J. Biol. Chem.* **265**, 13276–13282.

Keller, K., Strube, M., and Mueckler, M. (1989). *J. Biol. Chem.* **264**, 18884–18889.

Kern, M., Wells, J. A., Stephens, J. M., Elton, C. W., Friedman, J. E., Tapscott, E. B., Pekala, P. H., and Dohm, G. L. (1990). *Biochem. J.* **270**, 397–400.

Klip, A., Ramlal, T., Young, D. A., and Holloszy, J. O. (1987). *FEBS Lett.* **224**, 224–230.

Kono, T., Suzuki, K., Dansey, L. E., Robinson, F. W., and Blevins, T. L. (1981). *J. Biol. Chem.* **256**, 6400–6407.

Koranyi, L., James, D., Mueckler, M., and Permutt, M. A. (1990). *J. Clin. Invest.* **85**, 962–967.

Lawrence, J. C., Hiken, J. F., and James, D. E. (1990a). *J. Biol. Chem.* **265**, 2324–2332.

Lawrence, J. C., Hiken, J. F., and James, D. E. (1990b). *J. Biol. Chem.* **265**, 19768–19776.

Maiden, M. C. J., Davis, E. O., Baldwin, S. A., Moore, D. C. M., and Henderson, P. J. F. (1987). *Nature (London)* **325**, 641–643.

May, J. M., Buchs, A., and Carter-Su, C. (1990). *Biochemistry* **29**, 10393–10398.

McCall, A. L., Fixman, L. B., Fleming, N., Tornheim, K., Chick, W., and Ruderman, N. B. (1986). *Am. J. Physiol.* **251**, E442–F447.

McKusick, V. A. (1990). "Mendelian Inheritance in Man." Johns Hopkins Univ. Press, Baltimore and London.

Mueckler, M. (1990). *Diabetes* **39**, 6–11.

Mueckler, M., and Lodish, H. F. (1986). *Cell (Cambridge, Mass.)* **44**,629–637.

Mueckler, M., Caruso, C., Baldwin, S. A., Panico, M., Blench, I., Morris, H. R., Allard, W. J., Lienhard, G. E., and Lodish, H. F. (1985). *Science* **229**, 941–945.

Nagamatsu, S., Kornhauser, J., Seino, S., Mayo, K. E., Steiner, D. F., and Bell, G. I. (1991). Submitted for publication.

Ohta, T., Isselbacher, K. J., and Rhoads, D. B. (1990). *Mol. Cell. Biol.* **10**, 6491–6499.

Oka, Y., Asano, T., Shibasaki, Y., Kasuga, M., Kanazawa, Y., and Takaku, F. (1988). *J. Biol. Chem.* **263**, 13432–13439.

Oka, Y., Asano, T., Shibasaki, Y., Lin, J.-L., Tsukuda, K., Akanuma, Y., and Takaku, F. (1990a). *Diabetes* **39**, 441–446.

Oka, Y., Asano, T., Shibasaki, Y., Lin, J.-L., Tsukuda, K., Katagiri, H., Akanuma, Y., and Takaku, F. (1990b). *Nature (London)* **345**, 550–553.

Orci, L., Thorens, B., Ravazzola, M., and Lodish, H. F. (1989). *Science* **245**, 295–297.

Orci, L., Unger, R. H., Ravazzola, M., Ogawa, A., Komiya, I., Baetens, D., Lodish, H. F., and Thorens, B. (1990a). *J. Clin. Invest.* **86**, 1615–1622.

Orci, L., Ravazzola, M., Baetens, D., Inman, L., Amherdt, M., Peterson, R. G., Newgard, C. B., Johnson, J. H., and Unger, R. H. (1990b). *Proc. Natl. Acad. Sci. U.S.A.* **87**, 9953–9957.

Pardridge, W. M., Boada, R. J., and Farrell, C. R. (1990a). *J. Biol. Chem.* **265**, 18035–18040.

Pardridge, W. M., Triguero, D., and Farrell, C. R. (1990b). *Diabetes* **39**, 1040–1044.

Pedersen, O., Bak, J. F., Andersen, P. H., Lund, S., Moller, D. E., Flier, J. S., and Kahn, B. B. (1990). *Diabetes* **39**, 865–870.

Peerce, B. E. (1990). *J. Biol. Chem.* **265** 1737–1741.

Permutt, M. A., Koranyi, L., Keller, K., Lacy, P. E., Scharp, D. W., and Mueckler, M. (1989). *Proc. Natl. Acad. Sci. U.S.A.* **86**, 8688–8692.

Pessin, J. E., and Czech, M. P. (1985). *In* "The Enzymes of Biological Membranes" (A. N. Martinosi, ed.), pp. 497–522. Plenum, New York.

Pilch, P. F. (1990). *Endocrinology (Baltimore)* **126**, 3–5.

Reed, B. C., Shade, D., Alperovich, F., and Vang, M. (1990). *Arch. Biochem. Biophys.* **279**, 261–274.

Richardson, J. M., Treadway, J. L., Balon, T., and Pessin, J. E. (1991). Submitted for publication.

Rodnick, K. J., Holloszy, J. O., Mondon, C. E., and James, D. E. (1990). *Diabetes* **39**, 1425–1429.

Sadiq, F., Holtzclaw, L., Chundu, K., Muzzafar, A., and Devaskar, S. (1990). *Endocrinology (Baltimore)* **126**, 2417–2424.

Salter, D. W., Baldwin, S. A., Lienhard, G. E., and Weber, M. J. (1982). *Proc. Natl. Acad. Sci. U.S.A.* **79**, 1540–1544.

Sarkar, H. K., Thorens, B., Lodish, H. F., and Kaback, H. R. (1988). *Proc. Natl. Acad. Sci. U.S.A* **85**, 5463–5467.

Sauer, N., Friedlander, K., and Graml-Wicke, U. (1990). *EMBO J.* **9**, 3045–3050.

Simpson, I. A., and Cushman, S. W. (1986). *Annu. Rev. Biochem.* **55**, 1059–1089.

Sistonen, L., Holtta, E., Lehvaslaiho, H., Lehtola, L., and Alitalo, K. (1989). *J. Cell Biol.* **109**, 1911–1919.

Sivitz, W. I., DeSautel, S. L., Kayano, T., Bell, G. I., and Pessin, J. E. (1989a). *Nature (London)* **340**, 72–74.

Sivitz, W. I., DeSautel, S. L., Walker, P. S., and Pessin, J. E. (1989b). *Endocrinology (Baltimore)* **124**, 1875–1880.

Sivitz, W. I., DeSautel, S. L., Kayano, T., Bell, G. I., and Pessin, J. E. (1990). *Mol. Endocrinol.* **4**, 583–588.

Slot, J. W., Moxley, R., Geuze, H. J., and James, D. E. (1990). *Nature (London)* **346**, 369–371.

Sogin, D. C., and Hinkle, P. C. (1978). *J. Supramol. Struct.* **8**, 447–453.

Stevens, B. R., Fernandez, A., Hirayama, B., Wright, E. M., and Kempner, E. S. (1990). *Proc. Natl. Acad. Sci. U.S.A.* **87**, 1456–1460.

Su, T.-S., Tsai, T.-F., Chi, C.-W., Han, S.-H., and Chou, C.-K. (1990). *Hepatology* **11**, 118–122.

Suzue, K., Lodish, H. F., and Thorens, B. (1989). *Nucleic Acids Res.* **17**, 10099.

Suzuki, K., and Kono, T., (1980). *Proc. Natl. Acad. Sci. U.S.A.* **77**, 2542–2545.

Takata, K., Kasahara, T., Kasahara, M., Ezaki, O., and Hirano, H. (1990). *Biochem. Biophys. Res. Commun.* **173**, 67–73.

Takeda, J., Burant, C. F., and Bell, G. I. (1991). In preparation.

Tal, M., Schneider, D. L., Thorens, B., and Lodish, H. F. (1990). *J. Clin. Invest.* **86**, 986–992.

Thorens, B., Sarkar, H. K., Kaback, H. R., and Lodish, H. F. (1988). *Cell (Cambridge, Mass.)* **55**, 281–290.

Thorens, B., Flier, J. S., Lodish, H. F., and Kahn, B. B. (1990a). *Diabetes* **39**, 712–719.

Thorens, B., Cheng, Z.-Q., Brown, D., and Lodish, H. F. (1990b). *Am. J. Physiol.* **259**, C279–C285.

Thorens, B., Lodish, H. F., and Brown, D. (1990c). *Am. J. Physiol.* **259**, C286–C294.

Thorens, B., Weir, G. C., Leahy, J. L., Lodish, H. F., and Bonner-Weir, S. (1990d). *Proc. Natl. Acad. Sci. U.S.A.* **87**, 6492–6496.

Turk, E., Zabel, B., Mundlos, S., Dyer, J., and Wright, E. M. (1991). *Nature (London)* **350**, 354–356.

Vera, J. C., and Rosen, O. M. (1989). *Mol. Cell. Biol.* **9**, 4187–4195.

Warburg, O. (1956). *Science* **123**, 309–314.

Wardzala, L. J., and Jeanrenaud, B. (1981). *J. Biol. Chem.* **256**, 7090–7093.

Wardzala, L. J., and Jeanrenaud, B. (1983). *Biochim. Biophys. Acta* **730**, 49–56.

Watanabe, T., Smith, M. M., Robinson, F. W., and Kono, T. (1984). *J. Biol. Chem.* **259**, 13117–13122.

Weiler-Guttler, H., Zinke, H., Mockel, B., Frey, A., and Gassen, H. G. (1989). *Biol. Chem. Hoppe-Seyler* **370**, 467–473.

Werner, H., Adamo, M., Lowe, W. L., Roberts, C. T., and LeRoith, D. (1989). *Mol. Encodrinol.* **3**, 273–279.

Witters, L. A., Vater, C. A., and Lienhard, G. E. (1985). *Nature (London)* **315**, 777–778.

Yamada, Y., Seino, S., Takeda, J., Fukumoto, H., Yano, H., Inagaki, N., Fukuda, Y., Seino, S., and Imura, H. (1990). *Biochem. Biophys. Res. Commun.* **168**, 1274–1279.

Yamamoto, T., Seino, Y., Fukumoto, H., Koh, G., Yano, H., Inagaki, N., Yamada, Y., Inoue, K., Manabe, T., and Imura, H. (1990). *Biochem. Biophys. Res. Commun.* **170**, 223–230.

Yano, H., Seino, Y., Inagaki, N., Hinokio, Y., Yamamoto, T., Yasuda, K., Masuda, K., Someya, Y., and Imura, H. (1991). *Biochem, Biophys, Res. Commun.* **174**, 470–477.

Zeller, W. P., Sian, M. T., Sweet, M., Goto, M., Gottschalk, M. E., Hurley, R. M., Filkins, J. P., and Hoffman, C. (1991). *Biochem. Biophys. Res. Commun.* **176**, 535–540.

Zhang, R., Logee, K. A., and Verkman, A. S. (1990). *J. Biol. Chem.* **265**, 15375–15378.

Ziel, F. H., Venkatesan, N., and Davidson, M. B. (1988). *Diabetes* **37**, 885–890.

Zorzano, A., Wilkinson, W., Kotliar, N., Thoidis, G., Wadzinkski, B. E., Ruoho, A. E., and Pilch, P. F. (1989). *J. Biol. Chem.* **264**, 12358–12363.

DISCUSSION

F. Murdoch. It is my understanding that one of the acute effects of insulin on GLUT4, the insulin-regulated glucose transporter, is to change its subcellular distribution so that it becomes localized to the plasma membrane. Now that you have the sequences of all these glucose transporters, I would think that you could identify regions of the GLUT4 protein that are important for this regulation by insulin.

G. Bell. We are trying to do just that. However, it is not obvious from the sequence comparisons which region(s) of the protein confers "insulin regulation." The approach that we have taken to address this problem is to prepare chimeras between GLUT4 and other glucose transporter isoforms and then examine their subcellular distribution when expressed in heterologous systems, including insulin-responsive cell lines such as 3T3-L1 adipocytes. Preliminary studies suggest that the intracellular N-terminal domain may be involved in targeting GLUT4, thus conferring insulin regulatability.

S. McKnight. Is the ability of glucose transporters to translocate in response to insulin specific for GLUT4 or would other transporters also translocate if expressed in the appropriate tissue?

G. Bell. That is a very good question, but unfortunately I have only a partial answer. Primary rat adipocytes express both GLUT1 and GLUT4, and the levels of GLUT4 are 20- to 30-fold greater than those of GLUT1. Studies of the subcellular distribution of these two glucose transporters in adipocytes have indicated that GLUT1 is approximately equally distributed between the plasma membrane and an intracellular vesicle population. By contrast, GLUT4 is associated almost exclusively with an intracellular vesicle population, and the vesicles with which it is associated are different from those containing GLUT1. On insulin stimulation, there is an ~2-fold increase in the amount of GLUT1 in the plasma membrane and a 10- to 20-fold increase in the amount of GLUT4. Thus, there is a quantitative difference in the ability of the two transporters expressed in the same cell to translocate. This suggests that the ability of the transporter to translocate in response to insulin is to a large degree specific for the transporter isoform. Our working hypothesis is that translocation is a property of specific glucose transporters and that it only occurs in cells with the appropriate "cellular machinery" for translocation.

W. Bardin.   You discussed a mechanism for explaining insulin resistance in diabetes. If you think about other explanations for abnormal insulin sensitivity in diabetes or obesity, what is the relative importance of GLUT4 depletion in adipocytes vs other causes of insulin resistance?

G. Bell.   Specific depletion of GLUT4 mRNA and protein may explain, at least in part, the insulin resistance of adipose tissue of obese and diabetic subjects. However, in addition to having lower levels of glucose transporters, there may be other postinsulin receptor defects in cells of these individuals which impair their responses to insulin, and these are likely to be in the intracellular signaling pathway that transfers the signal from the insulin receptor to the cellular effectors of insulin action. In fact, since GLUT4 levels are only slightly decreased, if at all, in skeletal muscle of insulin-resistant individuals, depletion of GLUT4 cannot explain the insulin resistance of this tissue, which is the major site of insulin-stimulated glucose disposal. Thus, I would have to say that the molecular basis for insulin resistance is still largely not understood and that while depletion of GLUT4 as well as genetic variation in the sequence of the insulin receptor can lead to insulin resistance, they do not provide a complete explanation.

INDEX